D1000346

Plant Proteomics

Plant Proteomics

Edited by

CHRISTINE FINNIE
Biochemistry and Nutrition Group
Biocentrum-DTU
Technical University of Denmark
Kgs Lyngby
Denmark

Blackwell
Publishing

Editorial Offices:
Blackwell Publishing Ltd, 9600 Garsington Road, Oxford OX4 2DQ, UK
Tel: +44 (0)1865 776868
Blackwell Publishing Professional, 2121 State Avenue, Ames, Iowa 50014-8300, USA
Tel: +1 515 292 0140
Blackwell Publishing Asia Pty Ltd, 550 Swanston Street, Carlton, Victoria 3053, Australia
Tel: +61 (0)3 8359 1011

First published 2006 by Blackwell Publishing Ltd

ISBN-13: 978-1-4051-4429-2
ISBN-10: 1-4051-4429-7

Library of Congress Cataloging-in-Publication Data

Plant proteomics/edited by Christine Finnie.
p. cm. — (Annual plant reviews)
Includes bibliographical references and index.
ISBN-13: 978-1-4051-4429-2 (hardback: alk. paper)
ISBN-10: 1-4051-4429-7 (hardback: alk. paper)
1. Plant proteins. 2. Plant proteomics.
I. Finnie, Christine. II. Series
QK898.P8P52 2006
572′.62—dc22

2006009416

A catalogue record for this title is available from the British Library

Set in 10/12 pt, Times
by Charon Tec Ltd, Chennai, India
www.charontec.com
Printed and bound in India
by Replika Press Pvt. Ltd, Kundli

The publisher's policy is to use permanent paper from mills that operate a sustainable forestry policy, and which has been manufactured from pulp processed using acid-free and elementary chlorine-free practices. Furthermore, the publisher ensures that the text paper and cover board used have met acceptable environmental accreditation standards.

For further information on Blackwell Publishing, visit our web site:
www.blackwellpublishing.com

Annual Plant Reviews

A series for researchers and postgraduates in the plant sciences. Each volume in this series focuses on a theme of topical importance and emphasis is placed on rapid publication.

Titles in the series:

1. **Arabidopsis**
   Edited by M. Anderson and J.A. Roberts
2. **Biochemistry of Plant Secondary Metabolism**
   Edited by M. Wink
3. **Functions of Plant Secondary Metabolites and Their Exploitation in Biotechnology**
   Edited by M. Wink
4. **Molecular Plant Pathology**
   Edited by M. Dickinson and J. Beynon
5. **Vacuolar Compartments**
   Edited by D.G. Robinson and J.C. Rogers
6. **Plant Reproduction**
   Edited by S.D. O'Neill and J.A. Roberts
7. **Protein–Protein Interactions in Plant Biology**
   Edited by M.T. McManus, W.A. Laing and A.C. Allan
8. **The Plant Cell Wall**
   Edited by J.K.C. Rose
9. **The Golgi Apparatus and the Plant Secretory Pathway**
   Edited by D.G. Robinson
10. **The Plant Cytoskeleton in Cell Differentiation and Development**
    Edited by P.J. Hussey
11. **Plant–Pathogen Interactions**
    Edited by N.J. Talbot
12. **Polarity in Plants**
    Edited by K. Lindsey
13. **Plastids**
    Edited by S.G. Moller
14. **Plant Pigments and Their Manipulation**
    Edited by K.M. Davies
15. **Membrane Transport in Plants**
    Edited by M.R. Blatt
16. **Intercellular Communication in Plants**
    Edited by A.J. Fleming
17. **Plant Architecture and its Manipulation**
    Edited by C. Turnbull

18. Plasmodesmata
Edited by K.J. Oparka
19. Plant Epigenetics
Edited by P. Meyer
20. Flowering and Its Manipulation
Edited by C. Ainsworth
21. Endogenous Plant Rhythms
Edited by A. Hall and H. McWatters
22. Control of Primary Metabolism in Plants
Edited by W.C. Plaxton and M.T. McManus
23. Biology of the Plant Cuticle
Edited by M. Riederer
24. Plant Hormone Signaling
Edited by P. Hedden and S.G. Thomas
25. Plant Cell Separation and Adhesion
Edited by J.R. Roberts and Z. Gonzalez-Carranza
26. Senescence Processes in Plants
Edited by S. Gan
27. Seed Development, Dormancy and Germination
Edited by K.J. Bradford and H. Nonogaki
28. Plant Proteomics
Edited by C. Finnie
29. Regulation of Transcription in Plants
Edited by K. Grasser
30. Light and Plant Development
Edited by G. Whitelam and K. Halliday

Contents

Colour plate falls after page 110

Preface

The proteome comprises all protein species resulting from gene expression in a cell, organelle, tissue or organism. By definition, proteomics aims to identify and characterize the expression pattern, cellular location, activity, regulation, post-translational modifications (PTMs), molecular interactions, three-dimensional (3D) structures and functions of each protein in a biological system. Due to the highly dynamic nature of the proteome, proteomics studies tend to focus on more specific goals as part of approaches to understand the function and regulation of biological systems. Proteome analysis is applied on different levels: (i) to catalogue the proteins synthesized in an organism, tissue or organelle; (ii) to characterize changes occurring during a developmental process; (iii) to identify proteins differing between biological samples or (iv) to identify proteins on the basis of a functional property (e.g. ligand binding). These types of analysis are all illustrated in the different chapters of this book.

Proteomics techniques are increasingly applied over the entire spectrum of biological sciences. All approaches have in common that they involve the separation, characterization and analysis of many proteins at once. Since 1975, 20 years before the term 'proteome' was first coined, separation of proteins in complex mixtures has been achieved by 2D-gel electrophoresis. Gel-based proteomics studies, despite limitations, are still of great value since 2D-gels can provide a unique overview of the different protein forms in a sample. However, gel-free techniques based solely on chromatographic separations and mass spectrometry are used increasingly frequently and have enabled the identification and characterization of low abundance, hydrophobic, basic or otherwise elusive proteins that are not amenable to 2D-gel analysis. Technological advances in 2D-gel electrophoresis systems and particularly in mass spectrometry have meant that proteome analysis is now financially and technically feasible for many laboratories. A wide spectrum of technical approaches will be found throughout the volume.

In plant science, the number of proteome studies is rapidly expanding after the completion of the *Arabidopsis thaliana* genome sequence, and proteome analyses of other important or emerging model systems and crop plants are in progress or are being initiated. Proteome analysis in plants is subject to the same obstacles and limitations as in other organisms, however the nature of plant tissues, with their rigid cell walls and complex variety of secondary metabolites, means that extra challenges are involved that may not be faced when analysing other organisms.

In several cases, the technologies described in this volume have not yet been applied to plant proteins, but point the way for future plant proteome studies. In Chapter 1, some of the challenges involved are introduced together with an overview of the technical and bioinformatic resources available for plant proteome analysis.

Much of the complexity of the proteome is due to PTMs of proteins. Mass spectrometry has revolutionized analysis of PTMs and new types of modification continue to be identified. Approaches for identification and characterization of PTMs in proteomes are presented in Chapter 2. In Chapter 3 approaches to analyse multiprotein complexes and protein–protein interactions in plants are discussed. Redox regulation of proteins is of major importance for many aspects of plant physiology. Recently there have been an increasing number of functional proteomics approaches taken to identify proteins undergoing cysteine and disulphide oxidoreduction. An overview of this field is given in Chapter 4. High throughput structure determination of proteins can give important and often unexpected insights into the roles of previously uncharacterized proteins. An example of such a structural proteomics approach is described in Chapter 5.

An overview of proteome studies in cereal crop plants and the status for comprehensive analysis of rice proteomes is presented in Chapter 6. Proteomics approaches to identify proteins involved in specific developmental processes in plants, with particular reference to seed development and germination are covered in Chapter 7. Two chapters covering organelle proteomics highlight some of the technical challenges involved in these analyses and the type of data that can be obtained from them. Chapter 8 covers analyses of the plant cell wall and secreted proteins that are notoriously challenging to isolate and study. Plant mitochondria have been the subject of numerous proteomics studies, including analysis of both soluble and membrane proteins, post-translationally modified proteins and spanning expression and functional proteome approaches using a wide range of techniques. Chapter 9 provides a detailed overview of the new information about plant mitochondrial function that has been gained from these studies.

A few common themes emerge from plant proteome analyses, including those presented in this volume: many proteins are typically identified in proteomics studies that are not similar to any protein with known function. In fact, the functions of approximately a third of the annotated genes in the *Arabidopsis thaliana* genome are unknown, suggesting that we still have a great deal to learn about plant biology. Another common outcome of plant proteomics projects is the identification of plant-specific proteins, that is, where the only similar sequences also originate from plants. The functions of many of these are also unknown, but are likely to be crucial for our understanding of plant physiology. Thus, one of the challenges arising from plant proteome research is to understand the functions of the proteins that are identified. Structural proteomics initiatives combined with information obtained from interaction, expression and functional proteomics studies can guide biochemical studies to elucidate protein function. In-depth biochemical characterization at the level of individual proteins or genes is an important requirement to validate and understand the significance of the data produced by proteomics, transcriptomics and metabolomics studies. Currently, few studies combine proteome analysis with genomics, transcriptomics and/or metabolomics, but future integration of these more or less high throughput techniques will bring us closer to a 'systems biology' approach to plant science.

The aim of this volume is to highlight ways in which proteome analysis has been used to probe the complexities of plant biochemistry and physiology. It is aimed at researchers in plant biochemistry, genomics, transcriptomics, proteomics and metabolomics who wish to gain an up-to-date insight into plant proteomes, the information plant proteomics can yield and the directions plant proteome research is taking.

Christine Finnie

Contributors

David J. Aceti — Center for Eukaryotic Structural Genomics (CESG), University of Wisconsin-Madison, USA

Craig A. Bingman — Center for Eukaryotic Structural Genomics (CESG), University of Wisconsin-Madison, USA

Eduard Bitto — Center for Eukaryotic Structural Genomics (CESG), University of Wisconsin-Madison, USA

Hans-Peter Braun — Institut für Pflanzengenetik, Naturwissenschaftliche Fakultät, Universität Hannover, Herrenhäuser Str. 2, D-30419 Hannover, Germany

Natalia V. Bykova — Agriculture and Agri-Food Canada, Cereal Research Centre, 195 Dafoe Road/195, Winnipeg MB, Canada R3T 2M9

Christine Finnie — Biochemistry and Nutrition Group, Biocentrum-DTU, Building 224, Technical University of Denmark, DK-2800 Kgs Lyngby, Denmark

Brian G. Fox — Center for Eukaryotic Structural Genomics (CESG), University of Wisconsin-Madison, USA

Ronnie O. Frederick — Center for Eukaryotic Structural Genomics (CESG), University of Wisconsin-Madison, USA

Karine Gallardo — Unité de Génétique et Ecophysiologie des légumineuses INRA, domaine d'Epoisses, 21110 Bretenières, France

Albrecht Gruhler — Protein Research Group, Department of Biochemistry and Molecular Biology, University of Southern Denmark, Odense, Campusvej 55, DK-5230 Odense M, Denmark

Per Hägglund — Biochemistry and Nutrition Group, Biocentrum-DTU, Building 224, Technical University of Denmark, DK-2800 Kgs Lyngby, Denmark

Joshua L. Heazlewood — ARC Centre of Excellence in Plant Energy Biology, Molecular and Chemical Science Building M310, The University of Western Australia, Crawley 6009, WA, Australia

Tal Isaacson Department of Plant Biology, 228 Plant Science Building, Cornell University, Ithaca, NY 14853, USA

Ole N. Jensen Protein Research Group, Department of Biochemistry and Molecular Biology, University of Southern Denmark, Odense, Campusvej 55, DK-5230 Odense M, Denmark

Won Bae Jeon Center for Eukaryotic Structural Genomics (CESG), University of Wisconsin-Madison, USA

Claudette Job CNRS/Bayer CropScience joint Laboratory (UMR CNRS 2847), Bayer CropScience, 14-20 rue Pierre Baizet, 69263 Lyon, France

Dominique Job CNRS/Bayer CropScience joint Laboratory (UMR CNRS 2847), Bayer CropScience, 14-20 rue Pierre Baizet, 69263 Lyon, France

Setsuko Komatsu Department of Molecular Genetics, National Institute of Agrobiological Sciences, Tsukuba 305-8602, Japan

Kenji Maeda Biochemistry and Nutrition Group, Biocentrum-DTU, Building 224, Technical University of Denmark, DK-2800 Kgs Lyngby, Denmark

John L. Markley Center for Eukaryotic Structural Genomics (CESG), University of Wisconsin-Madison, USA

A. Harvey Millar ARC Centre of Excellence in Plant Energy Biology, Molecular and Chemical Science Building M310, The University of Western Australia, Crawley 6009, WA, Australia

Ian M. Møller Department of Agricultural Sciences, Thorvaldsensvej 40, The Royal Veterinary and Agricultural University, DK-1871 Frederiksberg C, Denmark

George N. Phillips Jr Center for Eukaryotic Structural Genomics (CESG), University of Wisconsin-Madison, USA

Loïc Rajjou CNRS/Bayer CropScience joint Laboratory (UMR CNRS 2847), Bayer CropScience, 14-20 rue Pierre Baizet, 69263 Lyon, France

Jocelyn K.C. Rose Department of Plant Biology, 228 Plant Science Building, Cornell University, Ithaca, NY 14853, USA

Udo K. Schmitz Institut für Pflanzengenetik, Naturwissenschaftliche Fakultät, Universität Hannover, Herrenhäuser Str. 2, D-30419 Hannover, Germany

Jikui Song Center for Eukaryotic Structural Genomics (CESG), University of Wisconsin-Madison, USA

Hassan K. Sreenath Center for Eukaryotic Structural Genomics (CESG), University of Wisconsin-Madison, USA

Birte Svensson Biochemistry and Nutrition Group, Biocentrum-DTU, Building 224, Technical University of Denmark, DK-2800 Kgs Lyngby, Denmark

Dmitriy A. Vinarov Center for Eukaryotic Structural Genomics (CESG), University of Wisconsin-Madison, USA

Frank C. Vojtik Center for Eukaryotic Structural Genomics (CESG), University of Wisconsin-Madison, USA

Russell L. Wrobel Center for Eukaryotic Structural Genomics (CESG), University of Wisconsin-Madison, USA

Zsolt Zolnai Center for Eukaryotic Structural Genomics (CESG), University of Wisconsin-Madison, USA

1 Plant proteomics: challenges and resources

Joshua L. Heazlewood and A. Harvey Millar

1.1 Introduction

The revolution in big biology that we are currently witnessing can be as readily attributed to our changing perceptions about biology as it can be credited to recent technical advances. The field of proteomics has exploded onto the scientific landscape in recent years as more and more biologists wield tools that were once the exclusive domain of the analytical chemist. The plant community has readily taken up the promise of this emerging field with the number of proteome articles containing 'plant proteome' consistently being ~5% of the total field from 1999 onwards (Figure 1.1).

Essentially the proteome refers to the protein complement of a particular biological system at any given point in time. By its nature, the proteome is thus a dynamic and changing entity, and our best efforts in analysis are usually only a snap-shot in the lifetime of the organisms we study. The term 'proteomics' has been widely used to describe any analysis of these complex sets of proteins by mass spectrometry (MS). This can range from analysis of a single isolated protein to characterization of protein–protein interactions to selecting proteins for analysis from a polyacrylamide gel

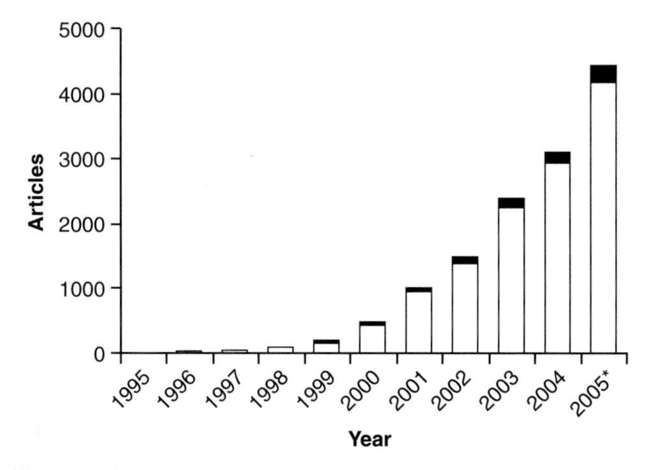

Figure 1.1 Number of articles using the term proteomics at ISI web of knowledge. Searches were conducted using the term 'proteom*' against the Science Citation Index Expanded for years 1995–2005 using the Topic field. The black box indicates the number of articles that also contained the term 'plant*'. The* indicates the extrapolation of the year of 2005 as searches only included January to October.

separation to large-scale multiple analyses of many hundreds or thousands of proteins from a complex biological system. Much of the recent progress in proteomics can be directly attributed to the many genome sequencing programs. This is because proteins are most commonly identified via MS through pattern matching of spectral information against large sets of predicted protein sequences derived from these genomic datasets. The completed genomes have provided us with a means to examine an organism's 'ORFeome', but this purely predictive set, while a valuable tool for pattern matching, can bare little resemblance to the actual proteome that can be experimentally observed.

While proteomics is theoretically the analysis of the full protein populations of complex biological systems, the technology we have today grossly restricts our ability to even come close to fulfilling this description experimentally. Proteomic analyses are normally restricted to studies of more discrete sub-systems which when assembled will form a mosaic that gives an overview of the system being investigated. As a result the experimental objectives of any proteomic study need to be well considered and the precise targets defined to ensure that technical limitations themselves are not the primary outcome of the research.

1.2 Challenges

Many of the challenges faced by plant researchers undertaking proteomics are the general problems shared by the wider field and relate to the matching of the broad physio-chemical properties of cellular protein complements to a narrow set of physical and chemical conditions dictated by each type of display and analysis technology. The following sections briefly outline some of the major challenges that the field of plant proteomics will be required to overcome to further advance this approach and assist in its integration into the analysis of plant systems.

1.2.1 Sample extraction

Sample preparation procedures for proteomic analysis require the reliable and consistent extraction of proteins from tissue homogenates. The resultant protein-rich extract must be substantially free of contaminating components associated with the tissue itself as well as being free from any components in the extraction solution that complicate downstream processing. Sample contaminants can affect every aspect of the subsequent analysis ranging from the modification of proteins by secondary metabolites (e.g. phenols), to effects on isoelectric focusing (IEF) of the sample (e.g. DNA and ionic strength) to suppression of ionization of the sample in the mass spectrometer (e.g. by salts and plastic residues) (Newton et al., 2004). The elimination or reduction of contaminating components can also lead to sample loss, modification or degradation. These problems are further compounded in plant tissues by the suite of plant-specific or plant-abundant compounds that can impede successful protein analysis techniques.

Plant tissues contain a wide variety of secondary metabolites many of which cause problems with protein extraction and proteomic procedures. These compounds can be specific to certain plant species or lineages while some compounds, such as quercetin, are relatively ubiquitous (Pierpoint, 2004). Several classes of secondary metabolites exist in plants that can interfere with proteins during homogenization, such as phenylpropanoids (e.g. lignins), terpenoids (e.g. gibberellins), flavanoids (e.g. pigments) and alkaloids (e.g. morphine). Problems often arise with specific tissue types such as fruit and flowers that can contain large amounts of these compounds. To further complicate matters, some tissues also contain large reserves of oil bodies (e.g. seeds) that can also impact on the success of protein extractions and subsequent proteomic analyses. Moreover, the abundance of complex carbohydrates in many plant samples also hampers protein extraction and analysis. Several reports have provided some direction when dealing with carbohydrate-rich material that may have wider applications (Andon et al., 2002; Østergaard et al., 2002). Finally, the presence of the plant cell wall provides a physical impediment to protein extraction requiring mechanical or enzymic removal; either method results in the production of large amounts of cell wall debris. The amount of contaminating components can be managed through the choice of tissue type or developmental stage (e.g. young leaves, root tips), but clearly in many cases such decisions are impractical due to the amounts of material available or the focus of the study being undertaken (Fido et al., 2004).

Most initial stages of sample preparation of plant tissue for proteomic analyses involve a mechanical disruption of material in a non-denaturing aqueous buffer that attempts to mimic the intracellular properties of the sample. Such a basic procedure would appear to favour the analysis of many plant tissues, although it is clearly not helpful if looking at proteins with poor solubility. Unfortunately, none of the issues regarding cellular contaminants such as lipids, DNA and secondary metabolites are addressed by such a simple approach. Many classic biochemical techniques to assist in the extraction process cannot be readily utilized due to the sensitivity of downstream proteomic analysis techniques. For example, many detergents cannot be used due to their effect on IEF, high-salt buffers will cause difficulties in the ionization of the sample and protease inhibitors can affect the digestion of samples prior to analysis (Newton et al., 2004). The choice of what can and cannot be used for the extraction process is dictated by the downstream analysis procedures that will be undertaken on the sample.

1.2.1.1 Two-dimensional gel electrophoresis

Many early studies on plant tissues assessed the compatibility of various extraction and precipitation procedures for reliable two-dimensional polyacrylamide gel electrophoresis (2D-PAGE) of samples (Hurkman and Tanaka, 1986; Granier, 1988). These studies tended to find that the precipitation of total protein by acetone or trichloroacetic acid after extraction produced the best resolution when analysed by 2D-PAGE. Not surprisingly, the addition of a precipitation step results in the removal of lipids, nucleic acids and a variety of metabolites thus producing a relatively clean protein sample. More recently the use of a precipitation step prior to

2D-PAGE analysis was also shown as an effective cleanup step for very problematic plant tissues such as banana meristem (Carpentier et al., 2005). Although a precipitation step undeniably results in a far better resolved 2D-PAGE array, there is no doubt that substantial protein loss occurs during this step, limiting the proteome that is displayed and further investigated.

One of the underlying objectives of 2D-PAGE is the reproducibility and consistency of protein extractions. This is often coupled to a desire to array the largest representative protein set that can be achieved. As a result, many attempts have been made to improve extraction reliability and performance utilizing a variety of detergents, chaotropes and mechanical disruption techniques (Giavalisco et al., 2003). Such optimization is often required due to the nature of different samples and the lack of complete solubilization in non-denaturing homogenizing buffers. Since the next step involves the solubilization of the sample in a harsh denaturing IEF solution, the opportunity exists to utilize IEF-compatible components in the extraction solution (Rabilloud et al., 1997). These typically include chaotropic agents like urea as well as zwitterionic detergents such as 3-[(3-Cholamidopropyl)dimethylammonio]-1-propanesulfonate (CHAPS). Poorly compatible compounds, such as the charged detergent SDS, can also be used in extraction buffers provided some cleanup step is incorporated prior to IEF (Hurkman and Tanaka, 1986).

1.2.1.2 Direct MS analysis of samples

Unlike the extraction of a sample for 2D-PAGE, the direct analysis of a sample by MS must be relatively free from contaminating compounds like salts and detergents. The presence of salt in the sample will dramatically affect ionization of the sample, can contaminate the mass spectrometer and can cause blockages in the liquid chromatography (LC) system. Detergents are also likely to suppress sample ionization and contaminate the mass spectrometer but in addition will affect the separation of the sample if an LC system is used prior to mass spectrometric analysis. The use of denaturing compounds such as urea or SDS for protein extraction are difficult to couple to direct MS analysis as they are hard to successfully eliminate from the sample. Sample cleanup via a solid-phase extraction (SPE) procedure could be used to remove these compounds, but this approach involves a considerable loss of sample. Some success has been achieved to minimize losses in SPE by undertaking this step as a coupled process with the analysis by MS (McDonald et al., 2002). The most straightforward approach is to precipitate the sample with acetone after solubilization in a MS-compatible buffer such as ammonium hydrogen carbonate or potassium phosphate prior to digestion and analysis (Link et al., 1999).

1.2.2 Sample preparation and arraying

The method chosen for sample analysis will be heavily dictated by the properties of the extract and by the objectives of the experiment (Figure 1.2). The following briefly outlines some of the major techniques that have been employed by plant researchers undertaking proteomic studies in recent years.

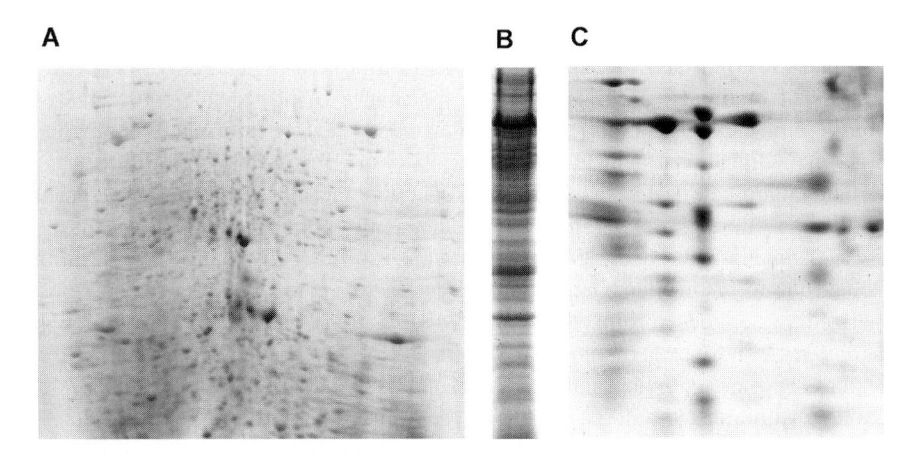

Figure 1.2 Various PAGE arraying techniques that are widely utilized in plant proteomics. **A:** 2D-PAGE of approximately 500 μg of whole-rice embryo lysate precipitated overnight in acetone prior to IEF. The IEF strip used was a 3–10 NL and the second dimension was run on a 12% gel. **B:** 1D-PAGE of approximately 50 μg of isolated mitochondria from rice coleoptiles run on a 12% gel. While mitochondria are likely to contain around 1500 proteins, less than this number of distinct bands is seen indicating that a high complexity level still exists. **C:** BN-PAGE of approximately 1 mg of *Arabidopsis* mitochondrial membrane proteins solubilized in 5% digitonin. The first dimension separated mitochondrial complexes on a 5–15% gradient gel, while the second dimension (shown) used a 10% tricine acrylamide gel. Components of the complexes are shown as vertical arrays of proteins.

1.2.2.1 Two-dimensional gel electrophoresis

The vast majority of recent proteomic studies undertaken on plant samples have employed 2D-PAGE for protein arraying of samples. This has had much to do with advances in IEF which now routinely utilize immobilized pH gradients (IPGs) in the first dimension (Görg et al., 1988). This technology has dramatically enhanced the ease and reproducibility of 2D-PAGE for proteomic analysis (Rabilloud, 1998). This arraying technique also provides a very clear picture for the researcher and can be easily used to assess similarities and differences between samples and treatments. It also provides information on two important physical properties of the arrayed proteins: the apparent molecular mass and the apparent isoelectric point (pI).

The basic procedures and techniques for undertaking 2D-PAGE have been well documented and reviewed in recent years (Herbert et al., 2001b; Rabilloud, 2002). For the past three decades many improvements have been made to the sample buffer used to solubilize the extracted sample prior to IEF. Many of these improvements have been targeted at increasing the efficiency of sample solubilization and the incorporation of new IEF-compatible compounds. The use of 9 M urea in the sample buffer still remains the chaotropic agent of choice for solubilizing and denaturing the sample prior to IEF (O'Farrell, 1975). Some care must be employed when using urea as it will readily degrade to ammonium and cyanate in solution with increased decomposition if the solution is heated above ~30°C. The isocyanic acid will subsequently react with amide groups in proteins (N-terminus and the side chains of arginine and lysine) as well as cysteine side chains resulting in protein carbamylation. This uncontrolled modification results in a protein population with varying degrees of modification and causes the appearance of horizontal reiterations of a protein on the 2D-PAGE. These are often interpreted as evidence of biological protein modifications such as phosphorylation or glycosylation. The effects of carbamylation can be limited through a reduction and alkylation step undertaken prior to IEF through the addition of iodoacetamide to the sample resulting in the *S*-carboxymethylation of cysteine (Herbert et al., 2001a). This controlled alkylation step also prevents uncontrolled alkylation by reactive components such as unpolymerized acrylamide (Bordini et al., 1999). Modifications to the urea concentration in sample buffer have been employed to improve solubility of other components in the buffer. The most common addition in recent times has been the combination of thiourea with urea (Rabilloud, 1998). The addition of thiourea (~2–3 M) requires a decrease in the concentration of urea in the sample buffer (down to ~6 M), but appears to improve solubilization of the sample resulting in many more arrayed proteins (Giavalisco et al., 2003).

The detergent CHAPS has been widely used in IEF sample buffers since it replaced Nonidet P-40 in the early 1980s (Perdew et al., 1983). Although CHAPS is still the detergent of choice, recently several surfactants have been developed in an attempt to improve the solubility of hydrophobic proteins in IEF. The addition of *N*-decyl-*N*,*N*-dimethyl-3-ammonio-1-propane sulphonate (SB 3–10) requires a lowering of the urea concentration to below 5 M in the sample buffer due to its poor solubility in high concentrations of urea (Rabilloud et al., 1997). While the addition of the more recently developed amidosulphobetaine 14 (ASB 14) is capable of withstanding urea concentrations of ~7–8 M (Chevallet et al., 1998; Herbert, 1999). The use of either surfactant in the sample buffer is likely to improve the solubility of hydrophobic proteins and allow for their separation on the 2D-PAGE. A recent 2D-PAGE examination of soluble proteins extracted from *Arabidopsis* aerial tissue found little visual difference when completely substituting CHAPS with ASB 14 (Giavalisco et al., 2003). These newer detergents are thus more likely to be better suited for more specifically solubilizing samples containing hydrophobic proteins (Rabilloud et al., 1997; Giavalisco et al., 2003). Lastly, tributylphosphine has been included in most IEF sample buffers as a powerful uncharged reductant (Herbert et al., 1998). Many protocols have now completely replaced dithiothrietol (DTT) or β-mercaptoethanol with this reducing agent as it can be used in lower concentrations and appears to increase protein solubility (Herbert et al., 2001a).

These adaptations to the classical IEF sample buffer have been driven by the desire to produce a generic solubilization solution that can reproducibly array as many proteins as possible from any given sample. Nonetheless, the largest disadvantage of the classical 2D-PAGE presentation technique still remains the selective loss of specific proteins due to physical properties incompatible with IEF. While there have been many attempts to overcome these shortcomings, hydrophobic proteins, basic proteins and high-molecular-mass proteins are often insoluble during IEF and do not readily appear in 2D gel presentations (Santoni et al., 1999b, 2000).

With the introduction of commercially available IEF strips with plastic supports and the accessibility of commercial-built 2D apparatus, the second dimension SDS-PAGE component has become relatively straightforward. After the IEF has completed, strips are equilibrated in an SDS loading buffer and placed on top of a precast acrylamide gel. Once the sample has successfully been arrayed in the second dimension acrylamide gel, sample analysis requires a successful digestion, and peptide extraction to be undertaken. The basic procedure has been widely reported and used with many minor modifications (Shevchenko et al., 1996a, b). More recently an examination of various parameters of the in-gel digestion and extraction process was able to suggest several useful modifications to the basic procedure with the added advantage of accelerating the whole procedure (Havlis et al., 2003).

1.2.2.2 One-dimensional gel electrophoresis

While the 2D-PAGE array provides unprecedented visualization of a wide range of proteins, ourselves and others have moved back to a one-dimensional polyacrylamide gel electrophoresis (1D-PAGE) to array and identify particular classes of proteins from our samples (Andon et al., 2002; Millar and Heazlewood, 2003). The use of such an approach can readily overcome the limitations of the 2D-PAGE as recalcitrant proteins such as hydrophobic or basic proteins are more easily arrayed, and this provides a mechanism to display and identify such proteins using MS (Wu and Yates, 2003). Unfortunately, unless some fractionation or sample purification approach is undertaken before 1D-PAGE, the band complexity observed can potentially reduce the effectiveness of this approach. Some assistance can be gained from using more extensive LC prior to mass spectrometric analysis to reduce this complexity.

1.2.2.3 Blue-native gel electrophoresis

This technique provides a means to array a protein complex using Coomassie Blue G-250 to bind and provide a net negative charge in the first dimension blue-native polyacrylamide gel electrophoresis (BN-PAGE), while a second dimension SDS-PAGE arrays individual polypeptides incorporated in or associated with protein complexes (Schägger and von Jagow, 1991; Jänsch et al., 1996). Such a technique has been readily used to array and identify components from the respiratory chain of mitochondria (Jänsch et al., 1996; Heazlewood et al., 2003a) and the photosystem complexes of the chloroplast (Heinemeyer et al., 2004). This approach can assist in preventing hydrophobic proteins from precipitating during the first dimension as they form the core of intact complexes during this phase of the separation. This method has its own particular set of conditions that need to be optimized to ensure successful solubilization of complexes and to minimize proteins streaking through the first dimension (Eubel et al., 2005).

1.2.2.4 Direct analysis of samples by MS

An alternative approach to arraying samples using a gel involves the direct analysis of intact proteins by the mass spectrometer. This technique was originally employed to analyse and characterize purified proteins or complexes but has more recently been used for the large-scale identification of hundreds of intact proteins from a single digested sample (Whitelegge et al., 1998; Wolters et al., 2001). Although still a useful tool for the global characterization of a protein or complex, whole-protein analysis can be problematic as it depends greatly on sample purity and detailed knowledge of the mature protein sequences and post-translational modifications. Undeniably the greatest advances in proteomics in recent years have been the successful analyses of complex mixtures of peptides formed by digestion of complex protein lysates. This method has actively exploited advances in hardware capabilities of mass spectrometers that have occurred in recent years. Specifically the method involves LC-MS/MS or LC/LC-MS/MS analysis of the sample employing chromatographic techniques to fractionate and concentrate peptides prior to MS. The latter technique employing multiple online chromatographic steps is more widely known as *Mu*lti*d*imensional *P*rotein *I*dentification *T*echnology (MudPIT) and has successfully identified thousands of peptides derived from many hundreds of proteins from a given sample (McDonald et al., 2002).

The direct analysis of complex peptide lysates provides a relatively quick means to assess a sample or preparation. The sample to be analysed can be readily digested using a protease such as trypsin or using a chemical cleavage method such as cyanogen bromide (CNBr). Digestion or cleavage of the sample occurs in solution allowing for control over conditions and providing greater accessibility to the substrate and subsequent release of peptides than occurs during in-gel digestions. Multiple strategies can be employed to target even highly insoluble materials since relatively soluble peptides will be released even from hydrophobic proteins (Millar and Heazlewood, 2003). The solution utilized for the digestion of the sample need not be completely compatible with the subsequent mass spectrometric analysis as soluble peptides can be cleaned offline using commercially available spin columns of reverse-phase LC material. Nonetheless, to avoid sample loss a direct analysis of the digested sample is often optimal (McDonald et al., 2002). Several digestion methods have been utilized that target insoluble materials such as membrane proteins (Wu and Yates, 2003). Such approaches are likely to provide a useful framework for the digestion of complex samples as they are often undertaken in a relatively denaturing environment. Samples have been successfully digested in solutions containing 0.5% SDS (Han et al., 2001) and high concentrations of solvents (80%) such as methanol and acetonitrile (Russell et al., 2001). By exploiting the high tolerance of the protease pepsin to acidic environments, samples have been solubilized in 70% formic acid with the subsequent digest at lower formic acid concentrations (Friso et al., 2004). The processes involved in the use of CNBr for protein cleavage involve solubilization of the sample in 70–90% formic acid (with samples diluted prior to analysis) greatly aiding cleavage efficiencies of insoluble samples (Skopp and Lane, 1989). The chemical digestive approach with CNBr tends to produce relatively large peptides and is often used in combination with a protease such as trypsin or Lys C (Washburn et al., 2001).

Although the multiple chromatographic steps provide a means of fractionating a complex sample, peptides resulting from the more abundant proteins will dominate the analysis masking less abundant peptides. Thus the complexity of many of these samples still challenges the current generation of mass spectrometers that have significant limitations in dynamic range. At this stage even the most successful analyses which have sought to characterize proteomes from a single sample representing a whole-cell lysate have identified less than 2000 proteins, far less than that required for the full analysis of a complex eukaryotic system (Washburn et al., 2001; Wu and Yates, 2003).

1.2.3 Mass spectrometry (MALDI and ESI)

The utilization of MS for the identification of proteins has become the primary technique for large-scale protein analysis in recent years. The choice of sample delivery methods and subsequent analysis are dictated by availability of hardware, costs and experimental objectives (Figure 1.3). The two most popular sample delivery devices currently being used in proteomics are *M*atrix-*A*ssisted *L*ASER *D*esorption/*I*onization (MALDI) and ElectroSpray Ionization (ESI). These sources are connected to various types of mass spectrometer, which further influence the type of data and quality produced. Both delivery methods share similarities as they rely on the processing of an isolated sample with a cleavage agent such as trypsin or CNBr, and the resulting peptides are introduced into the mass spectrometer as peptide ions in the gas phase.

1.2.3.1 MALDI

The most widely used method of sample delivery and analysis uses the MALDI source attached to a Time-Of-Flight mass spectrometer (MALDI-TOF). This combination has been successfully employed in the proteomic identification of proteins for over 10 years (James et al., 1994). This method of analysis is best suited for gel-separated proteins producing samples of relatively low complexity (three to four proteins). This approach uses the absolute mass of each peptide to produce a peptide mass fingerprint (PMF) of the protein, which is then used to search a database of theoretical peptide masses (Yates, 1998). The analysis of samples by MALDI-TOF to produce PMF data is a relatively high-throughput technique but produces data of poorer quality when compared to MS/MS-type analyses. This has a direct impact on pattern-matching confidence scores and creates more ambiguity when interrogating a large database for a match. Typically if six to eight peptide masses can be successfully matched with small error tolerances (e.g. 50–100 ppm), this technique can be very accurate for protein identifications; but when less than six peptides match, the possibility of false positives increases very sharply. Nonetheless, using this approach it is possible to analyse hundreds of samples a day (Figure 1.4A). The delivery process relies on aromatic acid matrices (e.g. dihydroxybenzoic acid or hydroxycinnamic acid) mixed with a digested sample. The matrix transforms the energy from the LASER to the sample leading to a release of peptides from the matrix. The LASER literally blasts the peptides from a solid dried sample into the gas phase for analysis by the mass spectrometer. No amino acid sequence information is obtained by basic MALDI-TOF analysis. Matching of data is based on the similarity of the masses to predicted patterns derived from an *in silico* digestion of the sequence databases (James et al., 1994). In most MALDI-TOF

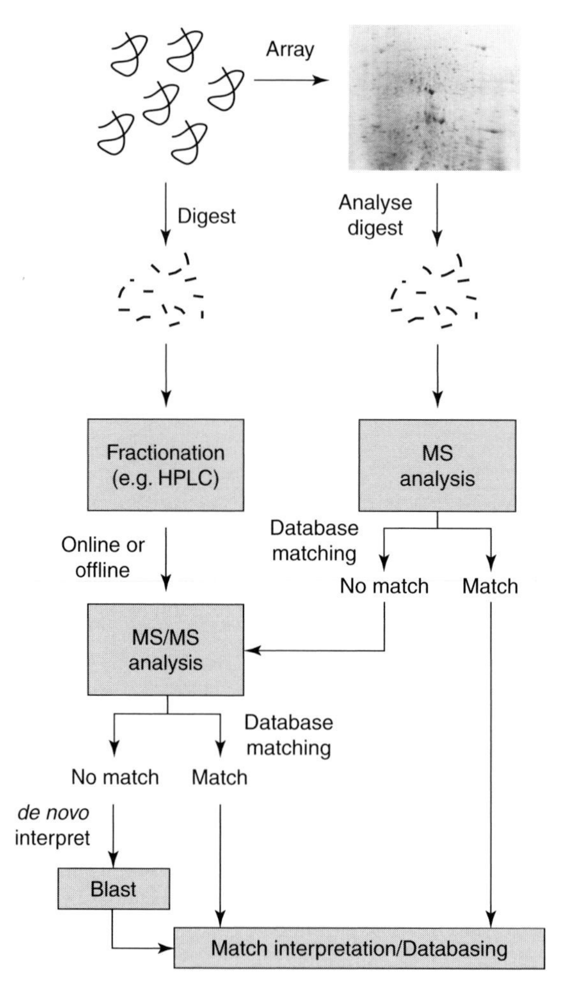

Figure 1.3 Flow chart of the basic proteomic analysis procedures. Proteins can be arrayed on a gel system (e.g. 2D-PAGE) analysed by a gel imaging system, differentially expressed proteins isolated and digested with a protease. Samples can initially be analysed by using a MALDI-TOF and data matched. If a match is unsuccessful, samples can be analysed using tandem MS and identified using data interrogation or *de novo* sequencing and BLAST. Alternatively samples can be directly digested with a protease, fractionated using an HPLC and analysed by tandem MS. Matching is undertaken using direct interrogation of MS/MS spectra with theoretical spectral data from DNA, EST or protein databases.

instruments there are now techniques that provide some sequence information by a post-source-decay (PSD) process (Patterson et al., 1996). However, PSD will generally only provide sequence tags for three to four amino acids from a 10–25 amino acid peptide. While these can be useful and provide an increase in matching confidences, they do not offer the same matching capabilities as authentic MS/MS spectra from a tandem mass spectrometer (Griffin et al., 1995).

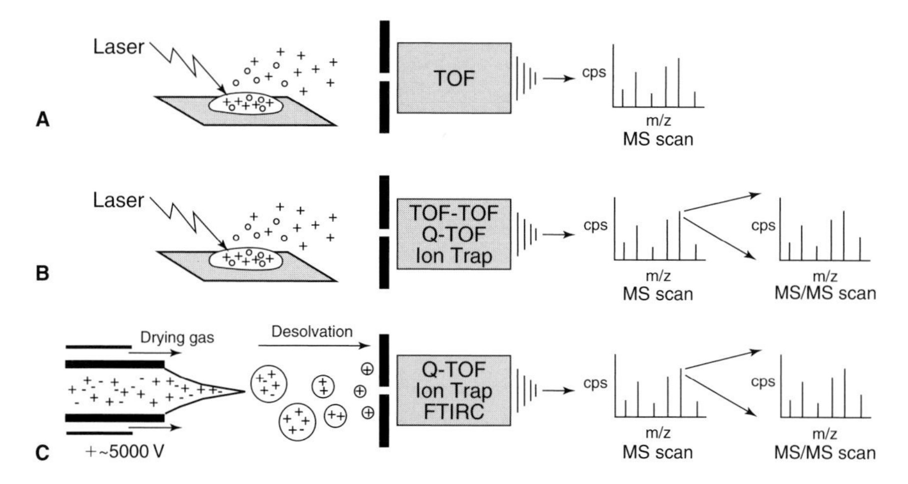

Figure 1.4 The principal ion delivery techniques and mass spectrometer combinations used in proteomics. A: The MALDI-TOF mass spectrometer is the proteomics workhorse primarily used for the analysis of 2D-PAGE arrayed proteins and produces PMF data for matching. It uses a LASER to literally blast the peptides out of the matrix into the gas phase. **B:** This setup also incorporating the MALDI source is also capable of producing MS/MS spectral data from ions observed in the initial scan. The TOF-based instruments have the added advantage of providing excellent mass accuracies. **C:** ESI ionization delivers peptides in a liquid spray using voltage and drying gas (e.g. nitrogen) to desolvate the charged droplets resulting in single ions entering the mass spectrometer. Instruments used with this delivery method undertake an initial scan resulting in an ion mass list, then select ions of interest for fragmentation producing MS/MS data.

More recently, MS/MS capabilities have been made available with the coupling of the MALDI source to tandem mass spectrometers allowing full-peptide fragmentation. The MALDI source has commonly been attached to a tandem mass spectrometer with a TOF component due to synergistic reasons involving timed points of ionization and the pulsing nature of the TOF (e.g. TOF-TOF and Quadrupole-TOF (Q-TOF)), but more recently has also been coupled to IonTrap (IT) mass spectrometers (Figure 1.4B). This arrangement of MS components provides the advantages of sample throughput and high-confidence matching of MS/MS spectra into a single system. Moreover, these systems now have the capacity to deal with complex lysates with the development of LC-MALDI applications allowing chromatographic separations to be directly spotted onto MALDI plates (Bodnar et al., 2003). One of the considerable advantages of this technique is the ability to lockdown samples providing the capacity to return to a plate for a subsequent detailed analysis of a particular spot of interest. The LC-MALDI setup appears to be providing an excellent means of obtaining proteomic depth and protein coverage when compared to other methods (Chen et al., 2005). These systems thus provide significant advances in the large-scale analysis of complex systems with unambiguous matching capabilities, high-throughput applications and the capability to analyse complex samples.

1.2.3.2 ESI

The ESI source is commonly used to analyse complex mixtures and is capable of interfacing with a wide range of mass spectrometers that usually provide MS/MS capabilities; these include the IT, the hybrid Q-TOF, and the Fourier Transform Ion Cyclotron Resonance system (FTICR). ESI is an excellent method for the ionization of a wide range of polar molecules and relies on the sample of interest being dissolved in a solvent and delivered to the source in a thin capillary. A high voltage is applied to the tip of the capillary, producing a strong electric current which causes the emerging sample to be dispersed into an aerosol of highly charged droplets. These droplets are vaporized by an inert gas such as nitrogen-releasing charged species for analysis by the mass spectrometer (Figure 1.4C).

Whatever the arrangement of MS components that is utilized, this approach relies on the fragmentation of each peptide in some sort of collision cell within the mass spectrometer yielding a fragmentation spectra or MS/MS spectra. Each MS/MS spectrum can then be interpreted individually to suggest a probable amino acid sequence (*de novo* sequencing) or can be used to search a database containing theoretical MS/MS spectra (pattern matching) to suggest both an amino acid sequence and a probable identity of the original protein (Yates et al., 1995). The quality of information obtained using these systems providing MS/MS spectra is superior to PMF data obtained through classic MALDI-TOF analyses. However, it is also more complicated and thus can be a limitation for researchers both in terms of time and finance.

The analysis of samples using ESI can readily be achieved manually using a syringe apparatus often provided with the system, but by far the most convenient and sensitive procedures involve a high-performance liquid chromatography (HPLC) connected to the ionization source. This provides sample automation and higher throughput but also has an added level of complexity. Samples are usually injected onto a reverse-phase HPLC column such as C_{18} allowing the concentration and separation of peptides prior to analysis by the mass spectrometer. This analysis produces MS/MS data from the fragmentation of a parent ion (peptide) in the process termed collision-induced dissociation (CID). These CID spectra are high-quality data for pattern matching and interpretation of peptides. The basic type of HPLC used for proteomic analyses over the past 10 years has been the capillary HPLC system that provides flow rates from 2 to 50 μL/min. More recently, however, a newer generation of ESI sources have been developed that are being coupled to LC systems and provide reliable flow stability at rates down to 50 nL/min and below. These ESI sources are often known as NanoSpray sources. These sources are becoming standard hardware options on the front-ends of most mass spectrometers and are being coupled to nano-flow HPLC systems. The upshot of this reduction in flow rate has been a substantial increase in sensitivity while maintaining a reasonable level of throughput through coupling to LC systems (Gatlin et al., 1998). While these gains in sensitivity are welcome, the trade-off is in the added technical complexity of the analysis. In order to load the sample onto a column, a second pump with higher flow capacities (e.g. a capillary flow pump) is usually required, before the nano-pump elutes the peptides into the mass spectrometer at low flow rates. These low flow rates mean that the internal tubing diameters used in the LC system become smaller and smaller (around 25 μm) with column internal diameter also becoming smaller (around 75 μm). There is a larger chance of blockages in such small diameter tubing which means samples must be more rigorously prepared to be free of particulate material prior to analysis. The analysis time per sample in nano-flow analyses increases dramatically, with a simple analysis on a nano-system taking many hours compared to less than 1 h using a capillary LC system alone. In complex sample analyses employing multiple chromatographic columns (MudPIT), a NanoSpray experiment can take days depending on the number of fractions and elution periods (Wolters et al., 2001).

1.2.4 Analysis depth

When initiating and undertaking a proteomic study of a complex system a major objective is to identify as many proteins as possible from the sample. In a complex sample such as a whole-cell lysate this could represent many thousands of proteins. A standard 2D-PAGE-based proteomic array technique will typically display a set of around 1000–2000 proteins. Several efforts have reported gels with over 8000 arrayed spots (Gauss et al., 1999). While these studies are useful at pushing the technical boundaries of arraying, they do not represent a normal outcome. From a more realistic gel, perhaps, 500 of the 1000–2000 proteins are abundant enough to be confidently assigned an identity using MS. If the sample analysed was derived from a whole cell or whole tissue there may be as many as 10,000 different gene products in the sample. Consequently this approach is likely to capture at best 5% of the expressed protein set from this sample. The direct analysis of a complex protein lysate by MudPIT has the potential to identify a larger dynamic range and could enable the identification of thousands of proteins from a single sample (McDonald et al., 2002). But in one of the most comprehensive proteomic identification attempts in plants to date, only around 2000 rice proteins were identified by extensive MS/MS from a variety of tissues using MudPIT (Koller et al., 2002).

The challenges faced by researchers attempting to obtain maximum proteomic depth from the sample are even further confounded by the dynamic range of proteins found with the cell (Patterson, 2004). Plant samples isolated from green tissue contain large quantities of the chloroplast protein RuBisCO, which can obscure less abundant proteins both on gel arrays and when directly analysing complex lysates (Komatsu et al., 1999). Furthermore, the majority of other proteins that are found in whole-cell or tissue lysates represent major metabolic enzymes and structural proteins, not the modulating and regulatory components such as transcription factors that are more often desired. Thus much of the current proteomic analyses are surveying the abundant cellular components, while the majority of proteins within the cell are being precluded from the analysis. As a consequence when proteomic techniques are being used to interpret phenomena such as the effects of certain hormonal regimes, developmental processes or defence against a pathogen only a small fraction of abundant housekeeping proteins are really being investigated. While observed differences in these proteins are likely real, they hardly explain the underlying mechanisms of signalling and regulation that are occurring within the cell under these situations. Furthermore, even when these data on changes in particular protein abundance can be identified from the experimental analysis, they tend to refer to only individual members of complex biochemical pathways and thus do not provide a complete picture of the response occurring in whole pathways. While disappointing, this is not really surprising given the fragmented nature of the display and analysis techniques.

The use of subcellular experimental approaches can allow the analysis of proteins that are co-localized and are often functionally interactive. This also decreases the complexity of protein samples, allowing a greater depth of analysis of the expressed proteome. These approaches commonly employ cellular fractionation, centrifugation-based purification of organelle or cellular compartment and MS to identify peptides (Millar, 2004). The chloroplast has been dissected through multiple studies that have explored its sub-organelle proteome including the thylakoid lumen (Peltier et al., 2002; Schubert et al., 2002), thylakoid membrane (Friso et al., 2004) and mixed envelope membranes (Ferro et al., 2003; Froehlich et al., 2003), and a whole-chloroplast analysis (Kleffmann et al., 2004). The *Arabidopsis* mitochondrial proteome was initially investigated through two studies using 2D-PAGE (Kruft et al., 2001; Millar et al., 2001). Subsequently further studies have targeted complexes and used a direct lysate analysis approach (Eubel et al., 2003; Heazlewood et al., 2003a, b, 2004). The proteome of nuclei has also been analysed (Bae et al., 2003; Calikowski et al., 2003; Pendle et al., 2005). The vacuole has been studied through analysis of the tonoplast membrane and the vacuole contents (Carter et al., 2004; Shimaoka et al., 2004; Szponarski et al., 2004). Only two small studies have been undertaken on the peroxisome isolated from cotyledons (Fukao et al., 2002, 2003). Several other studies have identified proteins from a variety of intracellular membrane systems including the Golgi and endoplasmic reticulum (ER; Prime et al., 2000) and plasma membrane fractions from *Arabidopsis* (Santoni et al., 1999a; Alexandersson et al., 2004). The proteomes of the cell wall (Chivasa et al., 2002; Mithoefer et al., 2002; Komatsu et al., 2004) and the apoplast (Haslam et al., 2003) have also been investigated. These studies provide a mechanism to penetrate deeper into cellular proteomes and can allow a more focused approach when undertaking comparative proteomic studies.

The model plant *Arabidopsis* probably represents the source of the largest number of such sub-proteomic studies and these data have been recently compiled in an online SUBcellular database for *Arabidopsis* proteins called SUBA (Heazlewood et al., 2005). This database provides subcellular localizations for approximately 2500 non-redundant *Arabidopsis* proteins that have been identified through MS and illustrates the capabilities of subcellular fractionation as a means to address proteomic depth (Table 1.1).

1.2.5 Data analysis

The success of most proteomic studies is often a function of data analysis and matching. There are two important aspects to this process: principally whether good quality

Table 1.1 Proteins identified through subcellular analysis of Arabidopsis*

Compartment	*Arabidopsis* proteins (non-redundant)	Number of publications
Mitochondria	547	18
Plastid	1017	8
Nucleus	367	3
Plasma membrane	534	5
Vacuole	378	3
Peroxisome	28	1

* These studies have identified over 2500 non-redundant proteins and provide evidence for their presence within the organism as well as some functional information. Data were obtained from the SUBA database (http://www.suba.bcs.uwa.edu.au/).

sequence data are available for the species being examined and secondly the format of the collected data (PMF or MS/MS).

1.2.5.1 Peptide mass fingerprints

The ability to rapidly screen many hundreds of gel-separated samples at relatively low costs still makes PMF-based analyses a practical route. Since no sequence data are obtained, (unless using PSD), data matching is heavily reliant upon mass accuracy and sequence availability. Most modern TOF mass spectrometers provide a high mass accuracy (±10–30 ppm) which is a crucial aspect to the analysis of PMF data. Thus the real limitation to using PMF in proteomic studies is sequence availability (Heazlewood and Millar, 2003). While several high-fidelity plant genomes have been made available in recent years (e.g. *Arabidopsis* and rice), many other well-studied plant species lack sufficient genomic data for confident matching assignments to be readily undertaken using PMF. This lack of genome data can lead some researchers to a loosening of search parameters in an attempt to produce 'a match'. Most of the interrogation programs such as Mascot (Matrix Sciences) or MSFit (Protein Prospector) will take these parameter shifts into account when scoring the matching process (Perkins et al., 1999). Nonetheless, it is still in the hands of the researcher to interpret the final result, as nearly all data are bound to match some protein from large databases if parameters are broadened enough. The lack of a successful species-specific positive match using PMF can also lead researchers to attempt using cross-species matching. Clearly genomic analysis has shown that there is a substantial degree of sequence conservation between many plant species. The problem faced for significant proteomic data matching is the requirement for identical masses of regions, specifically cleaved peptides under investigation, between proteins of divergent species. The alternatives are to attempt to allow for error tolerances to account for amino acid substitutions. This will lead to much less confidence in the match and greatly increases the number of false and misleading identifications. Some attempts have been made to examine the opportunities available for cross-species matching with PMF data, but it is certainly not an optimal avenue for large-scale proteomic studies (Liska and Shevchenko, 2003).

1.2.5.2 Peptide fragmentation data (MS/MS)

The use of MS/MS data produced by a mass spectrometer capable of targeted peptide fragmentation provides an unsurpassed method of data matching and analysis in species with both a well characterized genome and those lacking genomic data. The data format provides the mass of the analysed peptide as well as some fragmentation data which can either be used directly for interrogation with a database using available software (e.g. Mascot or SEQUEST) or can be deconvoluted to provide some sequence information (*de novo* sequencing), which can then be used to match to the sequence databases using BLAST (Altschul et al., 1997). The use of either method will provide a far greater confidence in the resulting match, but just as with PMF matching, the final interpretation still requires some level of assessment. The coupling of an LC system with the production of MS/MS data can provide substantial sample depth from both complex lysates and gel-arrayed samples. Since peptides can be concentrated and separated prior to analysis, far more data on each peptide can be obtained. This method of data analysis is commonly utilized for complex samples since significant matches to proteins can be obtained with only one or two MS/MS spectra. Furthermore, more stringent scoring procedures are being incorporated into interrogation software in an attempt to provide a tighter scoring procedure for use with complex lysates or MudPIT-type experiments (Mascot: http://www.matrixscience.com/help/results_help.html). Although MS/MS spectra contain some sequence information, the matching process still requires an accurate peptide mass for confident matching (Yates, 1998). Finally a significant benefit of using MS/MS data for protein identifications is the ability to more readily analyse post-translational modifications through the interrogation of the fragmentation patterns produced.

1.2.5.3 Analysis options

Based on the observations above, it would seem that PMF-based analyses are not a desirable procedure to employ in modern proteomic surveys. This is not entirely accurate since this technique has obvious financial advantages, the availability and ease of instrument operation needs to be considered as does the ability to process hundreds or thousands of samples a week. The advantages gained through MS/MS analysis need to be offset by the procedures scalability in large-scale proteomic studies. While a lack of complete genome information can potentially be viewed as inhibitive, there are many studies in plant species outside the model systems that have employed both PMF and MS/MS data analysis successfully (Neubauer et al., 1998; Mathesius et al., 2001). Such studies often exploit the availability of large-scale expressed sequence tags (EST) sequencing programs. The Institute for Genomic Research (TIGR: Quackenbush et al., 2001) provides non-redundant EST datasets or Gene Indices (GI) for over 30 commonly researched plant species (Table 1.2). These EST sets can be used to successfully interrogate either PMF or MS/MS data specifically for these species or from closely related species and can significantly assist proteomic studies outside the model plant species. But if the requirement is to readily identify proteins from a species with poor genomic data, then there is currently little

Table 1.2 Available plant species and total unique ESTs from the TIGR Gene Indices[a]

Species	Release version	Total unique ESTs	Proteomic potential[b]
Aquilegia	2.0	17,801	++
Arabidopsis	12.1	62,010	+++++
Barley	9.0	50,453	++++
Brassica napus	1.0	15,151	++
Chlamydomonas reinhardtii	5.0	31,608	++++
Cocoa	1.0	2539	+
Common bean	1.0	9484	+
Cotton	6.0	40,348	+++
Grape	4.0	23,871	++
Ice plant	4.0	8455	+
Lettuce	2.0	22,185	++
Lotus	3.0	28,460	+++
Maize	16.0	72,047	++++
Medicago truncatula	8.0	36,878	+++
Nicotiana benthamiana	2.0	7554	+
Onion	1.0	11,726	++
Pepper	2.0	13,003	++
Petunia	1.0	4466	+
Pinus	6.0	45,557	++++
Poplar	2.0	54,756	++++
Potato	10.0	38,239	+++
Rice	16.0	89,147	+++++
Rye	3.0	5,347	+
Sorghum bicolor	8.0	39,148	+++
Soya bean	12.0	63,676	++++
Spruce	1.0	27,194	+++
Sugar beet	1.0	13,618	++
Sugar cane	2.1	78,547	++++
Sunflower	3.0	20,520	++
Tobacco	2.0	21,107	++
Tomato	10.1	31,838	+++
Triphysaria	1.0	3847	+
Wheat	10.0	122,282	++++

[a]These datasets are available for download and can be incorporated into software to enable data matching for poorly represented plant species in genome databases.
[b]Based on the number of available ESTs for well sequenced organisms such as *Arabidopsis* and rice, a Proteomic Potential score was assigned to provide suitability of these sets to proteomic studies.

choice other than to employ an MS/MS data approach complemented with *de novo* sequencing and BLAST.

1.2.6 Quantitation

While the identification of proteins from a sample is a fundamental necessity in proteomic studies, it is often the ability to observe and quantify differences in proteomes that can provide the most significant findings from such work. While the field has developed and expanded its capabilities in the quantitation of gel-arrayed samples, it is in the area of direct MS analysis of samples that significant advances are taking shape. One of the major issues in the area of quantitation is the problem of dynamic range, because protein abundances can cover five to six orders of magnitude in cell extracts. The ability to accurately resolve abundance across this dynamic range is still well outside the scope of current technologies. Nonetheless, it is still possible to quantify a limited range with some accuracy.

1.2.6.1 Gel stains

For the last decade Colloidal Coomassie has been the dominant stain in the field of proteomics. It is a comparatively sensitive staining method with a detection limit of around 5–10 ng of protein. The stain is relatively inexpensive, it is compatible with MS, it is simple to remove through washing and importantly it does not result in any modifications to the protein prior to MS (Neuhoff et al., 1988). While silver staining is generally more sensitive with detection limits around 0.5–2 ng of protein, it has reproducibility issues and is not particularly compatible with MS (Lopez et al., 2000). An alternative to the more traditional staining methods is the new generation of fluorescent stains such as SYPRO Ruby, Red and Orange (Steinberg et al., 1996). These stains offer similar sensitivities to silver staining but provide a greater dynamic range and a far better level of compatibility with subsequent MS (Lauber et al., 2001). There are still some advantages to using either Colloidal Coomassie or silver staining as scanning the gels for the comparative analysis requires a standard flatbed scanner, while fluorescent stains will require a charge-coupled device (CCD) camera or commercial LASER scanner for visualization (Steinberg et al., 1996). Furthermore, the fluorescent stains are relatively expensive which can be a factor if undertaking a large-scale gel-based comparative proteomic study (Rose et al., 2004). Another problem with increasing levels of sensitivity offered by these newer stains is the inability to routinely identify the proteins they can stain at the low range. Although many mass spectrometers are marketed with promises of identifications at the femtomole or attomole level, this usually occurs under ideal conditions and is not a consistent outcome when generic settings are used in high-throughput MS protocols. It should also be noted that these sensitivity levels are calculated on the isolated peptides supplied at the ionization source and do not take into account the efficiencies of peptide extraction from gels, sample handling, etc.

For accurate comparative gel analysis, the scanned gels need to be analysed and some level of quantitation undertaken. There are a variety of commercially available software options as well as free packages to undertake these procedures (Young et al., 2004), but essentially this process is able to detect protein spots on the digitized gels and use the volume and greyscale intensities to provide an approximate measure of abundance based on some background subtraction algorithm. Several factors can affect the performance of these programs including: poor spot detection or the inability to discriminate between two valid spots, poor differentiation of proteins spots from background, and issues in matching spots between gels when undertaking a comparison building a replicate gel (Rosengren et al., 2003). Although most of these programs allow some user input to set optimal levels during the initial phases, manual corrections and interpretations are often required to correct problems which can be time-consuming, labour-intensive and subjective. Understanding every nuance of these programs is not a trivial exercise, with much time investment needed to allow the full exploitation of most of the features.

Clearly one of the major problems when undertaking comparative proteomic studies using 2D-PAGE is the reproducibility between replicates and between samples. Inconsistencies and procedural variation between IEF focusing and arraying of samples can cause real problems with statistical analysis and validation. More recently a fluorescent-labelling technique has been developed that attempts to overcome some of the technical discrepancies that invariably occur during 2D-PAGE. The system employs three different CyDye fluors with varying excitation wavelengths and is referred to as 2D Fluorescence DIfference Gel Electrophoresis or DIGE (Yan et al., 2002). This method labels the two samples to be analysed with two different CyDye fluors and significantly can utilize the third fluor to label a loading control. These three samples are mixed and analysed using IEF 2D-PAGE. The resultant gel is analysed using a fluorescent scanner with appropriate filters. The scanned images can be viewed independently to assess arraying of samples, but since the same protein in either of the samples being analysed will migrate to the same point in the gel after IEF and electrophoresis, the scanned images can be accurately overlayed, providing unsurpassed 2D-PAGE comparative proteomic capabilities (Rose et al., 2004).

1.2.6.2 Chemical labelling of sample

The movement of proteomics away from a gel-based environment to LC analysis of complex lysates has created some difficulties in reliable quantitation. It is difficult to make too many inferences about abundances of the protein identified, since the process of peptide ionization is not similar for each of the peptides for a given protein. As a consequence techniques have been developed to provide relative quantitation of the same peptide between samples. This method uses thiol-reactive isotope-coded affinity tags (ICAT) to label cysteine residues in peptides from each sample. The system employs two different ICAT tags to label the two samples for analysis. One tag contains eight hydrogen atoms (light tag) while the second contains eight deuterium atoms (heavy tag), providing a distinct mass difference (8 amu). The samples are differentially labelled and mixed in equal amounts, digested and tagged peptides purified using a biotin moiety built into the tags. The enriched ICAT-labelled peptides are then analysed using MS/MS, with identical differentially labelled peptides analysed simultaneously. The origins of the two peptides can be readily determined by the 8 amu mass difference between peaks. The relative intensity of each of the peptides can provide a comparative ratio to give relative quantitation of the protein identified (Gygi et al., 1999, 2000).

One of the problems that were soon identified with the ICAT approach is the requirement for the presence of cysteine in the protein for labelling and quantitation to occur. Unfortunately cysteine is not a very common amino acid with around 8% of proteins in yeast not containing this residue (Gygi et al., 2000). While several other labelling techniques have employed the use of stable isotopes such as ^{15}N (Oda et al., 1999), these methods have required propagation of the organism in media containing the isotope itself. More recently an amine-specific isotope labelling system was developed that could provide tagging of all peptides in a sample. The system employs four isobaric (same mass) tags that can be mixed to analyse four different samples simultaneously and is known as iTRAQ (Ross et al., 2004). The system uses a similar logic to ICAT, where identical peptides will be analysed by the mass spectrometer simultaneously, and it provides strong quantitation capabilities without being affected by the vagaries of peptide ionization between samples. Samples are differentially labelled, pooled and analysed by MS/MS. Relative or even absolute quantitation (if a standard is labelled and included) is obtained when peptides with the isobaric tags are fragmented, causing the release of a specific mass reporter for each tag (Ross et al., 2004). Since all peptides are labelled by this method, as opposed to the cysteine specificity of ICAT, the analysis process can be very complicated due to the increased abundance of ions in complex samples. Moreover, just like the analysis of complex lysates, there are limitations to the number of proteins that can be identified and thus quantified in these labelled samples. Furthermore, since there is often a significant reduction in the quality of MS/MS spectra when undertaking automated analysis of complex samples, the data required to produce significant quantitation of less abundant peptides is often inadequate.

1.2.7 Modifications

The number of predicted proteins estimated by many eukaryotic sequencing projects is around 30,000–40,000. Although recent estimates have suggested that the number of functionally distinct proteins that could be produced by post-translational modifications may be 10–100 times this base number (Rappsilber and Mann, 2002). Protein modifications leading to conformational changes of the target protein will have major influences on the proteins activation state, cellular location and/or function.

A wide range of techniques can be used to characterize protein modifications using proteomic analysis techniques. The 2D-PAGE array can be useful for the visualization of apparent post-translational modifications as many modifications alter the pI of the protein with only minor changes to the molecular mass (Rabilloud, 2002). This uncontrolled modification results in a protein population with a variety of alterations and causes the appearance of a series of horizontal repates of the protein on the 2D-PAGE. Often these putative modifications are in fact non-biological and are a result of degraded urea in the sample buffer causing carbamylation (Herbert et al., 2001a). However, clearly biological protein modifications are also present in these analyses and these moieties do in fact result in pI shifts in the 2D-PAGE array (Zhu et al., 2005). The difficulty is in locating the proteins on the gel and characterizing the modified residues. Several approaches have been applied to this problem and by far the most successful has been studies of protein phosphorylation. A combination of differential labelling with ^{32}P and the use of antibodies to phosphoproteins (Peck et al., 2001), as well as a phosphoprotein-specific stain (Wang et al., 2005) have been used to highlight phosphoproteins. While such approaches can provide a nice picture in the analysis process, increasingly mass spectrometric techniques rather than 2D-PAGE arrays are being utilized to confidently assign a plethora of biologically significant protein modifications including acetylation, phosphorylation, glycosylation and methylation (Wilkins et al., 1999; Mann and Jensen, 2003). These approaches utilize techniques like immobilized metal ion affinity chromatography (IMAC) to selectively recover and analyse phosphorylated peptides from complex lysates. Significantly such techniques can be used to enrich low-abundant modified proteins or peptides from complex lysates.

The study of post-translational modifications in proteomics represents one of the major issues in this field. Currently studies analysing complex samples using MudPIT-type analyses produce a significant number of unmatched ions or peptides that do not match any protein in the database (Washburn et al., 2001; Wolters et al., 2001). Many such studies are undertaken on model systems where excellent genome sequence data are available. Whether many of these unmatched ions turn out to be peptides of known proteins that have been modified beyond recognition in some way remains to be seen. The ability to identify modified proteins and to conclusively map modifications to amino acid residues will provide a significant resource in the functional interpretation of the proteome, although the area is still in early development.

1.2.8 Data

While technical issues of protein identification, arraying, dynamic range and advances in instrumentation tend to dominate the field of proteomics, there is also a growing desire to capture and archive the data that are currently being produced (Prince et al., 2004). This desire has developed from the realization that much of the data currently being produced is being lost. Currently the data presented in the majority of publications is represented in binary format, with protein identifications simply listed within the manuscript. Consequently the primary data that were used to make these identifications are essentially inaccessible to the wider community. Moreover, much of the detail of the analysis that was undertaken in the experiment is lost. There is a real fear that there are hidden gems in the primary data of many studies that will be simply discarded or just gather dust on computers around the world. With the development of better matching algorithms and the ongoing sequencing of more genomes, there is a possibility that much of these neglected data could be exploited by researchers in the future. Attempts are currently underway to standardize the capture and storage of proteomic data forms. Currently most of the hardware and software designers are involved in producing a standard data format (mzDATA), which can be readily examined and distributed for all major platforms (Orchard et al., 2004). There are also moves to produce a standardized management schema for proteomics, similar to the MIAME schema developed by the microarray community. The Proteome Experimental Data Repository (PEDRo: http://pedro.man.ac.uk/) attempts to standardize proteomic analyses by providing guidelines for methods and experiments as well as acting as a data repository and analysis facility for proteomic studies (Taylor et al., 2003; Garwood et al., 2004).

Recently there have been a series of editorials from prominent proteomic journals which now minimally require the inclusion of the data file that was used for matching as well as a variety of other basic information on how the matching was undertaken during submission of manuscripts (Beavis, 2005; Bradshaw, 2005). These decisions have come about because of the explosion in the use of proteomics by non-specialist researchers and the opinion that many applications of this technology were not stringent enough. Much of these concerns stem from the problem that too many of the identifications that are currently being reported in the literature are likely to be false positives (Carr et al., 2004). Such problems usually occur when data matching is pushed to its limit as in most circumstances mass spectral data will always match something in peptide databases. Such issues were not as serious when the majority of studies were carried out using 2D-PAGE which greatly limited the amount of data produced, and provided some empirical information about the protein being analysed (molecular weight and pI). The problem has become far more prevalent with the explosion in the use of direct analysis of complex samples where many thousands of spectra are produced and any physical relationship with the proteins being analysed is lost.

1.3 Resources

1.3.1 Proteomic databases

The development of the biological database has been driven by the genome sequencing projects of the last decade and as such the structures and search options for these databases reflect these origins. With the number of large-scale proteomic analyses that are currently being undertaken, many researchers have found themselves moving away from the ubiquitous spreadsheet and into database platforms that allow them to more readily analyse the data flows produced (Table 1.3). These databases largely grow out of the desire for researchers to manage their own data in-house, but many flow into the public arena.

The 2D-PAGE array was the primary analytical method used in proteomics for many years, and as such there are several 2D-PAGE-type databases that have attempted to provide reference maps of arrayed proteins from different species and different organs for the community at large. The best example of this is illustrated by the SWISS-2D-PAGE database (http://www.expasy.org/ch2d/). The problems with 2D-PAGE databases are that they are largely limited to the cell lines and species that are chosen to be displayed. The ability to match 2D-PAGE arrays is problematic even when they are produced as replicates from a similar tissue source, so attempting to use online digital maps as accurate matching tools to a users own protein gels is exceedingly frustrating. Such databases thus provide a rough means of comparing identified proteins between species or samples, rather than a complete and accurate reference

Table 1.3 Online plant proteomic databases

Database type	Name	Focus	Address
Subcellular	AMPDB	Mitochondria	http://www.ampdb.bcs.uwa.edu.au/
	AMPP	Mitochondria	http://www.gartenbau.uni-hannover.de/genetik/AMPP
	PPDB	Plastid	http://ppdb.tc.cornell.edu/
	PLPROT	Plastid	http://www.plprot.ethz.ch/
	PPMdb	Plasma membrane	http://sphinx.rug.ac.be:8080/
	AtNoPDB	Nuceolar	http://bioinf.scri.sari.ac.uk/cgi-bin/atnopdb/home
	SUBA	Subcellular	http://www.suba.bcs.uwa.edu.au/
	CWN	Cell wall	http://bioinfo.ucr.edu/projects/Cellwall/index.pl
2D-PAGE	SWISS-2DPAGE	*Arabidopsis*	http://www.expasy.org/ch2d/
	ANU-	*Medicago*, rice	http://semele.anu.edu.au/2d/2d.html
	2DPAGE	*Arabidopsis* seed	http://seed.proteome.free.fr/
	RPD	Rice	http://gene64.dna.affrc.go.jp/RPD/main_en.html
		Pine	http://www.pierroton.inra.fr/genetics/2D/
		Medicago	http://www.noble.org/2dpage
	Mt Proteomics	*Medicago*	http://www.mtproteomics.com/
	Gel bank	*Arabidopsis*	http://gelbank.anl.gov/

map utility. Although there have been several attempts to use 2D-PAGE reference maps to identify proteins across plant species, to date these have not been entirely successful (Mathesius et al., 2002). Furthermore, these databases are not equipped to deal with the newer proteomic techniques involving direct analysis of complex samples.

The proliferation of specific plant proteomic databases in recent years has probably reflected the movement of the community into more advanced proteomic studies. Much of this work has discarded the 2D-PAGE array and focused on specific sets of proteins within the cell. These subsets have invariably represented well characterized biochemical entities, such as the mitochondrion, the plastid and the nucleus (Table 1.3). These studies have not only defined distinct functional sets of proteins, but as many groups represent sub-fractions of the cellular milieu, they have also enabled proteomic depth of the cellular proteome. At this stage these databases generally provide the presence or absence of a protein from a given subset. Proteomic databases will need to become far more dynamic in the future and will need to provide user-annotation features as well as full integration with the genomic, transcriptomic and metabolomic datasets. Importantly they will need to allow tracking and archiving of experimental procedures and the ability to access the primary data. Such features will allow the current databases to integrate with other areas providing advanced search and data-mining abilities that will be necessary in order to interpret the vast amount of information currently being produced. While at the present stage there is no organized data storage program for primary data produced through plant proteomics, there are moves in other organisms to establish some consensus in proteomic data handling that are likely to show the way to the plant proteomic community (Taylor et al., 2003). As an example, The Open Proteomic Database has attempted to provide an online resource for yeast, humans, *Escherichia coli* and *Mycobacteria* communities (http://bioinformatics.icmb.utexas.edu/OPD/) and may provide a model for a plant proteomic database in the future (Prince et al., 2004).

Table 1.4 Online proteomic resources

Resource	Address
Tutorials/Resources	http://www.spectroscopynow.com http://www.ionsource.com/ http://www.abrf.org/ http://www.systemsbiology.org/ http://www.expasy.org/ http://www.proteomicworld.org/ http://fields.scripps.edu/
Data matching	http://www.matrixscience.com/ http://prospector.ucsf.edu/ http://prowl.rockefeller.edu/ http://wolf.bms.umist.ac.uk/mapper/ http://pubchem.ncbi.nlm.nih.gov/omssa/ http://phenyx.vital-it.ch/pwi/login/login.jsp
Data storage	http://bioinformatics.icmb.utexas.edu/OPD/ http://www.expasy.org/ch2d/ http://pedro.man.ac.uk/
Proteome journals	http://www.mcponline.org/ http://pubs.acs.org/journals/jprobs/index.html http://www.wiley-vch.de/publish/en/journals/alphabeticIndex/2120/ http://www.proteomesci.com/ http://www.bentham.org/cp/index.htm

1.3.2 Online proteomic tools and resources

A variety of resources exist online that can readily be accessed to provide technical assistance, data matching and a wide variety of utilities for proteomic research (Table 1.4). Recently there have been a wide variety of freeware utilities to undertake many of the techniques that were previously only available from commercial products. There are some inherent limitations of these free utilities, namely many are very early releases of preliminary techniques that may not have been tested as widely as many of the larger commercial products. Nonetheless, many of these new tools provide far more flexible tools for analysis or manipulation of data.

1.4 Future

The future for plant proteomics will most likely be to try to stay on top of technical advances made in MS and protein chemistry, and exploit and adapt new and developing techniques. Currently this will involve the adoption of quantitation methods that employ the mass spectrometer directly (Chelius et al., 2003; Liu et al., 2004) or utilizing the ICAT/iTRAQ-type approach (Gygi et al., 1999; Ross et al., 2004). The study of post-translational modifications and how these changes affect biological processes is still a largely unexplored area in plant proteomics. However, these techniques only exist in the context of biological questions and it is here that we need to maintain our understanding of issues of interest in plant biology in order to do interesting experiments and link our findings to those of our colleagues working with very different techniques. Much of our current work in plant proteomics tends to be binary identifications, and for this to link with other large data analysis technologies, adequate quantitation will be required. Such information will provide the ability to more closely link protein function with global protein abundances, overlaying them into regulatory and metabolic networks within the cell and ultimately play a part in explaining how plants grow, develop and react to their environment.

Acknowledgements

This work is supported through grants provided by the Australian Research Council Centres of Excellence Program and ARC Discovery Program, an ARC QEII Research Fellowship to A.H.M. and an ARC Postdoctoral Fellowship to J.L.H.

References

Alexandersson, E., Saalbach, G., Larsson, C. and Kjellbom, P. (2004) *Arabidopsis* plasma membrane proteomics identifies components of transport, signal transduction and membrane trafficking. *Plant Cell Physiol.*, **45**, 1543–1556.

Altschul, S.F., Madden, T.L., Schaffer, A.A., Zhang, J., Zhang, Z., Miller, W. and Lipman, D.J. (1997) Gapped BLAST and PSI-BLAST: a new generation of protein database search programs. *Nucleic Acid. Res.*, **25**, 3389–3402.

Andon, N.L., Hollingworth, S., Koller, A., Greenland, A.J., Yates III, J.R. and Haynes, P.A. (2002) Proteomic characterization of wheat amyloplasts using identification of proteins by tandem mass spectrometry. *Proteomics*, **2**, 1156–1168.

Bae, M.S., Cho, E.J., Choi, E.Y. and Park, O.K. (2003) Analysis of the *Arabidopsis* nuclear proteome and its response to cold stress. *Plant J.*, **36**, 652–663.

Beavis, R. (2005) The Paris consensus. *J. Proteome Res.*, **5**, 1475.

Bodnar, W.M., Blackburn, R.K., Krise, J.M. and Moseley, M.A. (2003) Exploiting the complementary nature of LC/MALDI/MS/MS and LC/ESI/MS/MS for increased proteome coverage. *J. Am. Soc. Mass Spectrom.*, **14**, 971–979.

Bordini, E., Hamdan, M. and Righetti, P.G. (1999) Matrix-assisted laser desorption/ionisation time-of-flight mass spectrometry for monitoring alkylation of beta-lactoglobulin B exposed to a series of N-substituted acrylamide monomers. *Rapid Commun. Mass Spectrom.*, **13**, 2209–2215.

Bradshaw, R.A. (2005) Revised draft guidelines for proteomic data publication. *Mol. Cell. Proteom.*, **4**, 1223–1225.

Calikowski, T.T., Meulia, T. and Meier, I. (2003) A proteomic study of the *Arabidopsis* nuclear matrix. *J. Cell. Biochem.*, **90**, 361–378.

Carpentier, S.C., Witters, E., Laukens, K., Deckers, P., Swennen, R. and Panis, B. (2005) Preparation of protein extracts from recalcitrant plant tissues: an evaluation of different methods for two-dimensional gel electrophoresis analysis. *Proteomics*, **5**, 2497–2507.

Carr, S., Aebersold, R., Baldwin, M., Burlingame, A., Clauser, K. and Nesvizhskii, A. (2004) The need for guidelines in publication of peptide and protein identification data: Working Group on Publication Guidelines for Peptide and Protein Identification Data, *Mol. Cell. Proteom.*, **3**, 531–533.

Carter, C., Pan, S., Zouhar, J., Avila, E.L., Girke, T. and Raikhel, N.V. (2004) The vegetative vacuole proteome of *Arabidopsis thaliana* reveals predicted and unexpected proteins. *Plant Cell*, **16**, 3285–3303.

Chelius, D., Zhang, T., Wang, G. and Shen, R.F. (2003) Global protein identification and quantification technology using two-dimensional liquid chromatography nanospray mass spectrometry. *Anal. Chem.*, **75**, 6658–6665.

Chen, H.S., Rejtar, T., Andreev, V., Moskovets, E. and Karger, B.L. (2005) Enhanced characterization of complex proteomic samples using LC-MALDI MS/MS: exclusion of redundant peptides from MS/MS analysis in replicate runs. *Anal. Chem.*, **77**, 7816–7825.

Chevallet, M., Santoni, V., Poinas, A., Rouquie, D., Fuchs, A., Kieffer, S., Rossignol, M., Lunardi, J., Garin, J. and Rabilloud, T. (1998) New zwitterionic detergents improve the analysis of membrane proteins by two-dimensional electrophoresis. *Electrophoresis*, **19**, 1901–1909.

Chivasa, S., Ndimba, B.K., Simon, W.J., Robertson, D., Yu, X.L., Knox, J.P., Bolwell, P. and Slabas, A.R. (2002) Proteomic analysis of the *Arabidopsis thaliana* cell wall. *Electrophoresis*, **23**, 1754–1765.

Eubel, H., Jansch, L. and Braun, H.P. (2003) New insights into the respiratory chain of plant mitochondria. Supercomplexes and a unique composition of complex II. *Plant Physiol.*, **133**, 274–286.

Eubel, H., Braun, H.P. and Millar, A.H. (2005) Blue-native PAGE in plants: a tool in analysis of protein–protein interactions. *Plant Method.*, **1**, 11.

Ferro, M., Salvi, D., Brugiere, S., Miras, S., Kowalski, S., Louwagie, M., Garin, J., Joyard, J. and Rolland, N. (2003) Proteomics of the chloroplast envelope membranes from *Arabidopsis thaliana*. *Mol. Cell. Proteom.*, **2**, 325–345.

Fido, R.J., Mills, E.N.C., Rigby, N.M. and Shewry, P.R. (2004) In: Cutler, P. (ed) *Methods in Molecular Biology*, Vol. 244. Humana Press Inc., Totowa, NJ, pp. 21–27.

Friso, G., Giacomelli, L., Ytterberg, A.J., Peltier, J.B., Rudella, A., Sun, Q. and Wijk, K.J. (2004) In-depth analysis of the thylakoid membrane proteome of *Arabidopsis thaliana* chloroplasts: new proteins, new functions, and a plastid proteome database. *Plant Cell*, **16**, 478–499.

Froehlich, J.E., Wilkerson, C.G., Ray, K., McAndrew, R.S., Osteryoung, K.W., Gage, D.A. and Phinney, B.S. (2003) Proteomic study of the *Arabidopsis thaliana* chloroplastic envelope membrane utilizing alternatives to traditional two-dimensional electrophoresis. *J. Proteome Res.*, **2**, 413–425.

Fukao, Y., Hayashi, M. and Nishimura, M. (2002) Proteomic analysis of leaf peroxisomal proteins in greening cotyledons of *Arabidopsis thaliana. Plant Cell Physiol.*, **43**, 689–696.

Fukao, Y., Hayashi, M., Hara-Nishimura, I. and Nishimura, M. (2003) Novel glyoxysomal protein kinase, GPK1, identified by proteomic analysis of glyoxysomes in etiolated cotyledons of *Arabidopsis thaliana. Plant Cell Physiol.*, **44**, 1002–1012.

Garwood, K., McLaughlin, T., Garwood, C., Joens, S., Morrison, N., Taylor, C.F., Carroll, K., Evans, C., Whetton, A.D., Hart, S., Stead, D., Yin, Z., Brown, A.J., Hesketh, A., Chater, K., Hansson, L., Mewissen, M., Ghazal, P., Howard, J., Lilley, K.S., Gaskell, S.J., Brass, A., Hubbard, S.J., Oliver, S.G. and Paton, N.W. (2004) PEDRo: a database for storing, searching and disseminating experimental proteomics data. *BMC Genom.*, **5**, 68.

Gatlin, C.L., Kleemann, G.R., Hays, L.G., Link, A.J. and Yates III, J.R. (1998) Protein identification at the low femtomole level from silver-stained gels using a new fritless electrospray interface for liquid chromatography–microspray and nanospray mass spectrometry. *Anal. Biochem.*, **263**, 93–101.

Gauss, C., Kalkum, M., Lowe, M., Lehrach, H. and Klose, J. (1999) Analysis of the mouse proteome. (I) Brain proteins: separation by two-dimensional electrophoresis and identification by mass spectrometry and genetic variation. *Electrophoresis*, **20**, 575–600.

Giavalisco, P., Nordhoff, E., Lehrach, H., Gobom, J. and Klose, J. (2003) Extraction of proteins from plant tissues for two-dimensional electrophoresis analysis. *Electrophoresis*, **24**, 207–216.

Görg, A., Postel, W. and Günther, S. (1988) The current state of two-dimensional electrophoresis with immobilized pH gradients. *Electrophoresis*, **9**, 531–546.

Granier, F. (1988) Extraction of plant proteins for two-dimensional electrophoresis. *Electrophoresis*, **9**, 712–718.

Griffin, P.R., MacCoss, M.J., Eng, J.K., Blevins, R.A., Aaronson, J.S. and Yates III, J.R. (1995) Direct database searching with MALDI-PSD spectra of peptides. *Rapid Commun. Mass Spectrom.*, **9**, 1546–1551.

Gygi, S.P., Rist, B., Gerber, S.A., Turecek, F., Gelb, M.H. and Aebersold, R. (1999) Quantitative analysis of complex protein mixtures using isotope-coded affinity tags. *Nat. Biotechnol.*, **17**, 994–999.

Gygi, S.P., Rist, B. and Aebersold, R. (2000) Measuring gene expression by quantitative proteome analysis. *Curr. Opin. Biotechnol.*, **11**, 396–401.

Han, D.K., Eng, J., Zhou, H. and Aebersold, R. (2001) Quantitative profiling of differentiation-induced microsomal proteins using isotope-coded affinity tags and mass spectrometry. *Nat. Biotechnol.*, **19**, 946–951.

Haslam, R.P., Downie, A.L., Raveton, M., Gallardo, K., Job, D., Pallett, K.E., John, P., Parry, M.A.J. and Coleman, J.O.D. (2003) The assessment of enriched apoplastic extracts using proteomic approaches. *Annal. Appl. Biol.*, **143**, 81–91.

Havlis, J., Thomas, H., Sebela, M. and Shevchenko, A. (2003) Fast-response proteomics by accelerated in-gel digestion of proteins. *Anal. Chem.*, **75**, 1300–1306.

Heazlewood, J.L. and Millar, A.H. (2003) Integrated plant proteomics – putting the green genomes to work. *Funct. Plant Biol.*, **30**, 471–482.

Heazlewood, J.L., Howell, K.A. and Millar, A.H. (2003a) Mitochondrial complex I form *Arabidopsis* and rice: orthologs of mammalian and fungal components coupled with plant-specific subunits. *Biochim. Biophys. Acta*, **1604**, 159–169.

Heazlewood, J.L., Whelan, J. and Millar, A.H. (2003b) The products of the mitochondrial orf25 and orfB genes are F_0 components in the plant F_1F_0 ATP synthase. *FEBS Lett.*, **540**, 201–205.

Heazlewood, J.L., Tonti-Filippini, J.S., Gout, A.M., Day, D.A., Whelan, J. and Millar, A.H. (2004) Experimental analysis of the *Arabidopsis* mitochondrial proteome highlights signalling and regulatory components, provides assessment of targeting prediction programs and points to plant specific mitochondrial proteins. *Plant Cell*, **16**, 241–256.

Heazlewood, J.L., Tonti-Filippini, J., Verboom, R.E. and Millar, A.H. (2005) Combining experimental and predicted datasets for determination of the subcellular location of proteins in *Arabidopsis*. *Plant Physiol.*, **139**, 598–609.

Heinemeyer, J., Eubel, H., Wehmhoner, D., Jansch, L. and Braun, H.P. (2004) Proteomic approach to characterize the supramolecular organization of photosystems in higher plants. *Phytochemistry*, **65**, 1683–1692.

Herbert, B. (1999) Advances in protein solubilisation for two-dimensional electrophoresis. *Electrophoresis*, **20**, 660–663.

Herbert, B.R., Molloy, M.P., Gooley, A.A., Walsh, B.J., Bryson, W.G. and Williams, K.L. (1998) Improved protein solubility in two-dimensional electrophoresis using tributyl phosphine as reducing agent. *Electrophoresis*, **19**, 845–851.

Herbert, B., Galvani, M., Hamdan, M., Olivieri, E., MacCarthy, J., Pedersen, S. and Righetti, P.G. (2001a) Reduction and alkylation of proteins in preparation of two-dimensional map analysis: Why, when, and how? *Electrophoresis*, **22**, 2046–2057.

Herbert, B.R., Harry, J.L., Packer, N.H., Gooley, A.A., Pedersen, S.K. and Williams, K.L. (2001b) What place for polyacrylamide in proteomics? *Trend. Biotechnol.*, **19**, S3–S9.

Hurkman, W.J. and Tanaka, C.K. (1986) Solubilization of plant membrane proteins for analysis by two-dimensional gel electrophoresis. *Plant Physiol.*, **81**, 802–806.

James, P., Quadroni, M., Carafoli, E. and Gonnet, G. (1994) Protein identification in DNA databases by peptide mass fingerprinting. *Protein Sci.*, **3**, 1347–1350.

Jänsch, L., Kruft, V., Schmitz, U.K. and Braun, H.P. (1996) New insights into the composition, molecular mass and stoichiometry of the protein complexes of plant mitochondria. *Plant J.*, **9**, 357–368.

Kleffmann, T., Russenberger, D., von Zychlinski, A., Christopher, W., Sjolander, K., Gruissem, W. and Baginsky, S. (2004) The *Arabidopsis thaliana* chloroplast proteome reveals pathway abundance and novel protein functions. *Curr. Biol.*, **14**, 354–362.

Koller, A., Washburn, M.P., Lange, B.M., Andon, N.L., Deciu, C., Haynes, P.A., Hays, L., Schlieltz, D., Ulaszek, R., Wei, J., Wolters, D. and Yates III, J.R. (2002) Proteomic survey of metabolic pathways in rice. *Proc. Natl Acad. Sci. USA*, **99**, 11969–11974.

Komatsu, S., Muhammad, A. and Rakwal, R. (1999) Separation and characterization of proteins from green and etiolated shoots of rice (*Oryza sativa* L.): towards a rice proteome. *Electrophoresis*, **20**, 630–636.

Komatsu, S., Kojima, K., Suzuki, K., Ozaki, K. and Higo, K. (2004) Rice proteome database based on two-dimensional polyacrylamide gel electrophoresis: its status in 2003. *Nucleic Acid. Res.*, **32**, D388–D392.

Kruft, V., Eubel, H., Jansch, L., Werhahn, W. and Braun, H.P. (2001) Proteomic approach to identify novel mitochondrial proteins in *Arabidopsis*. *Plant Physiol.*, **127**, 1694–1710.

Lauber, W.M., Carroll, J.A., Dufield, D.R., Kiesel, J.R., Radabaugh, M.R. and Malone, J.P. (2001) Mass spectrometry compatibility of two-dimensional gel protein stains. *Electrophoresis*, **22**, 906–918.

Link, A.J., Eng, J., Schieltz, D.M., Carmack, E., Mize, G.J., Morris, D.R., Garvik, B.M. and Yates III, J.R. (1999) Direct analysis of protein complexes using mass spectrometry. *Nat. Biotechnol.*, **17**, 676–682.

Liska, A.J. and Shevchenko, A. (2003) Expanding the organismal scope of proteomics: cross-species protein identification by mass spectrometry and its implications. *Proteomics*, **3**, 19–28.

Liu, H., Sadygov, R.G. and Yates III, J.R. (2004) A model for random sampling and estimation of relative protein abundance in shotgun proteomics. *Anal. Chem.*, **76**, 4193–4201.

Lopez, M.F., Berggren, K., Chernokalskaya, E., Lazarev, A., Robinson, M. and Patton, W.F. (2000) A comparison of silver stain and SYPRO Ruby Protein Gel Stain with respect to protein detection in two-dimensional gels and identification by peptide mass profiling. *Electrophoresis*, **21**, 3673–3683.

Mann, M. and Jensen, O.N. (2003) Proteomic analysis of post-translational modifications. *Nat. Biotechnol.*, **21**, 255–261.

Mathesius, U., Keijzers, G., Natera, S.H., Weinman, J.J., Djordjevic, M.A. and Rolfe, B.G. (2001) Establishment of a root proteome reference map for the model legume *Medicago truncatula* using the expressed sequence tag database for peptide mass fingerprinting. *Proteomics*, **1**, 1424–1440.

Mathesius, U., Imin, N., Chen, H., Djordjevic, M.A., Weinman, J.J., Natera, S.H., Morris, A.C., Kerim, T., Paul, S., Menzel, C., Weiller, G.F. and Rolfe, B.G. (2002) Evaluation of proteome reference maps for cross-species identification of proteins by peptide mass fingerprinting. *Proteomics*, **2**, 1288–1303.

McDonald, W.H., Ohi, R., Miyamoto, D.T., Mitchison, T.J. and Yates III, J.R. (2002) Comparison of three directly coupled HPLC MS/MS strategies for identification of proteins from complex mixtures: single-dimension LC-MS/MS, 2-phase MudPIT, and 3-phase MudPIT. *Int. J. Mass. Spectrom.*, **219**, 245–251.

Millar, A.H. (2004) Location, location, location: surveying the intracellular real estate through proteomics in plants. *Funct. Plant. Biol.*, **31**, 563–571.

Millar, A.H. and Heazlewood, J.L. (2003) Genomic and proteomic analysis of mitochondrial carrier proteins in *Arabidopsis*. *Plant. Physiol.*, **131**, 443–453.

Millar, A.H., Sweetlove, L.J., Giege, P. and Leaver, C.J. (2001) Analysis of the *Arabidopsis* mitochondrial proteome. *Plant Physiol*, **127**, 1711–1727.

Mithoefer, A., Mueller, B., Wanner, G. and Eichacker, L.A. (2002) Identification of defence-related cell wall proteins in *Phytophthora sojae*-infected soybean roots by ESI-MS/MS. *Mol. Plant Pathol.*, **3**, 163–166.

Neubauer, G., King, A., Rappsilber, J., Calvio, C., Watson, M., Ajuh, P., Sleeman, J., Lamond, A. and Mann, M. (1998) Mass spectrometry and EST-database searching allows characterization of the multi-protein spliceosome complex. *Nat. Genet.*, **20**, 46–50.

Neuhoff, V., Arold, N., Taube, D. and Ehrhardt, W. (1988) Improved staining of proteins in polyacrylamide gels including isoelectric focusing gels with clear background at nanogram sensitivity using Coomassie Brilliant Blue G-250 and R-250. *Electrophoresis*, **9**, 255–262.

Newton, R.P., Brenton, A.G., Smith, C.J. and Dudley, E. (2004) Plant proteome analysis by mass spectrometry: principles, problems, pitfalls and recent developments. *Phytochemistry*, **65**, 1449–1485.

Oda, Y., Huang, K., Cross, F.R., Cowburn, D. and Chait, B.T. (1999) Accurate quantitation of protein expression and site-specific phosphorylation. *Proc. Natl Acad. Sci. USA*, **96**, 6591–6596.

O'Farrell, P.H. (1975) High resolution two-dimensional electrophoresis of proteins. *J. Biol. Chem.*, **250**, 4007–4021.

Orchard, S., Hermjakob, H., Julian Jr. R.K., Runte, K., Sherman, D., Wojcik, J., Zhu, W. and Apweiler, R. (2004) Common interchange standards for proteomics data: public availability of tools and schema. *Proteomics*, **4**, 490–491.

Østergaard, O., Melchior, S., Roepstorff, P. and Svensson, B. (2002) Initial proteome analysis of mature barley seeds and malt. *Proteomics*, **2**, 733–739.

Patterson, S.D. (2004) How much of the proteome do we see with discovery-based proteomics methods and how much do we need to see? *Curr. Proteom.*, **1**, 3–12.

Patterson, S.D., Thomas, D. and Bradshaw, R.A. (1996) Application of combined mass spectrometry and partial amino acid sequence to the identification of gel-separated proteins. *Electrophoresis*, **17**, 877–891.

Peck, S.C., Nuhse, T.S., Hess, D., Iglesias, A., Meins, F. and Boller, T. (2001) Directed proteomics identifies a plant-specific protein rapidly phosphorylated in response to bacterial and fungal elicitors. *Plant Cell*, **13**, 1467–1475.

Peltier, J.B., Emanuelsson, O., Kalume, D.E., Ytterberg, J., Friso, G., Rudella, A., Liberles, D.A., Soderberg, L., Roepstorff, P., von Heijne, G. and van Wijk, K.J. (2002) Central functions of the lumenal and peripheral thylakoid proteome of *Arabidopsis* determined by experimentation and genome-wide prediction. *Plant Cell*, **14**, 211–236.

Pendle, A.F., Clark, G.P., Boon, R., Lewandowska, D., Lam, Y.W., Andersen, J., Mann, M., Lamond, A.I., Brown, J.W. and Shaw, P.J. (2005) Proteomic analysis of the *Arabidopsis* nucleolus suggests novel nucleolar functions. *Mol. Biol. Cell*, **16**, 260–269.

Perdew, G.H., Schaup, H.W. and Selivonchick, D.P. (1983) The use of a zwitterionic detergent in two-dimensional gel electrophoresis of trout liver microsomes. *Anal. Biochem.*, **135**, 453–455.

Perkins, D.N., Pappin, D.J., Creasy, D.M. and Cottrell, J.S. (1999) Probability-based protein identification by searching sequence databases using mass spectrometry data. *Electrophoresis*, **20**, 3551–3567.

Pierpoint, W.S. (2004) In: Cutler, P. (ed.) *Methods in Molecular Biology*, Vol. 244. Humana Press Inc., Totowa, NJ, pp. 65–74.

Prime, T.A., Sherrier, D.J., Mahon, P., Packman, L.C. and Dupree, P. (2000) A proteomic analysis of organelles from *Arabidopsis thaliana*. *Electrophoresis*, **21**, 3488–3499.

Prince, J.T., Carlson, M.W., Wang, R., Lu, P. and Marcotte, E.M. (2004) The need for a public proteomics repository. *Nat. Biotechnol.*, **22**, 471–472.

Quackenbush, J., Cho, J., Lee, D., Liang, F., Holt, I., Karamycheva, S., Parvizi, B., Pertea, G., Sultana, R. and White, J. (2001) The TIGR gene indices: analysis of gene transcript sequences in highly sampled eukaryotic species. *Nucleic Acid. Res.*, **29**, 159–164.

Rabilloud, T. (1998) Use of thiourea to increase the solubility of membrane proteins in two-dimensional electrophoresis. *Electrophoresis*, **19**, 758–760.

Rabilloud, T. (2002) Two-dimensional gel electrophoresis in proteomics: old, old fashioned, but it still climbs up the mountains. *Proteomics*, **2**, 3–10.

Rabilloud, T., Adessi, C., Giraudel, A. and Lunardi, J. (1997) Improvement of the solubilization of proteins in two-dimensional electrophoresis with immobilized pH gradients. *Electrophoresis*, **18**, 307–316.

Rappsilber, J. and Mann, M. (2002) What does it mean to identify a protein in proteomics? *Trends. Biochem. Sci.*, **27**, 74–78.

Rose, J.K., Bashir, S., Giovannoni, J.J., Jahn, M.M. and Saravanan, R.S. (2004) Tackling the plant proteome: practical approaches, hurdles and experimental tools. *Plant J.*, **39**, 715–733.

Rosengren, A.T., Salmi, J.M., Aittokallio, T., Westerholm, J., Lahesmaa, R., Nyman, T.A. and Nevalainen, O.S. (2003) Comparison of PDQuest and Progenesis software packages in the analysis of two-dimensional electrophoresis gels. *Proteomics*, **3**, 1936–1946.

Ross, P.L., Huang, Y.N., Marchese, J.N., Williamson, B., Parker, K., Hattan, S., Khainovski, N., Pillai, S., Dey, S., Daniels, S., Purkayastha, S., Juhasz, P., Martin, S., Bartlet-Jones, M., He, F., Jacobson, A. and Pappin, D.J. (2004) Multiplexed protein quantitation in *Saccharomyces cerevisiae* using amine-reactive isobaric tagging reagents. *Mol. Cell. Proteom.*, **3**, 1154–1169.

Russell, W.K., Park, Z.Y. and Russell, D.H. (2001) Proteolysis in mixed organic-aqueous solvent systems: applications for peptide mass mapping using mass spectrometry. *Anal. Chem.*, **73**, 2682–2685.

Santoni, V., Doumas, P., Rouquie, D., Mansion, M., Rabilloud, T. and Rossignol, M. (1999a) Large scale characterization of plant plasma membrane proteins. *Biochimie*, **81**, 655–661.

Santoni, V., Rabilloud, T., Doumas, P., Rouquie, D., Mansion, M., Kieffer, S., Garin, J. and Rossignol, M. (1999b) Towards the recovery of hydrophobic proteins on two-dimensional electrophoresis gels. *Electrophoresis*, **20**, 705–711.

Santoni, V., Kieffer, S., Desclaux, D., Masson, F. and Rabilloud, T. (2000) Membrane proteomics: use of additive main effects with multiplicative interaction model to classify plasma membrane proteins according to their solubility and electrophoretic properties. *Electrophoresis*, **21**, 3329–3344.

Schägger, H. and von Jagow, G. (1991) Blue native electrophoresis for isolation of membrane protein complexes in enzymatically active form. *Anal. Biochem.*, **199**, 223–231.

Schubert, M., Petersson, U.A., Haas, B.J., Funk, C., Schroder, W.P. and Kieselbach, T. (2002) Proteome map of the chloroplast lumen of *Arabidopsis thaliana*. *J. Biol. Chem.*, **277**, 8354–8365.

Shevchenko, A., Jensen, O.N., Podtelejnikov, A.V., Sagliocco, F., Wilm, M., Vorm, O., Mortensen, P., Boucherie, H. and Mann, M. (1996a) Linking genome and proteome by mass spectrometry: large-scale identification of yeast proteins from two-dimensional gels. *Proc. Natl Acad. Sci. USA*, **93**, 14440–14445.

Shevchenko, A., Wilm, M., Vorm, O. and Mann, M. (1996b) Mass spectrometric sequencing of proteins silver-stained polyacrylamide gels. *Anal. Chem.*, **68**, 850–858.

Shimaoka, T., Ohnishi, M., Sazuka, T., Mitsuhashi, N., Hara-Nishimura, I., Shimazaki, K., Maeshima, M., Yokota, A., Tomizawa, K. and Mimura, T. (2004) Isolation of intact vacuoles and proteomic analysis of tonoplast from suspension-cultured cells of *Arabidopsis thaliana*. *Plant Cell Physiol.*, **45**, 672–683.

Skopp, R.N. and Lane, L.C. (1989) Fingerprinting of proteins cleaved in solution by cyanogen bromide. *Appl. Theor. Electrophor.*, **1**, 61–64.

Steinberg, T.H., Jones, L.J., Haugland, R.P. and Singer, V.L. (1996) SYPRO orange and SYPRO red protein gel stains: one-step fluorescent staining of denaturing gels for detection of nanogram levels of protein. *Anal. Biochem.*, **239**, 223–237.

Szponarski, W., Sommerer, N., Boyer, J.C., Rossignol, M. and Gibrat, R. (2004) Large-scale characterization of integral proteins from *Arabidopsis* vacuolar membrane by two-dimensional liquid chromatography. *Proteomics*, **4**, 397–406.

Taylor, C.F., Paton, N.W., Garwood, K.L., Kirby, P.D., Stead, D.A., Yin, Z., Deutsch, E.W., Selway, L., Walker, J., Riba-Garcia, I., Mohammed, S., Deery, M.J., Howard, J.A., Dunkley, T., Aebersold, R., Kell, D.B., Lilley, K.S., Roepstorff, P., Yates III, J.R., Brass, A., Brown, A.J., Cash, P., Gaskell, S.J., Hubbard, S.J. and Oliver, S.G. (2003) A systematic approach to modeling, capturing, and disseminating proteomics experimental data. *Nat. Biotechnol.*, **21**, 247–254.

Wang, X., Goshe, M.B., Soderblom, E.J., Phinney, B.S., Kuchar, J.A., Li, J., Asami, T., Yoshida, S., Huber, S.C. and Clouse, S.D. (2005) Identification and functional analysis of *in vivo* phosphorylation sites of the *Arabidopsis* BRASSINOSTEROID-INSENSITIVE1 receptor kinase. *Plant Cell*, **17**, 1685–1703.

Washburn, M.P., Wolters, D. and Yates III, J.R. (2001) Large-scale analysis of the yeast proteome by multidimensional protein identification technology. *Nat. Biotechnol.*, **19**, 242–247.

Whitelegge, J.P., Gundersen, C.B. and Faull, K.F. (1998) Electrospray-ionization mass spectrometry of intact intrinsic membrane proteins. *Protein Sci.*, **7**, 1423–1430.

Wilkins, M.R., Gasteiger, E., Gooley, A.A., Herbert, B.R., Molloy, M.P., Binz, P.A., Ou, K., Sanchez, J.C., Bairoch, A., Williams, K.L. and Hochstrasser, D.F. (1999) High-throughput mass spectrometric discovery of protein post- translational modifications. *J. Mol. Biol.*, **289**, 645–657.

Wolters, D.A., Washburn, M.P. and Yates III, J.R. (2001) An automated multidimensional protein identification technology for shotgun proteomics. *Anal. Chem.*, **73**, 5683–5690.

Wu, C.C. and Yates III, J.R. (2003) The application of mass spectrometry to membrane proteomics. *Nat. Biotechnol.*, **21**, 262–267.

Yan, J.X., Devenish, A.T., Wait, R., Stone, T., Lewis, S. and Fowler, S. (2002) Fluorescence 2-D difference gel electrophoresis and mass spectrometry based proteomic analysis of *Escherichia coli*. *Proteomics*, **2**, 1682–1698.

Yates III, J.R. (1998) Database searching using mass spectrometry data. *Electrophoresis*, **19**, 893–900.

Yates III, J.R., Eng, J.K., McCormack, A.L. and Schieltz, D. (1995) Method to correlate tandem mass spectra of modified peptides to amino acid sequences in the protein database. *Anal. Chem.*, **67**, 1426–1436.

Young, N., Chang, Z. and Wishart, D.S. (2004) GelScape: a web-based server for interactively annotating, manipulating, comparing and archiving 1D and 2D gel images. *Bioinformatics*, **20**, 976–978.

Zhu, K., Zhao, J., Lubman, D.M., Miller, F.R. and Barder, T.J. (2005) Protein pI shifts due to posttranslational modifications in the separation and characterization of proteins. *Anal. Chem.*, **77**, 2745–2755.

2 Proteomic analysis of post-translational modifications by mass spectrometry

Albrecht Gruhler and Ole N. Jensen

2.1 Summary

Post-translational modifications (PTMs) are key regulators of protein structure and function, and their elucidation is a central goal of functional proteomics research in plant biology. Mass spectrometry (MS) based analytical strategies are already providing detailed insights into the molecular functions of PTMs, including phosphorylation, glycosylation and acylation. Emerging quantitative proteomic methods will soon enable detailed spatial and temporal studies of protein modifications, thereby revealing dynamic aspects of cellular regulatory networks in plants. We present an overview of recent analytical strategies for the systematic determination of PTMs by MS.

2.2 Introduction

Proteomics is a rapidly expanding field aimed at systematic studies of protein structure, function, interaction and dynamics. The pace of development is propelled by integration of novel computational techniques, ever more advanced and sensitive analytical methods and optimized genetic and biochemical approaches to study complex biological systems. In early proteomics efforts, the main tools were two-dimensional (2D) electrophoresis for protein separation and image analysis of protein spot intensities for protein quantitation. The 2D gel based methods allow separation and quantitation of complex protein mixtures and visualization of the heterogeneity at the intact protein level (Fey and Larsen, 2001; Görg et al., 2004). The emergence of soft ionization MS techniques (i.e. ESI-MS and MALDI-MS) in the late 1980s (Karas and Hillenkamp, 1988; Fenn et al., 1989) and the completion of model organism genome sequencing projects during the 1990s allowed efficient and robust protein identification by peptide mass fingerprinting and peptide sequencing by 2D gel electrophoresis, MS and sequence database searching (Shevchenko et al., 1996; Jensen et al., 1998). The combination of multidimensional capillary chromatography and tandem MS soon emerged as an increasingly robust approach in proteomics (Washburn et al., 2001). Whereas most initial efforts were in the field of 'expression proteomics', that is, the global analysis and cataloguing of the protein contents of organisms, it became clear that a more functionally focused approach for investigation of proteins would be appropriate.

PTMs are essential for protein function and activity, and their characterization is becoming the centre of attention in many proteomics experiments, including studies of plant biology. PTMs of proteins present a formidable analytical challenge. First of all, more than 200 different types of PTMs have been characterized (Krishna and Wold, 1993) and new ones are regularly reported (see e.g. Delta-Mass list at www.ABRF.org). PTMs come in many sizes and with a wide range of physicochemical properties (Table 2.1). PTMs generate a large diversity of gene products due to

Table 2.1 Examples of PTMs, their features, and examples of methods for their analysis at a proteomic scale*

PTM	Modified amino acid	Mass shift	Proteomics approach (examples)	References (examples)
Phosphorylation	Serine Threonine Tyrosine	80 Da	IMAC	Neville et al. (1997); Posewitz and Tempst (1999)
			Antibodies	Gronborg et al. (2002); Steen et al. (2003)
			TiO_2	Pinkse et al. (2004); Larsen et al. (2005b)
			β-elimination	Jaffe et al. (1998); Oda et al. (2001)
			SCX	Beausoleil et al. (2004); Gruhler et al. (2005a)
Glycosylation	Asparagine	>1000 Da	Lectin chromatography	Gabius et al. (2002)
	Serine	>162 Da	HILIC	Hagglund et al. (2004)
	Threonine	>162 Da	Hydrazide coupling	Zhang et al. (2003)
Proteolysis		Mass reduction	COFRADIC	Van Damme et al. (2005)
Ubiquitylation	Lysine	>8 kDa	Epitope-tagged ubiquitin	Peng et al. (2003)
GPI anchor	Small amino acids	>1200 Da	Phase partition and phospholipase C cleavage	Borner et al. (2003); Elortza et al. (2003)
Methylation	Lysine Arginine	14 Da	SILAC	Ong et al. (2004)
Sulphation	Tyrosine	80 Da	MS in negative mode	Budnik et al. (2001); Onnerfjord et al. (2004)
Nitrosylation	Cysteine	45 Da	*S*-Nitrosyl-specific biotinylation	Jaffrey et al. (2001)
Farnesylation	Cysteine	>200 Da	Tagging-via-substrate method	Kho et al. (2004)

*With proteomics approach we mean generally applicable approaches to study modifications on a system wide/large scale in contrast to the identification of PTMs on isolated proteins by MS, with the exception of sulphation. There are a multitude of relevant references for the different PTMs and we apologize to all the authors we could not cite due to the limited space.

multisite occupancy in proteins and frequently also heterogeneous or mutually exclusive PTM structures at distinct amino acid residues. Since PTMs alter the molecular weight of proteins, the mapping, identification and characterization of individual modifications are often achieved by MS analysis of intact proteins, when feasible, and then by MS analysis of the proteolytically derived peptides (Figure 2.1). In the last few years, significant progress in proteomics analysis of proteins and PTMs in plants has been made (Huber and Hardin, 2004; Laugesen et al., 2004; Peck, 2005). In the following sections, we will describe approaches to the detection and characterization of PTM proteins by MS. The discussion will cover not only plants, but also include examples from other species, where appropriate. Several recent reviews are recommended (Jensen, 2004, 2006; Kirkpatrick et al., 2005).

2.3 Considerations for the experimental design of PTM analysis by proteomics

In the vast majority of MS based proteomic studies, identification of PTMs occurs on the peptide level, requiring the proteolytic cleavage of proteins prior to analysis by MS. Peptide sequence and PTMs are then determined by MS-MS fragmentation and database searching of the product ions. This method, as effective as it is for the analysis of many low-complexity samples, poses serious challenges to the characterization of proteins and PTMs on a system-wide basis. One of the main reasons is the large number of distinct peptides generated from the protein content of any cell. The yeast genome as an example encodes approximately 6100 genes, giving rise to more than 300,000 tryptic peptides in an *in silico* digestion of the corresponding gene products with trypsin. Considering PTMs will increase this number several fold. Typically, in an LC-MS/MS experiment several hundred to a few thousand peptides are sequenced (depending on the sample complexity, the length of the gradient of the LC separation and the sensitivity and speed of the mass spectrometer), making it impossible to detect all but the most abundant peptides of a complex sample in a single experiment. In addition, some PTMs are substoichiometric and transient, meaning that a particular PTM attachment site is not modified on all proteins present in a cell at a given time. Other PTMs are heterogeneous, as is the case for glycosylation, where differential modification can give rise to several glycan structures with different masses occupying the same site of a protein. These characteristics of PTMs often result in low amounts of modified peptides that are available for analysis by MS and pose great demands on the sensitivity of mass spectrometers. Some of the PTMs (e.g. glycosylation, lipid attachment or phosphorylation) significantly change the physical properties of proteins and peptides, which can interfere with either their recovery during sample preparation or the efficiency of their analysis by MS. Therefore, the sensitivity of many proteomic approaches depends on the reduction of sample complexity, often achieved by fractionation or the enrichment of post-translationally modified proteins and peptides. Figure 2.1 outlines some of the general principles that can be used to achieve a larger coverage of PTMs in a proteomic study.

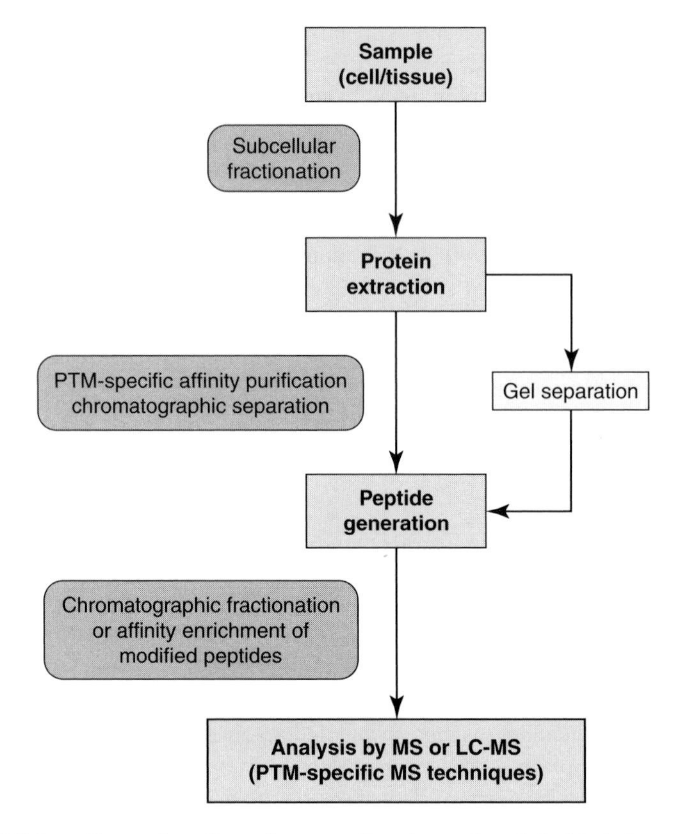

Figure 2.1 Schematic overview of sample preparation steps and analytical techniques commonly employed in proteomic experiments for PTM analysis. The combination of protein and peptide separation techniques with advanced MS provide the means to achieve high specificity, selectivity and sensitivity in proteomics experiments. Basic steps of a proteomics experiment (marked in bold) are the cell lysis, protein extraction, peptide generation and analysis by MS. Chromatographic fractionation and affinity based enrichment techniques are very useful for the analysis of post-translationally modified proteins and peptides. They can be employed during different stages of sample preparation.

The most direct proteomic approach to analyse a cell or tissue sample by MS comprises only a few steps (Figure 2.1). These include homogenization of the tissue, the disruption of the cells, the extraction of proteins (often aided by detergents or denaturing buffer systems) and the generation of peptides by proteolytic cleavage of proteins. For an LC-MS analysis, the peptide mixture is further fractionated by reversed phase chromatography prior to injection into the mass spectrometer. This 'in-solution' approach (and variations thereof) relies on chromatographic steps and does not involve gel electrophoresis. In addition to the reversed phase separation of peptides, additional orthogonal chromatography steps often serve to further segregate the peptide mixture into different fractions that can be analysed separately by LC-MS. It has been widely applied to the large-scale identification of proteins and PTMs (Issaq et al., 2005). In gel based approaches, proteins are separated by gel electrophoresis, often with SDS-PAGE or by 2D electrophoresis, where in principle each protein spot contains a single or only few protein species that can be analysed individually by MS. Protein amounts are commonly determined by densitometry of Coomassie and silver stained gels or by Western blotting. More recently, fluorescent dyes have been developed that allow more accurate quantitation over a larger linear range (Patton, 2002).

Fractionation and enrichment strategies can be employed at all levels of the sample preparation and can often be combined with each other. One way to reduce the number of proteins is the isolation of individual subcellular compartments such as cell organelles, membranes or protein complexes (Brunet et al., 2003; Warnook et al., 2004). However, the most efficient way to systematically investigate PTMs is the affinity enrichment of either modified proteins or peptides. Methods have been developed that are based on different techniques such as affinity chromatography, PTM specific antibodies or PTM-directed chemical modification of proteins or peptides. A number of these applications are described below.

Many PTMs are regulated by the cell and in order to investigate their dynamics, quantitative proteomic studies are desirable. Advances in MS based technologies within the last few years have tremendously enlarged the repertoire of methods that can be used for quantitation of proteins and PTMs. Most of them rely on stable isotope labelling of a control sample, which is mixed and analysed together with the sample of interest (Julka and Regnier, 2005). Thereby relative amounts of the unlabelled and labelled peptide pair – which have similar physical properties – can be determined from the same spectrum. Introduction of the isotopic label can occur at different stages of the proteomic experiment, dependent on the type of labelling method used. Cell cultures can be labelled *in vivo* either by growing them within the presence of an ^{15}N-containing nitrogen source (Oda et al., 1999; Krijgsveld et al., 2003) or by the replacement of certain amino acids with isotope labelled counterparts such as $^{13}C_6$-arginine, $^{13}C_6$-lysine or $^{2}H_3$-leucine (Ong et al., 2002; Zhu et al., 2002). The latter method termed SILAC (Stable Isotope Labelling by Amino acids in Cell culture) has been used in mammalian, yeast and *Arabidopsis thaliana* cells (Ong et al., 2003; Gruhler et al., 2005a, b). Alternatively, stable isotopes can be introduced by chemical modifications of proteins and peptides (Leitner and Lindner, 2004). There exists a broad variety of reagents that have affinity for different reactive groups, for example, thiols, free amines or carboxylic groups. Commercially available reagents include isotope coded affinity tags (ICAT) reagents that covalently bind to cysteines (Gygi et al., 1999), iTRAQ (Ross et al., 2004) and Mass-tag (Peters et al., 2001), which react with amino groups.

Alternatively, MS experiments of different samples can be compared to each other by aligning and matching identical peptides and comparing their relative intensities (Peters et al., 2001). This requires standardized conditions for sample preparation, liquid chromatography and mass spectrometric analysis, in order to achieve good reproducibility between individual samples. The combination of high peptide mass accuracy (accurate mass tags) and highly reproducible chromatographic separation technologies may eventually provide the means for routine, comparative analysis of complex peptide samples (Pasa-Tolic et al., 2004).

2.4 Analysis of PTMs by proteomic approaches

Proteomics has been defined as the analysis of all the proteins expressed by a given genome under given conditions, and thereby lays claim to completeness of the analysis. So far, this has not been (and maybe never will be) achieved, since even the most comprehensive studies of cells or subproteomes only identify a fraction of the total number of proteins present. A recent large-scale analysis of yeast proteins, for example, identified almost 1500 proteins comprising about 22% of the known yeast genome (Washburn et al., 2001). In higher organisms, the percentage of identified proteins drops, due to the larger complexity of their genomes, not even taking into account that many genes give rise to splice variants, which are not yet completely annotated, and that most proteins bear PTMs. This lack in completeness is due partly to the widely varying physical properties of distinct proteins and PTMs and the vast differences in protein abundance within cells and tissues. In addition, current proteomics methods are limited, both with regards to sample preparation and to the sensitivity of the MS instrumentation. In a typical LC-MS experiment of a complex sample, only the most abundant peptides are selected for fragmentation and only part of the product ion spectra yield an identification of a peptide by database searching (in the order 30–50%). Some PTMs (e.g. glycosylation and phosphorylation), change the fragmentation behaviour of peptides, thereby giving rise to spectra that contain only few or weak fragment ions that can be used to identify the amino acid sequence. For these reasons, identification of modified proteins and peptides and characterization of the PTMs are not comprehensive in complex samples, but even with these limitations, the present technologies provide powerful tools to study PTMs, which allow for the detection of hundreds of modified peptides in a single experiment.

In the following paragraphs, we will give examples for proteomics approaches that have been used to analyse classes of PTMs. There will be references to plant studies; however, not all of the methods have been applied to plants on a system-wide basis yet.

2.4.1 Phosphorylation

Protein phosphorylation is one of the most important regulatory PTM modulating enzyme activity, interaction with other proteins, subcellular localization and protein stability. Cell differentiation, propagation and responses to environmental clues are governed by cell signalling pathways that involve protein phosphorylation. For this reason, the study of protein phosphorylation has long been the focus of many investigations and numerous methods have been employed for the detection and modification of phosphoproteins and phosphorylation sites.

A widely used technique for the visualization of phosphoproteins is radioactive labelling of cells with ^{32}P in the form of phosphate salts or [γ-^{32}P]-ATP. Phosphoproteins are then usually separated by gel electrophoresis and detected by autoradiography. Classical methods for the identification of phosphoamino acid and phosphorylation sites are hydrolysis of proteins and identification of radioactive phosphoserine, -threonine or -tyrosine by thin layer chromatography or enzymatic digestion of proteins and Edman sequencing of peptides (Sickmann and Meyer, 2001). Recently, a fluorescent dye called Pro-Q diamond was introduced that specifically stains phosphoproteins in 1- or 2D gels and has the potential to replace the more hazardous ^{32}P labelling (Martin et al., 2003). Alternatively, phosphorylation site specific antibodies can be employed for the detection and characterization of protein phosphorylation, either by use of Western blotting or in immunoprecipitation experiments.

With the advent of sensitive and phosphorylation specific techniques, MS has become the preferred method for the characterization of protein phosphorylation, because in addition to the detection of phosphate groups also sequence information is obtained that in many cases allows pinpointing the phosphorylation site by observation of a mass increment of 80 Da for the modified amino acid residue ($+HPO_3$). A number of different methods can be found in the literature, which have been implemented on different types of mass spectrometers (Mann et al., 2002). They centre on the detection of different phosphoamino acid specific fragments generated during the analysis either by post-source decay or during intentional peptide fragmentation in both positive and negative ion modes. Examples are the neutral loss of H_3PO_4 from phosphoserine and phosphothreonine, the immonium ion of phosphotyrosine or the loss of PO_3^- from all phosphoamino acids detected in negative ion mode at -79 amu.

However, the intrinsic chemical properties of the phosphate groups, namely their low pK_a value and their relatively labile phosphoester bonds, present challenges to their analysis by MS. Hydrophilic phosphopeptides might not bind to the reversed phase material commonly used for sample purification or LC-MS (Larsen et al., 2004). In addition, their ionization is believed to be less efficient than that of unphosphorylated peptides, leading to their suppression in complex sample mixtures. And finally, during peptide sequencing, the loss of phosphoric acid from phospho-serine and -threonine is favoured (Figure 2.2), sometimes interfering with efficient peptide backbone fragmentation, thereby aggravating the identification of the amino acid sequence. Combined with the fact, phosphorylated proteins often are of low abundance, many efforts have been made to specifically enrich phosphopeptides prior to their analysis by MS.

Immobilized metal-affinity chromatography (IMAC), with different metal ions such as Fe^{3+}, Ga^{3+} or ZrO^{2+} bound by nitrilotriacetic acid (NTA) to a solid support, has been used extensively for large-scale analyses of phosphorylation in different organisms (Neville et al., 1997). Ficarro and co-workers applied IMAC to the analysis of protein phosphorylation in yeast by on-line LC-MS. IMAC columns prepared in fused silica capillaries were fused to a reversed phase (RP) precolumn and subsequent analytical column fitted with a micro-emitter tip for ESI-MS (Ficarro et al., 2002). Starting from a whole cell lysate, more than 200 phosphopeptides containing close to 400 phosphorylation sites could be identified in a single mass spectrometric experiment on an ion trap mass spectrometer. IMAC binds preferentially phosphopeptides, but has also affinity towards acidic peptides, which decreases the specificity of purification of phosphopeptides due to concomitant binding of unphosphorylated

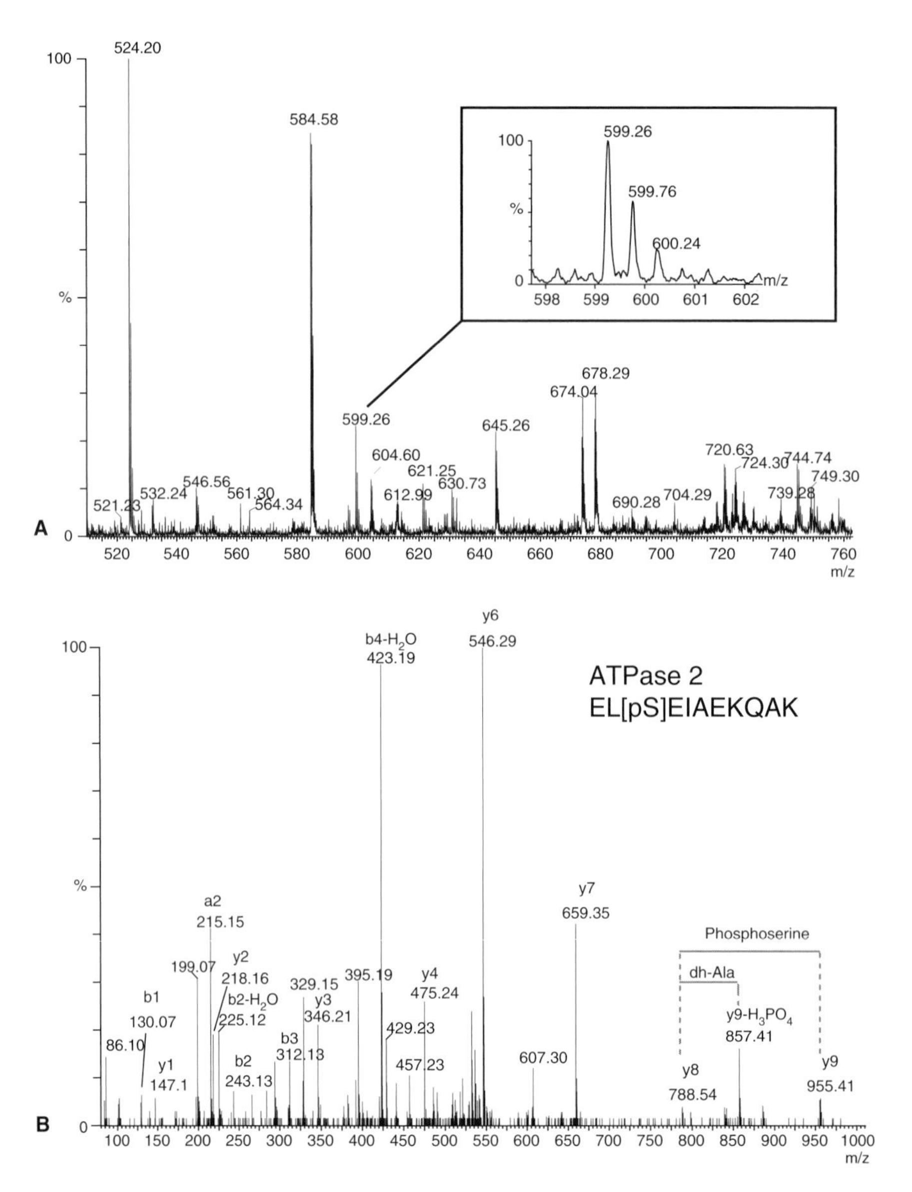

Figure 2.2 Identification by MS and MS/MS of a phosphopeptide from *Arabidopsis thaliana* ATPase2 (SwissProt Accession P19456). Plasma membrane fractions from *Arabidopsis thaliana* were treated with trypsin and the generated peptide mixture was fractionated by strong anion exchange chromatography (SAX). Peptides eluted from the SAX column with 120 mM NaCl were further subjected to IMAC affinity purification of phosphopeptides and analysed by LC-MS/MS on a Q-TOF Ultima instrument (Waters). For experimental details and further information see Nühse et al. (2003, 2004). **A:** Mass spectrum of peptide eluate that contained a phosphopeptide at *m/z* 599.26. Insert shows isotopic distribution of the precursor peptide ion that was automatically selected for MS/MS sequencing. **B:** MS/MS spectrum of peptide reveals the amino acid sequence. Protein database searching using the MASCOT software identified the *m/z* 599.26 peptide as a phosphopeptide originating from ATPase2 and containing a phosphoserine in position 3. Fragment ion signals corresponding to phosphoserine and to dehydro-alanine (Dha) after neutral loss of H_3PO_4, respectively, were observed.

acidic peptides. Therefore, methylation of carboxyl groups was performed prior to enrichment of phosphopeptides on the IMAC column. Methyl estrification of peptides reduces unspecific binding to IMAC beads, however additional steps of sample handling, including desalting and lyophilization are necessary, which require relatively large sample amounts and might entail sample loss, partial methylation of peptides and side reaction such as deamidation of glutamines and aspartames. Methyl esterification may also be the reason for the large number of doubly or multiply phosphorylated peptides observed in this study, since in the absence of free carboxylic groups peptides with multiple phosphates are expected to bind stronger to the Fe–NTA complexes than singly phosphorylated peptides.

Fe(III)-IMAC and LC-MS/MS was employed to study protein phosphorylation of *Arabidopsis* membrane proteins using a shave-and-conquer approach (Nuhse et al., 2003). Plasma membrane fractions were prepared from *Arabidopsis* suspension cells by phase partitioning, washed with Na_2CO_3 in the presence of the detergent Brij-58 that facilitates the formation of inside-out vesicles, where the cytoplasmic moieties of membrane proteins are accessible on the outside of the vesicles. These vesicles were treated with trypsin and released peptides were fractionated by strong anion exchange (SAX) chromatography. Phosphopeptides in these fractions were further enriched by IMAC without prior methylation leading to up to 75% phosphopeptides. Peptide sequencing by LC-MS/MS identified almost 300 phosphopeptides in a MASCOT search. The combination of SAX and IMAC increased sensitivity of the analysis and thereby the number of detected phosphopeptides. Bioinformatic analysis of this large-scale dataset revealed a series of novel features of plant plasma membrane phosphoproteins, including receptor-like kinases (Nuhse et al., 2004).

A similar phosphoproteomic approach has been used to quantitatively investigate the activation of an MAP kinase pathway in yeast in response to alpha-factor, the mating hormone of yeast (Gruhler et al., 2005a). Auxotrophic yeast cells were labelled by growing them in the presence of [^{13}C]-arginine and [^{13}C]-lysine. Equal amounts of isotope encoded cells and pheromone treated cells were then mixed and analysed together. Phosphopeptides were enriched from whole cell lysates using a combination of strong cation exchange chromatography (SCX) and IMAC. It has been shown previously that at pH 2.7 most tryptic phosphopeptides have a net charge of +1, due to the low pK_a value of the phosphate group, whereas most unmodified tryptic peptides have higher net charges. Therefore, the first fractions of an SCX gradient are enriched for phosphopeptides (Beausoleil et al., 2004). Peptides from these fractions were subjected to IMAC for further enrichment of phosphopeptides prior to analysis by LC-MS on an LTQ-FT instrument using a neutral loss directed MS/MS/MS protocol. This technique facilitates the isolation of abundant (MH–$H_3PO_4)^{n+}$ precursor ions from a product ion scan and triggers their subsequent fragmentation by another round of collision induced dissociation (CID). This increased the number of identified phosphopeptides by 40% as compared to MS/MS fragmentation and led to the identification and quantitation of more than 700 phosphopeptides. Regulated phosphorylation was demonstrated for many known members of the MAP kinase signalling pathway, including the alpha-factor receptor Ste2, the

MAP kinase Fus3, Fus3 substrates and other downstream components involved in key aspects of the pheromone response such as transcriptional activation, cell cycle arrest, morphological changes and feedback regulation. In this study, it has been demonstrated for the first time, that quantitative proteomics is capable to characterize a prototypic signalling pathway.

Recently, TiO_2 has been described as a new solid support for the purification of phosphopeptides, which exhibits less unwanted binding of acidic peptides than IMAC, but still retains high affinity for phosphorylated peptides (Pinkse et al., 2004; Meno et al., 2005). Pinkse and co-workers demonstrated the potential of TiO_2 columns for the affinity purification of phosphopeptides with on-line multidimensional LC-MS by coupling a TiO_2 column to reversed phase pre- and analytical columns. From a tryptic digest of autophosphorylated cGMP protein kinase, phosphopeptides were selectively retained on the TiO_2 column and thereby separated from other peptides, which were analysed by LC-MS. In a subsequent step, the bound phosphopeptides were eluted with NH_4HCO_3, pH 9 onto the precolumn and identified by LC-MS. Larsen and co-workers modified the method for off-line analysis of phosphopeptides with TiO_2 micro-columns. The specificity of phosphopeptide retention was increased by the use of a binding buffer containing 80% acetonitrile, 0.5% Trifluoroacetic acid (TFA) and 300 mg/mL 2,5-dihydroxy benzoic acid (DHB). The low pH value of this buffer decreases the unspecific binding of acidic peptides to TiO_2 as compared with the acetic acid based binding buffer used by Pinkse et al. This improvement in selectivity at lower pH has subsequently also been reported for the enrichment of phosphopeptides with IMAC (Kokubu et al., 2005). Likewise, the addition of DHB improves markedly the specificity of phosphopeptide purification. The authors explain this phenomenon by the existence of two different types of binding sites on the surface of TiO_2 beads, which are preferentially occupied by either phosphate groups or acidic amino acids via their carboxyl groups. DHB is thought to compete with acidic peptides for binding sites leaving the phosphate binding unaffected. Although a direct comparison between IMAC and TiO_2 for affinity purification of phosphopeptides from complex samples is still lacking, TiO_2 is a promising alternative to IMAC.

The majority of phosphorylation occurs on serine and threonine residues, whereas tyrosine phosphorylation only accounts for 0.05%. Nevertheless, tyrosine phosphorylation is essential in many cell signalling pathways in mammalian cells and deregulation often contributes to oncogenesis (Blume-Jensen and Hunter, 2001). In particular, receptor tyrosine kinases (e.g. the epidermal growth factor (EGF) receptor), and MAP kinases are regulated by tyrosine phosphorylation. Tyrosine phosphorylation in plant cells is at present largely uncharacterized, with the exception of MAP kinases that have been shown to participate in various stress responses (Zhang and Klessig, 2001). Since phosphotyrosine-containing peptides comprise a minor fraction of all phosphopeptides, generally only few have been detected in large-scale phosphoproteomics studies that have employed enrichment strategies, which did not discriminate between different types of phosphorylation. In contrast, efficient enrichment of phosphotyrosine-containing proteins can be achieved with the help of antibodies with high affinity for phosphotyrosine residues.

In most cases, phosphotyrosine antibodies have been used to precipitate tyrosine-phosphorylated proteins, which then were further characterized by LC-MS. Two recent elegant studies combined immunoprecipitation of tyrosine-phosphorylated proteins with *in vivo* stable isotope labelling to follow the kinetics and regulation of tyrosine phosphorylation (Blagoev et al., 2004; Kratchmarova et al., 2005). In HeLa cells, the analysis of five different time points of EGF stimulation could follow the up- and downregulation of tyrosine phosphorylation of the EGF receptor and many of its known interacting proteins (Blagoev et al., 2004). Likewise, analysis of EGF and platelet derived growth factor (PDGF) signalling in mesenchymal stem cells identified not only proteins activated by both growth factors, but also differences between the two pathways. Since only EGF, but not PDGF, triggers the formation of bone-forming cells, these results may help to determine the factors necessary for stem cell differentiation (Kratchmarova et al., 2005).

However, purification of phosphotyrosine proteins and analysis by MS often only identifies the protein, but not the site of phosphorylation. Therefore, IMAC has been used to isolate phosphopeptides from proteins immunopurified with phosphotyrosine antibodies (Salek et al., 2003; Brill et al., 2004; Zhang et al., 2005). With this strategy, a large number of phosphotyrosine peptides could be identified by LC-MS in a single analysis. Zhang and co-workers used iTRAQ labelling for the time-resolved analysis of EGF receptor signalling (Zhang et al., 2005). They identified 78 phosphorylation sites from 58 proteins, 52 of which had been previously associated with EGF signalling. Phosphorylation of the EGF receptor on tyrosines 1172 and 1092 was strongly increased 5 min after stimulation of the cells followed by slow dephosphorylation. Cluster analysis with self-organizing maps allowed the identification of proteins with similar phosphorylation profiles.

Recently, phosphotyrosine specific antibodies have also been applied to the affinity purification of phosphotyrosine peptides (Rush et al., 2005). Investigating tyrosine phosphorylation in three different cancer cell lines, the authors prepared protein extracts that were enzymatically digested and subsequently partitioned into three fractions by reversed phase solid phase extraction. Peptides from each fraction were incubated with immobilized phosphotyrosine antibody. Captured phosphopeptides were eluted with 0.1% TFA and subsequently detected by LC-MS on an ion trap mass spectrometer. Employment of a panel of four different proteases, trypsin, chymotrypsin, endoproteinase GluC and elastase, increased phosphopeptide yield by a factor of more than two and increased the confidence of a positive identification, because in many cases several distinct peptides were detected for a specific phosphorylation site. With this approach, close to 700 phosphotyrosine peptides were identified from three cell lines, making it the largest study of tyrosine phosphorylation to date.

2.4.2 Protein glycosylation

Glycosylation of proteins affects protein solubility, stability, localization and turnover. The three major types of protein glycosylation include (i) *N*-linked glycosylation of asparagine (Asn) residues, (ii) *O*-linked glycosylation of Ser/Thr residues and (iii) glycosylphosphatidyl-inositol (GPI) anchored proteins (AP), where the GPI anchor is attached to the processed C-terminus of the protein (ω-site). In plants, protein glycosylation plays a role in metabolism and cell remodelling mediated by, for example, plasma membrane proteins with *N*-linked glycan structures or GPI anchors, and cytosolic proteins with *O*-linked glycans.

Oligosaccharides are very diverse with regards to their structure and composition and their complete characterization by MS is, therefore, not trivial, usually requiring purification of the glycoprotein. Determination of the glycan structure can be achieved by chromatographic separation of the different molecular species, followed by a combination of successive treatment with exoglycosidases and MS or by direct fragmentation MS analysis (Zaia, 2004). In many cases however, the sites of carbohydrate attachment are of interest. Isolation of glycoproteins or glycopeptides and the subsequent removal of the oligosaccharide chain allow the identification of glycosylation sites from complex samples by MS. Different strategies to enrich glycoproteins or glycopeptides have been described in the literature. These include the use of lectins, which bind more or less specifically certain classes of oligosaccharide chains, hydrophilic (or normal phase) chromatography and chemical modification with subsequent attachment of affinity tags. In the following, we will give examples for these methods.

Lectins are defined as carbohydrate binding proteins that do not modify the oligosaccharides, a property which propagated their widespread use in the biochemical study of glycoproteins. Lectins are found in many organisms, including plants and mammals, where they have important roles in cell–cell interactions, innate immunity, angiogenesis and modulation of tumour progression. They exhibit diverse affinities towards different carbohydrate structures (see Figure 30.4 in Menon, 1994) and have, therefore, found widespread use in the analysis of glycoproteins. In several studies, lectin-enriched glycopeptides have been deglycosylated with peptide *N*-glycosidase F (PNGase F) in the presence of $H_2{}^{18}O$ (Xiong and Regnier, 2002; Kaji et al., 2003). During the enzymatic cleavage, the Asn residue to which the glycan is attached is converted to aspartic acid. If the reaction occurs in the presence of $H_2{}^{18}O$, aspartic acid incorporates two ^{18}O atoms shifting the molecular weight of the peptide by 4 Da. This mass shift facilitates the identification of formerly *N*-glycosylated peptides by MS and their discrimination from non-enzymatic deamidation of asparagines that can occur either *in vivo* or during sample preparation. Kaji and co-workers used Concanavalin A to capture glycoproteins from *C. elegans* and could identify more than 400 glycosylation sites originating from 250 glycoproteins (Xiong and Regnier, 2002).

A strategy to capture glycoproteins by covalent binding/capture of glycans to a solid support has been reported (Zhang et al., 2003). Cis-diol-groups in oligosaccharides are first oxidized with peroxide to form free aldehyde groups, and then incubated with immobilized hydrazide. Thereby, glycopeptides are specifically and covalently linked to the hydrazide resin and can be released by enzymatic deglycosylation with PNGase F. Optionally, for quantitative analyses, the α-amino groups of captured glycopeptides can be labelled with stable isotopes, for example, by

D_4-containing succinic anhydrate after blocking the ε-amino groups of lysines. Application of this method to a membrane fraction of a prostate cancer epithelial cell line led to the identification of 104 glycopeptides from more than 60 unique proteins (Zhang et al., 2003).

Most carbohydrate chains are hydrophilic due to the presence of polar groups such as hydroxyl, carboxyl, aminoacyl or sulphate groups. This property can be exploited for their enrichment by hydrophilic interaction liquid chromatography (HILIC). Hägglund and co-workers observed, that glycopeptides specifically bound to HILIC columns under conditions were most unmodified tryptic peptides which were not retained (Hägglund et al., 2004). Applying a mixture of tryptic peptides from several glycoproteins to an HILIC column made it possible to selectively enrich glycopeptides on the column and to subsequently analyse them by MS. This method was then applied to the characterization of *N*-glycosylation sites from human plasma. Plasma glycoproteins were enriched on a Concanvalin A column, separated on a 1D SDS PAGE and cleaved with trypsin by means of in-gel proteolysis. Glycopeptides were bound to an HILIC column and subsequently deglycosylated with a mixture of endo-β-*N*-acetylglucosaminidases (Endo H and Endo D), which cleave the glycosidic bond between the two proximal GlcNac residues in the chitobiose core, leaving one GlcNac residue attached to the glycopeptide. This one residual carbohydrate group serves as a marker for *N*-glycosylation that provides unambiguous evidence for the position of the oligosaccharide attachment site but at the same time allows the identification of the peptide sequence by MS fragmentation. The analysis of the deglycosylated peptides from the HILIC fractions resulted in the identification of 37 *N*-glycosylated peptides. Strikingly, no glycopeptides were found in the flow-through fractions from the HILIC column, demonstrating the high efficiency of this method to isolate glycopeptides.

The previously described approaches aim at the identification of glycosylation sites rather than the characterization of the carbohydrates. The latter task is challenging, because of the heterogeneity of glycosylation, where often many structurally and composition-wise different oligosaccharides occupy one glycosylation site, thereby leading to a spreading of the signal into several glycopeptide peaks during analysis of by MS. In addition to this reduction in signal intensity, oligosaccharides may interfere with peptide sequencing by MS complication, the concomitant determination of glycosylation sites. Therefore, these analyses generally require high amounts of purified protein. Larsen and co-workers recently introduced a strategy that allowed them to characterize glycan structure and glycosylation sites from gel separated proteins in the low picomole range (Larsen et al., 2005a). Glycoproteins were first digested with trypsin, whereupon a small aliquot was used to identify the protein by MALDI-MS. The remaining part was further digested with protease K, an unspecific proteinase that further cleaves the tryptic peptides. However, the bulky glycan group sterically hinders complete cleavage of glycopeptides, rendering a few amino acids around the glycosylation site protected. Since the protein has been identified, these small peptides are sufficient to pinpoint the glycosylation site. The sample is then further purified by successive desalting on micro-columns with reversed phase and graphite, respectively. The reversed phase step removes remaining uncleaved and unmodified tryptic peptides, but does not retain glycopeptides due to their hydrophilicity, which are subsequently collected on the graphite column. Elution from the graphite column allows the analysis of the glycopeptides by MS. Using this strategy, the authors identified 13 different glycans from two *N*-glycosylation sites of ovalbumin and characterized the *N*-glycosylation of two contaminating proteins of commercial ovalbumin.

2.4.3 GPI-AP

GPI-AP constitute an important class of cell surface biomolecules that are attached to the membrane by means of a lipid moiety. The basic structure of a GPI group comprises a phosphoethanolamine, a core tetrasaccharide consisting of three mannose and one glucosamine and the phosphatidylinositol group with the diacylglycerol that provides the membrane anchorage. There exist a variety of modifications to this core structure including alterations of the fatty acid groups and addition of carbohydrates or ethanolamine (Menon, 1994). Modification of proteins with GPI takes place in the lumen of the endoplasmic reticulum (ER), where a GPI transamidase cleaves at the carboxyterminus of proteins and catalyses an amide bond formation between the phosphoethanolamine of the GPI anchor and the new C-terminal amino acid. The attachment site – also called ω-site – and the two following amino acids contain typically small side chains and are located around 10 amino acids upstream of a hydrophobic stretch (Eisenhaber et al., 1998). Using these sequence constraints for prediction of GPI-AP, close to 250 potential GPI-AP have been identified from the genome of *Arabidopsis thaliana* (Borner et al., 2002, 2003).

GPI-AP can be selectively released from the membrane by the action of the GPI specific phospholipase C that cleaves between the phosphoinositol and the diacylglycerol. This enzyme has been successfully employed in proteomic studies for the enrichment of GPI-AP. The lipid moiety of the GPI anchor is firmly inserted into one leaflet of the membrane and confers the characteristics of an integral membrane protein to the GPI-AP, such as resistance to alkaline extraction by Na_2CO_3. Due to their hydrophobicity, GPI-APs and other membrane proteins will accumulate in the detergent phase during a partitioning experiment with the detergent Triton X-114. Treatment of this fraction with phospholipase C will cleave the diacylglycerol and release the GPI-APs from the membranes, so that they will be extracted into the soluble phase in a subsequent partitioning experiment. This combination of Triton X-114 partitioning and phospholipase C treatment has been used by several groups to isolate and identify GPI-APs in proteomic studies (Fivaz et al., 2000; Borner et al., 2003; Elortza et al., 2003). Recently, phospholipase D was also found useful for the proteomic investigation of GPI-AP in plants (Elortza et al., 2006).

Two large-scale studies performed with *Arabidopsis* suspension cells identified a number of novel potential GPI-APs. Dupree and co-workers enriched GPI-AP from total *Arabidopsis* membranes and separated proteins by 2D gel electrophoresis using fluorescent DIGE staining (Borner et al., 2003). Soluble fractions from a mock treated control sample and GPI-AP enriched, phospholipase C treated fractions were labelled with two different Cy fluorophores and analysed on the same gel, allowing to distinguish between proteins specifically released by phospholipase C and background binding. In addition, GPI-APs were analysed by 1D SDS PAGE to identify proteins not detectable by 2D gel electrophoresis. This combined analysis led to the identification of 30 potential GPI-APs belonging to several protein families with predicted GPI anchors. Unexpectedly, two of these proteins contained hydrophilic or charged residues in their hydrophobic termini. This information was used to modify the search algorithm, which led to the prediction of nine novel candidate GPI-APs from *Arabidopsis*.

Elortza and co-workers purified GPI-AP from *Arabidopsis* microsomes also by the combination of Triton X-114 partitioning and phospholipase C treatment. The fraction enriched for GPI-AP was compared to a mock sample by means of 1D SDS PAGE. Phospholipase C specific proteins were excised from the gel and identified by automated LC-MS analysis. In this study 44 putative GPI-APs were detected. Bioinformatic analysis of these proteins by three different search algorithms categorized all of these proteins as putative GPI-APs by at least one of the programs.

The proteomic approach applied in these two cases proved very efficient for isolating and identifying a large number of GPI-APs. However, proof of modification of the detected GPI-proteins was acquired indirectly by phospholipase C sensitivity and prediction of a GPI attachment sequence motif. For none of these proteins was the cleavage site or the structure of the GPI moiety determined by MS, although this is feasible when microgram amounts of GPI-protein are available (Omaetxebarria et al., 2006).

Another study focused on short arabinogalactan (AG) peptides from *Arabidopsis*, a subgroup of the AG family of proteins, which are plasma membrane anchored or part of the extracellular matrix in plants and have been implicated functions in plant growth, embryogenesis and apoptosis. Arabinogalactan proteins (AGPs) show the hallmarks of GPI-AP, namely a signal peptide and an ω cleavage site followed by a hydrophobic C-terminus. The mature part of the proteins is rich in hydroxyproline, alanine, serine and threonine, and contains large AG oligosaccharide chains attached to hydroxyproline and additional shorter glycans attached either to hydroxyproline or serine. Schultz and co-workers precipitated AGP from solubilized *Arabidopsis* membranes by covalent linkage to β-glucosyl Yariv, an AG specific reagent (Schultz et al., 2004). Precipitated AGPs were chemically deglycosylated with anhydrous hydrogen fluoride, which cleaves between the phosphoethanolamine and the carbohydrate moiety of the GPI anchor. AG-peptides were separated by reversed phase high performance liquid chromatography (HPLC) from the larger AGPs and analysed by Edman sequencing and MS. The sequence of 8 out of the 12 predicted AG-peptides could be characterized including N- and C-terminal processing sites.

2.4.4 *Farnesylation*

Protein farnesylation involves the attachment of a C15 prenyl group to the side chain of a cysteine via formation of a thioether. The consensus sequence for protein farnesylation is a C-terminal CAAX box where A is an aliphatic and X any amino acid.

The three last amino acids are subsequently removed and the cysteine becomes methylated. Farnesylated proteins are sequestered to membranes, a process which in some cases is associated with their activation, a prominent example being Ras. Recently, a proteomics technique designated Tagging-via-substrate (TAS) has been applied to the isolation of farnesylated proteins (Kho et al., 2004). African green monkey derived kidney cells (COS-1) were grown in the presence of either azido farnesyl diphosphate or azido farnesyl alcohol, which were taken up by the cells and attached to proteins. The azido group is small and relatively inert under physiological conditions and is, therefore, not expected to interfere with protein farnesylation. This was demonstrated by the relocalization of azido farnesylated Ras from the cytosol to the plasma membrane followed by activation of the MAP kinase pathway detected by the phosphorylation of MEK. Addition of lovastatin, an inhibitor of the farnesyl biosynthesis, increased the efficiency of azido farnesylation. The azido group undergoes a specific conjugation with phosphines via the Staudinger reaction. Kho and co-workers used a biotinylated phosphine reagent to capture azido farnesylated proteins, which subsequently were purified on streptavidin beads and analysed by LC-MS. They could identify 17 CAAX box containing proteins.

2.4.5 *N-terminally modified proteins*

Recently, a method termed combined fractional diagonal chromatography (COFRADIC) has been introduced, which facilitates the selective enrichment of amino terminal peptides or peptides containing specific amino acids (Gevaert et al., 2002, 2003). In a variation of this method, N-terminally modified peptides can be isolated from a complex sample and analysed separately from the bulk of tryptic peptides. COFRADIC is based on two subsequent chromatography analyses under identical conditions, where peptides are modified between runs so that the retention time of selected (modified) peptides changes and they are shifted away from the remainder of (unmodified) peptides. This leads to a reduction of sample complexity and, therefore, increases the probability to identify peptides from low abundance proteins. As a drawback, sequence coverage is also reduced; in the case of isolation of N-terminal peptides to one peptide per protein.

For the isolation of amino terminal peptides, cysteines of intact proteins are alkylated and free amino groups (both on the amino terminus and on lysines) are acetylated prior to digestion with trypsin. The peptide mixture is then fractionated by reversed phase chromatography in a primary run. Free amino groups generated on the N-termini of tryptic peptides are then blocked by 2,4,6-trinitrobenzenesulfonic acid (TNBS), which strongly increases their hydrophobicity. Each modified fraction of the primary run is then analysed again under identical conditions and fractions in the same retention window, which now contain the alkylated N-terminal peptides that have not been modified by TNBS, are collected for analysis by LC-MS. Applying this procedure to cytosolic and membrane skeleton fraction of human thromobocytes, 264 proteins could be identified via their N-terminal peptides (Gevaert et al., 2003). Peptides with amino terminal proline or pyroglutamine residues did not react very well with TNBS and were, therefore, also recovered with the amino terminal peptides. Likewise, an estimated 2% of internal peptides did not react completely with TNBS and was, therefore, also detected in the MS analysis.

Omission of the protein acetylation step can be used to identify *in vivo* blocked amino terminal tryptic peptides that contain arginine at their C-terminus (lysine-containing peptides will react with TNBS and be shifted away during the secondary chromatographic fractionation). TNBS treatment will modify all peptides except N-terminally blocked arginyl peptides that can thereby be isolated and analysed by MS. Gevaert and co-workers identified 78 *in vivo* acetylated proteins from human thrombocytes (Gevaert et al., 2003).

2.5 Conclusions and perspectives

PTMs are being discovered at a rapid pace by use of advanced and sensitive analytical technology, including MS. The introduction of increasingly accurate and robust methods for quantitative analysis of proteins by MS will facilitate detailed investigations of plant cells and tissues. Signal propagation from receptors at the plasma membrane into the cell nucleus can be revealed in an unbiased fashion by quantitative measurements of protein phosphorylation in model organisms. Plasma membrane proteins and cell wall molecules important for plant cell growth and pathogen resistance can be studied by modern proteomics technologies. We expect that in the next few years, large-scale studies will provide accurate and temporally resolved quantitative information of protein modifications in *Arabidopsis thaliana*, rice and other plant species. This will eventually facilitate simulation and modelling of complex cellular signalling networks and molecular interactions in various plant species and also aid in developing plant based systems for biotechnology applications.

Acknowledgements

We thank past and present members of the Protein Research Group at the University of Southern Denmark and our collaborators for their contributions to proteomics research described in this chapter. The Protein Research Group is supported by grants from the Danish Research Councils, the Carlsberg Foundation and from the European Union.

References

Beausoleil, S.A., Jedrychowski, M., Schwartz, D., Elias, J.E., Villen, J., Li, J., Cohn, M.A., Cantley, L.C. and Gygi, S.P. (2004) Large-scale characterization of HeLa cell nuclear phosphoproteins. *Proc. Natl Acad. Sci. USA*, **101**, 12130–12135.

Blagoev, B., Ong, S.E., Kratchmarova, I. and Mann, M. (2004) Temporal analysis of phosphotyrosine-dependent signaling networks by quantitative proteomics. *Nat. Biotechnol.*, **22**, 1139–1145.

Blume-Jensen, P. and Hunter, T. (2001) Oncogenic kinase signalling. *Nature*, **411**, 355.

Borner, G.H.H., Sherrier, D.J., Stevens, T.J., Arkin, I.T. and Dupree, P. (2002) Prediction of glycosylphosphatidylinositol-anchored proteins in *Arabidopsis*. A genomic analysis. *Plant Physiol.*, **129**, 486–499.

Borner, G.H., Lilley, K.S., Stevens, T.J. and Dupree, P. (2003) Identification of glycosylphosphatidylinositol-anchored proteins in *Arabidopsis*. A proteomic and genomic analysis. *Plant Physiol.*, **132**, 568–577.

Brill, L.M., Salomon, A.R., Ficarro, S.B., Mukherji, M., Stettler-Gill, M. and Peters, E.C. (2004) Robust phosphoproteomic profiling of tyrosine phosphorylation sites from human T cells using immobilized metal affinity chromatography and tandem mass spectrometry. *Anal. Chem.*, **76**, 2763–2772.

Brunet, S., Thibault, P., Gagnon, E., Kearney, P., Bergeron, J.J.M. and Desjardins, M. (2003) Organelle proteomics: looking at less to see more. *Trend. Cell Biol.*, **13**, 629.

Budnik, B.A., Haselmann, K.F. and Zubarev, R.A. (2001) Electron detachment dissociation of peptide di-anions: an electron–hole recombination phenomenon. *Chem. Phys. Lett.*, **342**, 299.

Eisenhaber, B., Bork, P. and Eisenhaber, F. (1998) Sequence properties of GPI-anchored proteins near the omega-site: constraints for the polypeptide binding site of the putative transamidase. *Protein Eng.*, **11**, 1155–1161.

Elortza, F., Nuhse, T.S., Foster, L.J., Stensballe, A., Peck, S.C. and Jensen, O.N. (2003) Proteomic analysis of glycosylphosphatidylinositol-anchored membrane proteins. *Mol. Cell. Proteom.*, **2**, 1261–1270.

Elortza, F., Mohammed, S., Bunkenborg, J., Foster, L.J., Nühse, T.S., Brodbeck, U., Peck, S.C. and Jensen, O.N. (2006) Modification-specific proteomics of plasma membrane proteins: identification and characterization of glycosylphosphatidylinositol-anchored proteins released upon phospholipase D treatment. *J. Proteome Res*, **5**, 935–943.

Fenn, J.B., Mann, M., Meng, C.K., Wong, S.F. and Whitehouse, C.M. (1989) Electrospray ionization for mass spectrometry of large biomolecules. *Science*, **246**, 64–71.

Fey, S.J. and Larsen, P.M. (2001) 2D or not 2D. Two-dimensional gel electrophoresis. *Curr. Opin. Chem. Biol.*, **5**, 26–33.

Ficarro, S.B., McCleland, M.L., Stukenberg, P.T., Burke, D.J., Ross, M.M., Shabanowitz, J., Hunt, D.F. and White, F.M. (2002) Phosphoproteome analysis by mass spectrometry and its application to *Saccharomyces cerevisiae*. *Nat. Biotechnol.*, **20**, 301–305.

Fivaz, M., Vilbois, F., Pasquali, C. and van der Goot, F.G. (2000) Analysis of glycosyl phosphatidylinositol-anchored proteins by two-dimensional gel electrophoresis. *Electrophoresis*, **21**, 3351–3356.

Gabius, H.J., Andre, S., Kaltner, H. and Siebert, H.C. (2002) The sugar code: functional lectinomics. *Biochim. Biophys. Acta*, **1572**, 165–177.

Gevaert, K., Van Damme, J., Goethals, M., Thomas, G.R., Hoorelbeke, B., Demol, H., Martens, L., Puype, M., Staes, A. and Vandekerckhove, J. (2002) Chromatographic isolation of methionine-containing peptides for gel-free proteome analysis: identification of more than 800 *Escherichia coli* proteins. *Mol. Cell. Proteom.*, **1**, 896–903.

Gevaert, K., Goethals, M., Martens, L., Van Damme, J., Staes, A., Thomas, G.R. and Vandekerckhove, J. (2003) Exploring proteomes and analyzing protein processing by mass spectrometric identification of sorted N-terminal peptides. *Nat. Biotechnol.*, **21**, 566–569.

Görg, A., Weiss, W. and Dunn, M.J. (2004) Current two-dimensional electrophoresis technology for proteomics. *Proteomics*, **4**, 3665–3685.

Gronborg, M., Kristiansen, T.Z., Stensballe, A., Andersen, J.S., Ohara, O., Mann, M., Jensen, O.N. and Pandey, A. (2002) A mass spectrometry-based proteomic approach for identification of serine/threonine-phosphorylated proteins by enrichment with phospho-specific antibodies: identification of a novel protein, Frigg, as a protein kinase A substrate. *Mol. Cell. Proteom.*, **1**, 517–527.

Gruhler, A., Olsen, J.V., Shabaz, M., Mortensen, P., Faergman, N., Mann, M. and Jensen, O.N. (2005a) Quantitative phosphoproteomics applied to the yeast pheromone signaling pathway. *Mol. Cell. Biol.*, **4**, 310–327.

Gruhler, A., Schulze, W.X., Matthiesen, R., Mann, M. and Jensen, O.N. (2005b) Stable isotope labeling of *Arabidopsis thaliana* cells and quantitative proteomics by mass spectrometry. *Mol. Cell. Proteom.*, **4**, 1697–1709.

Gygi, S.P., Rist, B., Gerber, S.A., Turecek, F., Gelb, M.H. and Aebersold, R. (1999) Quantitative analysis of complex protein mixtures using isotope-coded affinity tags. *Nat. Biotechnol.*, **17**, 994–999.

Hägglund, P., Bunkenborg, J., Elortza, F., Jensen, O.N. and Roepstorff, P. (2004) A new strategy for identification of *N*-glycosylated proteins and unambiguous assignment of their glycosylation sites using HILIC enrichment and partial deglycosylation. *J. Proteome Res.*, **3**, 556–566.

Huber, S.C. and Hardin, S.C. (2004) Numerous posttranslational modifications provide opportunities for the intricate regulation of metabolic enzymes at multiple levels. *Curr. Opin. Plant Biol.*, **7**, 318–322.

Issaq, H.J., Chan, K.C., Janini, G.M., Conrads, T.P. and Veenstra, T.D. (2005) Multidimensional separation of peptides for effective proteomic analysis. *J. Chromatogr. B*, **817**, 35.

Jaffe, H., Veeranna and Pant, H.C. (1998) Characterization of serine and threonine phosphorylation sites in beta-elimination/ethanethiol addition-modified proteins by electrospray tandem mass spectrometry and database searching. *Biochemistry*, **37**, 16211–16224.

Jaffrey, S.R., Erdjument-Bromage, H., Ferris, C.D., Tempst, P. and Snyder, S.H. (2001) Protein *S*-nitrosylation: a physiological signal for neuronal nitric oxide. *Nat. Cell Biol.*, **3**, 193–197.

Jensen, O.N. (2004) Modification-specific proteomics: characterization of post-translational modifications by mass spectrometry. *Curr. Opin. Chem. Biol.*, **8**, 33–41.

Jensen, O.N. (2006) Interpreting the protein language using proteomics. *Nat. Rev. Mol. Cell. Biol.* (in press).

Jensen, O.N., Larsen, M.R. and Roepstorff, P. (1998) Mass spectrometric identification and microcharacterization of proteins from electrophoretic gels: strategies and applications. *Proteins*, Suppl 2, 74–89.

Julka, S. and Regnier, F.E. (2005) Recent advancements in differential proteomics based on stable isotope coding. *Brief Funct. Genom. Proteom.*, **4**, 158–177.

Kaji, H., Saito, H., Yamauchi, Y., Shinkawa, T., Taoka, M., Hirabayashi, J., Kasai, K., Takahashi, N. and Isobe, T. (2003) Lectin affinity capture, isotope-coded tagging and mass spectrometry to identify *N*-linked glycoproteins. *Nat. Biotechnol.*, **21**, 667–672.

Karas, M. and Hillenkamp, F. (1988) Laser desorption ionization of proteins with molecular masses exceeding 10,000 daltons. *Anal. Chem.*, **60**, 2299–2301.

Kho, Y., Kim, S.C., Jiang, C., Barma, D., Kwon, S.W., Cheng, J., Jaunbergs, J., Weinbaum, C., Tamanoi, F., Falck, J. and Zhao, Y. (2004) A tagging-via-substrate technology for detection and proteomics of farnesylated proteins. *Proc. Natl Acad. Sci.*, **101**, 12479–12484.

Kirkpatrick, D.S., Denison, C. and Gygi, S.P. (2005) Weighing in on ubiquitin: the expanding role of mass-spectrometry-based proteomics. *Nat. Cell Biol.*, **7**, 750–757.

Kokubu, M., Ishihama, Y., Sato, T., Nagasu, T. and Oda, Y. (2005) Specificity of immobilized metal affinity-based IMAC/C18 tip enrichment of phosphopeptides for protein phosphorylation analysis. *Anal. Chem.*, **77**, 5144–5154.

Kratchmarova, I., Blagoev, B., Haack-Sorensen, M., Kassem, M. and Mann, M. (2005) Mechanism of divergent growth factor effects in mesenchymal stem cell differentiation. *Science*, **308**, 1472–1477.

Krijgsveld, J., Ketting, R.F., Mahmoudi, T., Johansen, J., Artal-Sanz, M., Verrijzer, C.P., Plasterk, R.H. and Heck, A.J. (2003) Metabolic labeling of *C. elegans* and *D. melanogaster* for quantitative proteomics. *Nat. Biotechnol.*, **21**, 927–931.

Krishna, R.G. and Wold, F. (1993) Post-translational modification of proteins. *Adv. Enzymol. Relat. Area. Mol. Biol.*, **67**, 265–298.

Larsen, M.R., Graham, M.E., Robinson, P.J. and Roepstorff, P. (2004) Improved detection of hydrophilic phosphopeptides using graphite powder microcolumns and mass spectrometry: evidence for *in vivo* doubly phosphorylated dynamin I and dynamin III. *Mol. Cell. Proteom.*, **3**, 456–465.

Larsen, M.R., Hojrup, P. and Roepstorff, P. (2005a) Characterization of gel-separated glycoproteins using two-step proteolytic digestion combined with sequential microcolumns and mass spectrometry. *Mol. Cell. Proteom.*, **4**, 107–119.

Larsen, M.R., Thingholm, T.E., Jensen, O.N., Roepstorff, P. and Jorgensen, T.J. (2005b) Highly selective enrichment of phosphorylated peptides from peptide mixtures using titanium dioxide microcolumns. *Mol. Cell. Proteom.*, **4**, 873–886.

Laugesen, S., Bergoin, A. and Rossignol, M. (2004) Deciphering the plant phosphoproteome: tools and strategies for a challenging task. *Plant Physiol. Biochem.*, **42**, 929–936.

Leitner, A. and Lindner, W. (2004) Current chemical tagging strategies for proteome analysis by mass spectrometry. *J. Chromatogr. B*, **813**, 1.

Mann, M., Ong, S.-E., Gronborg, M., Steen, H., Jensen, O.N. and Pandey, A. (2002) Analysis of protein phosphorylation using mass spectrometry: deciphering the phosphoproteome. *Trend. Biotechnol.*, **20**, 261.

Martin, K., Steinberg, T.H., Cooley, L.A., Gee, K.R., Beechem, J.M. and Patton, W.F. (2003) Quantitative analysis of protein phosphorylation status and protein kinase activity on microarrays using a novel fluorescent phosphorylation sensor dye. *Proteomics*, **3**, 1244–1255.

Meno, K., Thorsted, P.B., Ipsen, H., Kristensen, O., Larsen, J.N., Spangfort, M.D., Gajhede, M. and Lund, K. (2005) The crystal structure of recombinant proDer p 1, a major house dust mite proteolytic allergen. *J. Immunol.*, **175**, 3835–3845.

Menon, A.K. (1994) Structural analysis of glycosylphosphatidylinositol anchors. *Method. Enzymol.*, **230**, 418–442.

Neville, D.C., Rozanas, C.R., Price, E.M., Gruis, D.B., Verkman, A.S. and Townsend, R.R. (1997) Evidence for phosphorylation of serine 753 in CFTR using a novel metal-ion affinity resin and matrix-assisted laser desorption mass spectrometry. *Protein Sci.*, **6**, 2436–2445.

Nuhse, T.S., Stensballe, A., Jensen, O.N. and Peck, S.C. (2003) Large-scale analysis of *in vivo* phosphorylated membrane proteins by immobilized metal ion affinity chromatography and mass spectrometry. *Mol. Cell. Proteom.*, **2**, 1234–1243.

Nuhse, T.S., Stensballe, A., Jensen, O.N. and Peck, S.C. (2004) Phosphoproteomics of the *Arabidopsis* plasma membrane and a new phosphorylation site database. *Plant Cell*, **16**, 2394–2405.

Oda, Y., Huang, K., Cross, F.R., Cowburn, D. and Chait, B.T. (1999) Accurate quantitation of protein expression and site-specific phosphorylation. *Proc. Natl Acad. Sci. USA*, **96**, 6591–6596.

Oda, Y., Nagasu, T. and Chait, B.T. (2001) Enrichment analysis of phosphorylated proteins as a tool for probing the phosphoproteome. *Nat. Biotechnol.*, **19**, 379–382.

Omaetxebarria, M., Hägglund, P., Elortza, F., Hooper, N.M., Arizmendi, J.M., and Jensen O.N. (2006) Isolation and characterization of GPI-anchored peptides by hydrophilic interaction chromatography (HILIC) and MALDI tandem mass spectrometry. *Anal. Chem.*, **78**, 3335–3341.

Ong, S.-E., Blagoev, B., Kratchmarova, I., Kristensen, D.B., Steen, H., Pandey, A. and Mann, M. (2002) Stable istotope labeling by amino acids in cell culture, SILAC, as a simple and accurate approach to expression proteomics. *Mol. Cell. Proteom.*, **1**, 376–386.

Ong, S.-E., Kratchmarova, I. and Mann, M. (2003) Properties of ^{13}C-substituted arginine in stable isotope labeling by amino acids in cell culture (SILAC). *J. Proteome Res.*, **2**, 173–181.

Ong, S.E., Mittler, G. and Mann, M. (2004) Identifying and quantifying *in vivo* methylation sites by heavy methyl SILAC. *Nat. Method.*, **1**, 119–126.

Onnerfjord, P., Heathfield, T.F. and Heinegard, D. (2004) Identification of tyrosine sulfation in extracellular leucine-rich repeat proteins using mass spectrometry. *J. Biol. Chem.*, **279**, 26–33.

Pasa-Tolic, L., Masselon, C., Barry, R.C., Shen, Y. and Smith, R.D. (2004) Proteomic analyses using an accurate mass and time tag strategy. *Biotechniques*, **37**, 621–624, 626–633, 636 passim.

Patton, W.F. (2002) Detection technologies in proteome analysis. *J. Chromatogr. B: Anal. Technol. Biomed. Life Sci.*, **771**, 3.

Peck, S.C. (2005) Update on proteomics in *Arabidopsis*. Where do we go from here? *Plant Physiol.*, **138**, 591–599.

Peng, J., Schwartz, D., Elias, J.E., Thoreen, C.C., Cheng, D., Marsischky, G., Roelofs, J., Finley, D. and Gygi, S.P. (2003) A proteomics approach to understanding protein ubiquitination. *Nat. Biotechnol.*, **21**, 921–926.

Peters, E.C., Horn, D.M., Tully, D.C. and Brock, A. (2001) A novel multifunctional labeling reagent for enhanced protein characterization with mass spectrometry. *Rapid Commun. Mass Spectrom.*, **15**, 2387–2392.

Pinkse, M.W., Uitto, P.M., Hilhorst, M.J., Ooms, B. and Heck, A.J. (2004) Selective isolation at the femtomole level of phosphopeptides from proteolytic digests using 2D-NanoLC-ESI-MS/MS and titanium oxide precolumns. *Anal. Chem.*, **76**, 3935–3943.

Posewitz, M.C. and Tempst, P. (1999) Immobilized gallium(III) affinity chromatography of phosphopeptides. *Anal. Chem.*, **71**, 2883–2892.

Ross, P.L., Huang, Y.N., Marchese, J.N., Williamson, B., Parker, K., Hattan, S., Khainovski, N., Pillai, S., Dey, S., Daniels, S., Purkayastha, S., Juhasz, P., Martin, S., Bartlet-Jones, M., He, F., Jacobson, A. and Pappin, D.J. (2004) Multiplexed protein quantitation in Saccharomyces cerevisiae using amine-reactive isobaric tagging reagents. *Mol. Cell. Proteom.*, **3**, 1154–1169.

Rush, J., Moritz, A., Lee, K.A., Guo, A., Goss, V.L., Spek, E.J., Zhang, H., Zha, X.M., Polakiewicz, R.D. and Comb, M.J. (2005) Immunoaffinity profiling of tyrosine phosphorylation in cancer cells. *Nat. Biotechnol.*, **23**, 94–101.

Salek, M., Alonso, A., Pipkorn, R. and Lehmann, W.D. (2003) Analysis of protein tyrosine phosphorylation by nanoelectrospray ionization high-resolution tandem mass spectrometry and tyrosine-targeted product ion scanning. *Anal. Chem.*, **75**, 2724–2729.

Schultz, C.J., Ferguson, K.L., Lahnstein, J. and Bacic, A. (2004) Post-translational modifications of arabinogalactan-peptides of *Arabidopsis thaliana*. Endoplasmic reticulum and glycosylphosphatidylinositol-anchor signal cleavage sites and hydroxylation of proline. *J. Biol. Chem.*, **279**, 45503–45511.

Shevchenko, A., Jensen, O.N., Podtelejnikov, A.V., Sagliocco, F., Wilm, M., Vorm, O., Mortensen, P., Shevchenko, A., Boucherie, H. and Mann, M. (1996) Linking genome and proteome by mass spectrometry: large-scale identification of yeast proteins from two-dimensional gels. *Proc. Natl Acad. Sci. USA*, **93**, 14440–14445.

Sickmann, A. and Meyer, H.E. (2001) Phosphoamino acid analysis. *Proteomics*, **1**, 200–206.

Steen, H., Fernandez, M., Ghaffari, S., Pandey, A. and Mann, M. (2003) Phosphotyrosine mapping in Bcr/Abl oncoprotein using phosphotyrosine-specific immonium ion scanning. *Mol. Cell. Proteom.*, **2**, 138–145.

Van Damme, P., Martens, L., Van Damme, J., Hugelier, K., Staes, A., Vandekerckhove, J. and Gevaert, K. (2005) Caspase-specific and nonspecific *in vivo* protein processing during Fas-induced apoptosis. *Nat. Method.*, **2**, 771–777.

Warnock, D.E., Fahy, E. and Taylor, S.W. (2004) Identification of protein associations in organelles, using mass spectrometry-based proteomics. *Mass Spectrom. Rev.*, **23**, 259–280.

Washburn, M.P., Wolters, D. and Yates III, J.R. (2001) Large-scale analysis of the yeast proteome by multidimensional protein identification technology. *Nat. Biotechnol.*, **19**, 242–247.

Xiong, L. and Regnier, F.E. (2002) Use of a lectin affinity selector in the search for unusual glycosylation in proteomics. *J. Chromatogr. B*, **782**, 405.

Zaia, J. (2004) Mass spectrometry of oligosaccharides. *Mass Spectrom. Rev.*, **23**, 161–227.

Zhang, H., Li, X.J., Martin, D.B. and Aebersold, R. (2003) Identification and quantification of *N*-linked glycoproteins using hydrazide chemistry, stable isotope labeling and mass spectrometry. *Nat. Biotechnol.*, **21**, 660–666.

Zhang, S. and Klessig, D.F. (2001) MAPK cascades in plant defense signaling. *Trends Plant Sci.*, **6**, 520.

Zhang, Y., Wolf-Yadlin, A., Ross, P.L., Pappin, D.J., Rush, J., Lauffenburger, D.A. and White, F.M. (2005) Time-resolved mass spectrometry of tyrosine phosphorylation sites in the epidermal growth factor receptor signaling network reveals dynamic modules. *Mol. Cell. Proteom.*, **4**, 1240–1250.

Zhu, H., Pan, S., Gu, S., Bradbury, E.M. and Chen, X. (2002) Amino acid residue specific stable isotope labeling for quantiative proteomics. *Rapid Commun. Mass Spectrom.*, **16**, 2115–2123.

3 Strategies for the investigation of protein–protein interactions in plants

Hans-Peter Braun and Udo K. Schmitz

3.1 Summary

Stable or dynamic protein–protein interactions are crucial for molecular physiology. Thousands of protein types are present within single cells, and the number of specific interactions between them is estimated to be even much higher. Methodologically, the investigation of protein–protein interactions is difficult, especially in the case of more dynamic interactions. The characterization of stable protein complexes can be achieved by biochemical purifications, which are based on column chromatography, sucrose gradient ultracentrifugation, native gel electrophoresis or immunoprecipitations. Genetic strategies like the 'Yeast two-hybrid', the 'Split-ubiquitin', the 'Bimolecular fluorescence' (BiFC) and the 'Förster resonance energy transfer' (FRET) systems are applied for the analysis of more labile protein–protein interactions or those that take place between proteins of extremely low abundance. In this chapter, we give an overview on available methods and their applications in plant biology.

3.2 Introduction

Genomes of higher plants encode more than 25,000 protein types, many of which are differentially modified by post-translational processes. The function of all these proteins depends on their specific interactions with defined molecules. Protein–protein interactions are of key importance for cellular processes. Many proteins form stable protein complexes, which offer several advantages compared to singular proteins, like substrate channelling, catalytic enhancement, protection of reactive metabolic intermediates, protein stabilization, integration of hydrophilic proteins into membranes and the regulation and coordination of metabolic pathways. Other proteins are part of dynamic protein complexes which assemble and de-assemble in response to physiological states of cells. Finally, many proteins form part of signal transduction pathways which are based on defined short-term interactions of proteins to transfer molecular information.

The characterization of protein–protein interactions is of great importance in proteomics. However, experimentally their investigation often is difficult. As a general rule, stable protein complexes are easier to handle than dynamic, soluble protein complexes easier than membrane-bound and small protein complexes easier than large ones. In the past decades, several methods were developed to define the so-called 'quaternary' and 'quinternary' structure of proteins. Here we review available biochemical, genetic and cytological procedures and summarize studies especially successful in applying these methods to investigate protein–protein interactions in plants.

3.3 Biochemical procedures to characterize protein–protein interactions

3.3.1 Chromatographic purifications

Column chromatography is one of the classical strategies to purify stable protein complexes. Since protein complexes usually are much larger than singular proteins, chromatographic methods based on size resolutions are often carried out, for example gel filtration chromatography. However, resolution by size is hardly sufficient in protein complex purification and most of the time is combined with at least one further type of column chromatography, which can be based on a separation by charge or hydrophobicity. Very successful strategies include affinity chromatography. For example, plastocyanin-, ferredoxin- or cytochrome c-affinity chromatography was employed to purify specifically interacting proteins and protein complexes in plants (Molnar et al., 1987; Braun and Schmitz, 1992; Sakihama et al., 1992). Recently, proteomic approaches based on affinity chromatography were designed to systematically identify proteins interacting with regulatory proteins, for example thioredoxin-affinity chromatography (Balmer et al., 2003). Combined with a pre-resolution based on size, these approaches also allow to systematically define protein complexes interacting with regulatory proteins.

3.3.2 Sucrose gradient ultrafiltration

Ultracentrifugation using sucrose gradients represents another classical strategy to purify and resolve protein complexes of plants, especially the photosystems (van Wijk et al., 1995; Muller and Eichacker, 1999) and the complexes of the oxidative phosphorylation system (Leterme and Boutry, 1993). This procedure is considered to be biochemically gentle and often leads to better resolutions than gel filtration chromatography. Recently, sucrose gradient ultracentrifugation became important for the intact isolation of labile respiratory supercomplexes in plants (Dudkina et al., 2005a, b; Perales et al., 2005).

3.3.3 Native gel electrophoresis

Polyacrylamide gel electrophoresis (PAGE) carried out under native conditions proved to be one of the most powerful systems to purify and characterize protein complexes. Usually, two-dimensional (2D) gel systems are used, which are based on (i) a native gel dimension for protein complex separation and (ii) a second gel dimension carried out in the presence of the anionic detergent sodium dodecyl sulphate (SDS) for the separation of subunits of protein complexes. The first such system used

SDS for both gel dimensions, but at a very low concentration in the first and a much higher in the second dimension (Andersson et al., 1982). 2D gels based on this system allowed to resolve the photosystems of plants. Later, SDS of the first gel dimension was replaced by Deriphat 160, a zwitterions surfactant (Peter and Thornber, 1991; references within). However, resolution capacity of these gel systems was still limited. As an alternative, native isoelectric focussing (IEF) can be combined with SDS PAGE to separate protein complexes. This strategy was successfully employed to characterize the clp protease complexes of chloroplasts and mitochondria of higher plants (Peltier et al., 2004). However, resolution capacity of this 2D gel system is also limited with respect to protein complexes.

In 1991, Schägger and von Jagow suggested a completely new 2D gel electrophoresis system, which is based on a pre-treatment of protein samples with the anionic wool dye Coomassie-blue (Schägger, 2003). Coomassie tightly binds to proteins and carefully introduces negative charge into proteins and protein complexes. Subsequently, protein complexes are electrophoretically separated by size on polyacrylamide gradient gels. The advantage of Coomassie-blue in comparison to SDS is that it does not lead to the denaturation of proteins and dissection of protein complexes. Blue-native PAGE has a high resolution capacity to separate protein complexes in the size range between 50 and 5000 kDa. It is usually combined with SDS PAGE for the second gel dimension to separate the subunits of protein complexes.

Today, Blue-native/SDS PAGE is a biochemical procedure of basic importance that is widely used for protein complex characterizations in plants. Applications include the characterization of mitochondrial protein complexes (Jänsch et al., 1996; Eubel et al., 2003; Giegé et al., 2003; Heazlewood et al., 2003; Sabar et al., 2003; Dudkina et al., 2005a), protein complexes of plastids (Kügler et al., 1997; Rexroth et al., 2003; Heinemeyer et al., 2004; Ossenbühl et al., 2004; Ciambella et al., 2005) those of the plasma membrane (Kjell et al., 2004), the apoplast (Fecht-Christoffers et al., 2003) and of various biochemically defined fractions (Rivas et al., 2002; Drykova et al., 2003; Piotrowski et al., 2003; Salomon et al., 2004; Boldt et al., 2005). In combination with *in-organello* phosphorylation experiments 2D Blue-native/SDS PAGE was also used to systematically identify phosphorylations of protein complexes (Bykova et al., 2003a, b). Experimental aims addressed with Blue-native/SDS PAGE range from characterizations of specific protein complexes on one side to proteomic approaches to systematically characterize protein complexes of cellular or subcellular fractions on the other (see Plate 3.1 in Plate Section). Blue-native PAGE was successfully employed to analyse stable protein complexes like the photosystems, and more dynamic protein complexes like the preprotein translocase of the outer mitochondrial membrane, the so-called TOM complex (Werhahn et al., 2001).

Recently, a modification of the Blue-native/SDS PAGE system was suggested, which is based on Blue-native PAGE on both gel dimensions (Schägger and Pfeiffer, 2000). Normally, if the same electrophoretic procedure is used for both dimensions, proteins are localized on a diagonal line on resulting 2D gels. However, for Blue-native/Blue-native PAGE, the first gel dimension is carried out under most gentle conditions, whereas the second gel dimension is carried out under slightly less gentle conditions (Sunderhaus et al., 2006a). Protein complexes specifically destabilized in the presence of the conditions of the second gel dimension are dissected on this 2D system into subcomplexes of enhanced electrophoretic mobility, which are visible beneath the diagonal line. Blue-native/Blue-native PAGE nicely allows to investigate the subcomplex-structure of protein complexes or the protein complex composition of supercomplexes in plastids and mitochondria of plants (Eubel et al., 2003, 2004; Heinemeyer et al., 2004; Krause et al., 2004; Millar et al., 2004; Sunderhaus et al., 2006b). An advantage of this gel system is that it can be combined with *in gel* enzyme activity staining procedures (e.g. Zerbetto et al., 1997; Eubel et al., 2004; Sunderhaus et al., 2005b).

Further modifications of Blue-native PAGE include three-dimensional gel electrophoresis systems, which either combine Blue-native PAGE, IEF and SDS PAGE (Werhahn and Braun, 2002) or Blue-native PAGE, another Blue-native PAGE and SDS PAGE (Schägger and Pfeiffer, 2000). The Blue-native/IEF/SDS PAGE system allows to separate isoforms of subunits of protein complexes. The native gel dimension is employed to separate protein complexes, which subsequently are electroeluted and purified from bound Coomassie-blue. Finally, the subunits of the protein complexes are further separated on the basis of IEF and SDS PAGE (Werhahn and Braun, 2002). The Blue-native/Blue-native/SDS system allows to separate subunits of protein complexes forming part of supercomplexes.

3.3.4 Immunoprecipitations

Immunoprecipitation represents another powerful approach to investigate stable and also less stable protein–protein interactions. Protein mixtures are supplemented with an antibody directed against a subunit of a putative protein complex. Next, the antigen–antibody conjugate is specifically precipitated by its binding to Protein A- or Protein G-Sepharose. This strategy was successfully applied to investigate a large number of protein–protein interactions in plants (e.g. Perron et al., 2004; Tetlow et al., 2004).

3.4 Genetic procedures to characterize protein–protein interactions

During the last two decades a wealth of exciting genetic methods for the study of protein–protein interactions has emerged. All these methods are based on recombinant DNA technology and require genetic engineering of partial or complete protein coding sequences into suitable vectors. The recombinant proteins are then expressed in the organism of choice or in a model organism (e.g. yeast) and their localization and interaction with other proteins is monitored with elaborated physical or chemical methods. In this context, reporter proteins having natural fluorescence or proteins producing a coloured dye in the presence of a certain substrate (e.g. β-galactosidase, β-glucuronidase) play an important role. Some of the genetic methods for the investigation of protein–protein interactions even allow monitoring the activity and interaction of a certain protein *in vivo*. With FRET, for example, a protein can be analysed in its natural environment within the cell.

Generally, one way to characterize the function of a plant protein is to identify interacting protein partners. If the function of a partner protein is known, the function of an unknown protein can often be proposed. Therefore, genetically based high throughput technologies for the study of protein–protein interactions in plants have strongly enhanced a better understanding of plant differentiation and development. Especially within the last 10 years plant signal transduction pathways and regulatory networks have been characterized using genetic systems for the study of protein–protein interactions.

3.4.1 Yeast two-hybrid system

The first and best-known genetic system for the analysis of protein–protein interactions within a living cell is the yeast two-hybrid system. It was published in 1989 by Fields and Song and several interesting variants of it (yeast three-hybrid system, yeast one-hybrid system) have been developed since then. The yeast two-hybrid system is based on the separation of two functional domains of the yeast transcription factor Gal4. In a typical yeast two-hybrid screen, a hybrid protein consisting of a DNA-binding domain (DBD) and a protein of interest ('bait') is assayed against a ('prey') library of proteins expressed as fusion with a transcriptional activation domain (TA). If a 'prey' protein is interacting with the protein of interest a downstream reporter gene (mostly β-galactosidase) is transcribed and the yeast colony turns blue in the presence of 5-bromo-4-chloro-3-indolyl-β-D-galactoside (X-Gal).

To date the yeast two-hybrid system has been used in *Arabidopsis* in more than 400 Medline listed publications. The screens mainly focused on gene regulation and plant signalling pathways. For example, analysis of interacting transcription factors elucidated different control levels in plant development (e.g. Calderon-Villalobos et al., 2005). Signalling pathways analysed include resistance gene signalling (Innes, 1998), phytochrome and cryptochrome interacting factors (Quail, 2000; Sang et al., 2005) as well as stress and hormone responses (Mizoguchi et al., 2000).

Apart from investigations focusing on the above-mentioned fields, the yeast two-hybrid system has been used to study protein–protein interactions in intracellular protein trafficking (Latijnhouwers et al., 2005), protein maturation (Meyer et al., 2005) and protein degradation (Dieterle et al., 2005).

In the post-genomic era traditional genetic and biochemical approaches are struggling to keep up with the growing amount of sequencing data. Thus, the accumulation of large amounts of genomic sequence data has prompted high throughput approaches also in plants. From the roughly 26,000 genes in *Arabidopsis* the function of less than 3000 genes has been experimentally proven. Recently, the feasibility of large scale yeast two-hybrid screens in *Arabidopsis* has been demonstrated (Bürkle et al., 2005). From this study, it became very clear that only elaborated shuttle vectors (e.g. the Gateway compatible pENTR) and highly representative cDNA libraries with full-length cDNA clones are useful for functional screens. Large scale

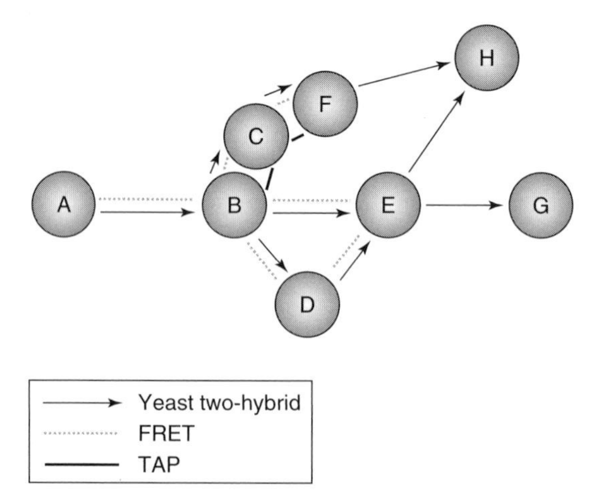

Figure 3.1 Mapping networks of interacting proteins. The map was elaborated by iterative interactor hunts using the yeast two-hybrid system (grey arrows). The starting protein 'A' was used as 'bait' which led to the isolation of the interacting protein 'B'. This 'prey' was converted into a 'bait' which identified three interacting proteins 'C', 'D' and 'E' in the subsequent hunt. From these proteins converted into 'baits' only 'C' and 'E' led to the isolation of further interacting proteins. The temporal and spatial interaction of all proteins shown was further analysed by FRET (dotted lines) and TAP (black bars). FRET confirmed the interaction of proteins 'A', 'B', 'C', 'D', 'E' and 'F' as illustrated above. It did neither proof the interaction of 'G' and 'H' with 'E' nor the interaction between 'F' and 'H'. TAP identified proteins 'B', 'C' and 'F' as subunits of the same protein complex.

two-hybrid screens are simplified and facilitated if the AD and DBD fusion proteins are cloned in two different haploid yeast strains with opposite mating types. To test for interaction the proteins are brought together by mating and positive diploids exhibit reporter gene activation. The system gives a high flexibility because the cDNA of a true positive prey protein may be subsequently cloned into the DBD containing vector which may be used in the next round as 'bait' for fishing further interacting proteins. Thereby protein interaction networks are created by exhaustive and sequential screening of AD fusion libraries (Figure 3.1). New AD fusion proteins isolated in each hunt are converted to 'baits' for the subsequent hunt. This approach has been used to create interaction maps not only in yeast (Uetz et al., 2000) and humans (Lehner and Fraser, 2004) but also in plants (Favaro et al., 2002). Recently, a comprehensive interaction map of the *Arabidopsis* MADS box transcription factors has been presented (de Folter et al., 2005). As interactions between MADS domain proteins are conserved between *Arabidopsis*, rice, *Petunia* and *Antirrhinum* (Immink and Angenent, 2002) it seems worth to validate interaction data by comparing interactions of orthologous proteins from different taxa.

3.4.2 Yeast three-hybrid system

Expanding the yeast two-hybrid system to a multiple-hybrid system facilitates the analysis of more complex protein networks established by the formation of protein complexes. For example, in a ternary protein complex, there are cases that protein X does not directly contact protein Y, but rather requires protein Z to mediate the interaction. The yeast three-hybrid system has been designed to introduce a third component in the traditional two-hybrid system; of course it is only applied, if there is circumstantial evidence that more than two components are involved in complex formation. Two cDNA sequences of proteins that do not directly interact but form part of a protein complex are fused in the normal way with AD and DBD of the Gal4 protein. If the coding sequence of a bridging protein having contact with both 'bait' proteins is then introduced into the yeast cell, the colonies will turn blue due to the activity of the reporter. This approach has been successfully used to characterize interaction partners of the *Arabidopsis* B_{sister} (ABS) protein which is involved in integument differentiation and contains a MADS box (Kaufmann et al., 2005). ABS was found to form a higher order complex with other proteins (SEP3, STK, SHP1 and SHP2) controlling flower development.

3.4.3 Yeast one-hybrid system

Another interesting derivative of the yeast two-hybrid system is the yeast one-hybrid system. This system rests on the same principal as the other yeast hybrid systems except that it is asexual and the 'bait' is in its classical design a promoter that binds the transcription factor(s). The yeast one-hybrid system has been used in *Arabidopsis* for isolating novel genes encoding transcription factors binding to the drought-responsive cis-element in the 'early responsive to dehydration stress 1' (ERD1) gene (Tran et al., 2004).

3.4.4 Limitations of yeast two-hybrid systems

Although being the most often used method for the study of protein–protein interactions in plants the yeast two-hybrid system has some limitations. One technical limitation is the occurrence of many false positives. Some baits activate the reporter genes present in the yeast strain in the absence of a protein–protein interaction. To alleviate this shortcoming, dual bait systems or dual hybrid systems based on two different reporter genes have been introduced and are nowadays commercially available. The most often used DBD-X fusion protein ('bait') is the bacterial protein lexA; it is transformed into an appropriate strain of *Saccharomyces cerevisiae* containing a dual reporter system. The dual reporters contain lexA operator sites (DNA-binding sites) upstream of the lacZ gene and a gene that confers the ability to grow in selective media lacking specific amino acids (mostly leucine or lysine). After transformation of the AD-cDNA library ('prey') into such yeast, bait–prey interactions are revealed by colonies that can grow on selective media and are blue when assayed with the chromogenic substrate X-Gal. For even higher reliability there are double dual bait yeast two-hybrid systems which use lexA and Gal4 operator sites, β-galactosidase and β-glucuronidase reporter genes as well as leucine and lysine synthesizing genes.

Still even these systems have some drawbacks. They are not suitable for several DNA-binding proteins, transcriptional activators or repressors and some proteins having intrinsic influence on transcription without being a transcription factor. Also cell cycle proteins interfere with the system, if they are expressed in the nucleus. Furthermore, it is difficult or impossible to analyse proteins which require cytoplasmic factors for proper folding or for post-translational modifications. Additionally, highly hydrophobic proteins are not suitable for yeast two-hybrid screens. For all theses cases another system the so-called split-ubiquitin system has been introduced.

3.4.5 Split-ubiquitin system

The split-ubiquitin system is a protein fragment complementation assay that is based on conditional proteolysis. The coding sequence of the C-terminal half of ubiquitin is fused to a reporter gene (mostly haemagglutinin-tagged dihydrofolate reductase, haDHFR) and to a gene encoding a protein Y of interest. The coding sequence of the N-terminal half of ubiquitin is fused to a cDNA library encoding potential interaction partners X. Both types of vectors are transformed into yeast cells containing ubiquitin specific proteases. These do not recognize the N- or C-terminal half of ubiquitin alone but only the complete protein. When proteins X and Y interact it leads to the formation of native-like ubiquitin and subsequent cleavage which is monitored on a Western blot by size variation of the protein detected by a monoclonal antibody against haemagglutinin. In contrast to the yeast two-hybrid system which is restricted to the nucleus the split-ubiquitin system works not only in the cytosol but also works in the hydrophobic environment of a membrane. This important advantage has been exploited in several recent investigations including an analysis of interactions between *Arabidopsis* sucrose transporters located in the plasma membrane (Schulze et al., 2003), K^+ channel interactions (Obrdlik et al., 2004) and an analysis of the hydrophobic thylakoid membrane proteins (Pasch et al., 2005). The split-ubiquitin system is also useful in host pathogen research (Deslandes et al., 2003) and was successfully applied to study the interactions between proteins controlling flower development in *Arabidopsis* (Chanvivattana et al., 2004).

3.4.6 Bimolecular fluorescence complementation (BiFC)

Technically, the BiFC approach is similar to the yeast two-hybrid and split-ubiquitin systems as again a protein is split into two halves to which proteins of interest are fused by genetic engineering. However, the BiFC approach is much more elegant as it employs a protein with intrinsic fluorescence that is only restored when the two halves of the reporter protein are joined together by interaction of proteins X and Y. The system is superior to the above-mentioned methods as it allows to monitor the spatial and temporal patterns of protein interactions *in vivo*. On the other hand the method is not suited for high throughput analysis.

Only very few proteins have intrinsic fluorescence like the green fluorescent protein (GFP) and its derivatives (e.g. the yellow fluorescent protein (YFP)). BiFC was introduced by Hu et al. (2002) and first applied to plants in 2004. Walter et al. (2004) generated several complementary sets of expression vectors which enable protein interaction studies in transiently or stably transformed plant cells. Using a split YFP Bracha-Drori et al. (2004) were able to determine which of the candidate proteins involved in endosperm development interact in *Nicotiana* and *Arabidopsis*. Functional heterodimerization of enzymes involved in ethylene biosynthesis was studied by Tsuchisaka and Theologis (2004) with BiFC.

3.4.7 Förster resonance energy transfer (FRET)

FRET is named after the German scientist Theodor Förster who described an energy transfer mechanism between two fluorescent molecules more than half a century ago. A fluorescent donor is excited at its specific fluorescence excitation wavelength. By a long-range dipole–dipole coupling mechanism, this excited state is then nonradiatively transferred to a second molecule, the acceptor. The donor returns to the electronic ground state. The described energy transfer mechanism is sometimes termed 'fluorescence resonance energy transfer', although the energy is not actually transferred by fluorescence. The FRET efficiency is determined by the distance between the donor and the acceptor (which has to be less than 10 nm), the spectral overlap of the donors emission spectrum and the acceptors absorption spectrum and the relative orientation of the donors emission dipole moment and the acceptors absorption dipole moment. As FRET only occurs, if the distance between the donor and the acceptor is less than 10 nm, the method may be used as an optical 'nanometre ruler' in plant biochemistry and cell biology. FRET is also a useful tool to quantify protein–protein interactions, protein–DNA interactions or protein conformational changes. For monitoring the complex formation between two proteins, one of them is labelled with a donor and the other with an acceptor. The most popular FRET pair is a cyan fluorescent protein (CFP) and YFP pair. Both are colour variants of GFP and can be easily fused genetically to any presumptive pair of interacting proteins. After transformation of the plant species of choice the interaction of the proteins may be monitored at different developmental stages using confocal microscopy. One of the first investigations using FRET in plants was the paper of Huang et al. (2001) who analysed the interaction of *Arabidopsis* proteins with the cauliflower mosaic virus movement protein. At that time FRET had already found broad application in the animal field for detection of changes or differences in calcium, membrane voltage, pH, metals or enzyme activity but only little attention in the plant field (reviewed in Hanson and Köhler, 2001). However, since then FRET has not only been used for studies on intracellular defense regulators and their interacting partners in plants (Feys et al., 2005) but also for analysis of plastid division by combinatorial assembly of plastid division proteins (Maple et al., 2005). It proved also to be useful for the investigation of subunit topology in protein complexes. Seidel et al. (2005) mapped the C-termini of V-ATPase subunits by *in vivo* FRET measurements in protoplasts transfected with CFP-donor and YFP-acceptor fluorophores.

3.4.8 Tagging technologies for the purification of protein complexes

While the above mentioned techniques for analysis of protein–protein interactions are either biochemical or genetic methods, the tagging technologies employ genetic engineering to facilitate the biochemical purification of protein complexes. Affinity tags are added to one or more subunits of a protein complex which can be subsequently, purified with appropriate high specificity resins. Some tags even enhance the solubility or proper folding of a protein but no tag is ideal with respect to all features affecting fusion protein expression and purification (e.g. low-metabolic burden, high specificity, mild elution conditions, etc.). Advantages and disadvantages of different tags like His_6, STREP, FLAG and others have been discussed in a recent review (Waugh, 2005). Usually, one appropriate tag is sufficient for purification of a known protein complex. However, in a high throughput setting designed to identify new interacting proteins, it is advisable to use at least two different tags in tandem which are either added to the N- or C-termini of all the proteins to be analysed. Tandem-affinity purification (TAP) was first introduced by Rigaut et al. (1999) who described its use for proteome exploration. The TAP tag secures high purity and rapid isolation of protein complexes from a relatively small number of cells without prior knowledge of complex composition or function. In plants the TAP technology was used for purification of the CF-9 disease resistance protein from tomato. The TAP-tagged protein turned out to be present in an approximately 420 kDa heteromultimeric membrane-associated complex at one molecule per complex (Rivas et al., 2002). Recently, Rubio et al. (2005) described an alternative tandem-affinity purification (TAPa) strategy applied to *Arabidopsis* protein complex isolation. In their system termed TAPa they replaced the tobacco etch virus (TEV) protease cleavage site with the more specific and low-temperature active rhinovirus 3C protease site. The system was successfully applied to characterize different light signalling pathway regulators. Interestingly, Witte et al. (2004) report that the eight amino acid StrepII epitope is faster but equally good for the purification of protein complexes from plant material.

3.5 Cytological procedures to characterize protein–protein interactions

The list of procedures for the characterization of protein–protein interactions so far described in this chapter is far from complete. For example, protein–protein interactions also can be monitored directly on a cytological level by the use of enzyme co-localizations on electron micrographs. This approach is based on immunolabelling of fixed cell sections with two different primary antibodies recognizing proteins speculated to physically interact. Primary antibody-specific secondary antibodies which are either labelled with small (15 nm) or large (25 nm) gold particles are used to simultaneously visualize the two proteins in ultrathin sections. Finally, statistical evaluation of protein distances by ‘nearest neighbour analyses’ allows to distinguish between proteins independently present in cellular compartments and those that physically interact. This method nicely allowed to monitor interaction of Calvin cycle enzymes (Anderson et al., 2003, 2005).

3.6 Outlook

In summary, quite a large number of very powerful methods are available to characterize protein–protein interactions. Recently, large scale projects were initiated to systematically explore the total number of protein–protein interactions in model organisms, the so-called 'Interactome' (summarized in von Mering et al., 2002). These projects are based on the two-hybrid technology, the TAP technology in combination with mass spectrometry or on 2D Blue-native/SDS PAGE. Large interactome datasets were published for yeast, *C. elegans* and man (Gavin et al., 2002; Ho et al., 2002; Camacho-Carvajal et al., 2004; Li et al., 2004; Rual et al., 2005; Stelzl et al., 2005). These datasets are complemented by transcriptome analyses, which use RNA expression profile correlations for the prediction of protein–protein interactions (Huges et al., 2000). At the same time, new databases are established to store experimental protein–protein interaction data (Han et al., 2004), which will be very helpful for future research. Interactome projects meanwhile were also initiated for plants. Broad-scale information on protein interaction networks will be of key importance for the understanding of the molecular basis of life.

Acknowledgements

Work of our laboratory is supported by the Deutsche Forschungsgemeinschaft (grant Br 1829 – 7/1) and by the Deutscher Akademischer Austauschdienst (DAAD).

References

Anderson, J.B., Carol, A.A., Brown, V.K. and Anderson, L.E. (2003) A quantitative method for assessing co-localization in immunolabelled thin section electron micrographs. *J. Struct. Biol.*, **143**, 95–106.

Anderson, L.E., Gatla, N. and Carol, A.A. (2005) Enzyme co-localization in pea leaf chloroplasts: glyceraldehydes-3-P dehydrogenase, triose-P isomerase and sedoheptulose biphosphatase. *Photosynth. Res.*, **83**, 317–328.

Andersson, B., Andersson, J.M. and Ryrie, I.J. (1982) Transbilayer organization of the chlorophyll-proteins of spinach thylakoids. *Eur. J. Biochem.*, **123**, 465–472.

Balmer, Y., Koller, A., del Val, G., Manieri, W., Schurmann, P. and Buchanan, B.B. (2003) Proteomics gives insight into the regulatory function of chloroplast thioredoxins. *Proc. Natl Acad. Sci. USA*, **100**, 370–375.

Boldt, A., Fortunato, D., Conti, A., Petersen, A., Ballmer-Weber, B., Lepp, U., Reese, G. and Becker, W.M. (2005) Analysis of the composition of an immunoglobulin E reactive high molecular weight protein complex of peanut extract containing Ara h 1 and Ara h 3/4. *Proteomics*, **5**, 675–686.

Bracha-Drori, K., Shichrur, K., Katz, A., Oliva, M., Angelovici, R., Yalovsky, S. and Ohad, N. (2004) Detection of protein–protein interactions in plants using bimolecular fluorescence complementation. *Plant J.*, **40**, 419–427.

Braun, H.P. and Schmitz, U.K. (1992) Affinity purification of cytochrome c reductase from plant mitochondria. *Eur. J. Biochem.*, **208**, 761–767.

Bürkle, L., Meyer, S., Dortay, H., Lehrach, H. and Heyl, A. (2005) *In vitro* recombination cloning of entire cDNA libraries in *Arabidopsis thaliana* and its application to the yeast two-hybrid system. *Funct. Integr. Genom.*, **5**, 175–183.

Bykova, N.V., Egsgaard, H. and Moller, I.M. (2003a) Identification of 14 new phosphoproteins involved in important plant mitochondrial processes. *FEBS Lett.*, **540**, 141–146.

Bykova, N.V., Stensballe, A., Egsgaard, H., Jensen, O.N. and Moller, I.M. (2003b) Phosphorylation of formate dehydrogenase in potato tuber mitochondria. *J. Biol. Chem.*, **278**, 26021–26030.

Calderon-Villalobos, L.I.A., Kuhnle, C., Dohmann, E.M.N., Li, H., Bevan, M. and Schwechheimer, C. (2005) The evolutionarily conserved TOUGH protein is required for proper development of *Arabidopsis thaliana*. *Plant Cell*, **17**, 2473–2485.

Camacho-Carvajal, M.M., Wollscheid, B., Aebersold, R., Steimle, V. and Schamel, W.W. (2004) Two-dimensional Blue native/SDS gel electrophoresis of multi-protein complexes from whole cellular lysates: a proteomics approach. *Mol. Cell. Proteom.*, **3**, 176–182.

Chanvivattana, Y., Bishopp, A., Schubert, D., Stock, C., Moon, Y.H., Sung, Z.R. and Goodrich, J. (2004) Interaction of Polycomb-group proteins controlling flowering in *Arabidopsis*. *Comp. Biologist*, **131**, 5263–5276.

Ciambella, C., Roepstorff, P., Aro, E.M. and Zolla, L. (2005) A proteomic approach for investigation of photosynthetic apparatus in plants. *Proteomics*, **5**, 746–757.

De Folter, S., Immink, R.G.H., Kieffer, M., Parenicova, L., Henz, S.R., Weigel, D., Busscher, M., Kooiker, M., Colombo, L., Kater, M.M., Davies, B. and Angenent, G.C. (2005) Comprehensive interaction map of the *Arabidopsis* MADS box transcription factors. *Plant Cell*, **17**, 1424–1433.

Deslandes, L., Olivier, J., Peeters, N., Feng, D.X., Khounlotham, M., Boucher, C., Somssich, I., Genin, S. and Marco, Y. (2003) Physical interaction between RRS1-R, a protein conferring resistance to bacterial wilt, and PopP2, a type III effector targeted to the plant nucleus. *Proc. Natl Acad. Sci. USA*, **100**, 8024–8029.

Dieterle, M., Thomann, A., Renou, J.P., Parmentier, Y., Cognat, V., Lemonnier, G., Muller, R., Shen, W.H., Kretsch, T. and Genschik, P. (2005) Molecular and functional characterization of *Arabidopsis* Cullin 3A. *Plant J.*, **41**, 386–399.

Drykova, D., Cenklova, V., Sulimenko, V., Volc, J., Draber, P. and Binarova, P. (2003) Plant gamma-tubulin interacts with alphabeta-tubulin dimers and forms membrane-associated complexes. *Plant Cell*, **15**, 465–480.

Dudkina, N.V., Eubel, H., Keegstra, W., Boekema, E.J. and Braun, H.P. (2005a) Structure of a mitochondrial supercomplex formed by respiratory chain complexes I and III. *Proc. Natl Acad. Sci. USA*, **102**, 3225–3229.

Dudkina, N.V., Heinemeyer, H., Keegstra, W., Boekema, E.J. and Braun, H.P. (2005b) Structure of dimeric ATP synthase from mitochondria: an angular association of monomers induces the strong curvature of the inner membrane. *FEBS Lett.*, **579**, 5769–5772.

Eubel, H., Jänsch, L. and Braun, H.P. (2003) New insights into the respiratory chain of plant mitochondria: supercomplexes and a unique composition of complex II. *Plant Physiol.*, **133**, 274–286.

Eubel, H., Heinemeyer, J. and Braun, H.P. (2004) Identification and characterization of respirasomes in potato mitochondria. *Plant Physiol.*, **134**, 1450–1459.

Favaro, R., Immink, R.G., Ferioli, V., Bernasconi, B., Byzova, M., Angenent, G.C., Kater, M. and Colombo, L. (2002) Ovule-specific MADS-box proteins have conserved protein–protein interactions in monocot and dicot plants. *Mol. Genet. Genom.*, **268**, 152–159.

Fecht-Christoffers, M.M., Braun, H.P., Guillier, C., VanDorssealer, A. and Horst W.J. (2003) Effect of manganese toxicity on the proteome of the leaf apoplast in cowpea (*Vigna unguiculata*). *Plant Physiol.*, **133**, 1935–1946.

Feys, B.J., Wiermer, M., Bhat, R.A., Moisan, L.J., Medina-Escobar, N., Neu, C., Cabral, A. and Parker, J.E. (2005) *Arabidopsis* SENESCENCE-ASSOCIATED GENE101 stabilizes and signals within an ENHANCED DISEASE SUSCEPTIBILITY1 complex in plant innate immunity. *Plant Cell*, **17**, 2601–2613.

Fields, F. and Song, O. (1989) A novel genetic system to detect protein–protein interactions. *Nature*, **340**, 245–246.

Gavin, A.C. et al. (2002) Functional organization of the yeast proteome by systematic analysis of protein complexes. *Nature*, **415**, 141–147.

Giegé, P., Sweetlove, L.J. and Leaver, C. (2003) Identification of mitochondrial protein complexes in *Arabidopsis* using two-dimensional blue-native polyacrylamide gel electrophoresis. *Plant Mol. Biol. Rep.*, **21**, 133–144.

Han, K., Park, B., Kim, H., Hong, J. and Park, J. (2004) HPID: the human protein interaction database. *Bioinformatics*, **20**, 2466–2470.

Hanson, M.R. and Köhler, R.H. (2001) GFP imaging: methodology and application to investigate cellular compartmentation in plants. *J. Exp. Bot.*, **356**, 529–539.

Heazlewood, J.L., Howell, K.A., Whelan, J. and Millar, A.H. (2003) Towards an analysis of the rice mitochondrial proteome. *Plant Physiol.*, **132**, 230–242.

Heinemeyer, J., Eubel, H., Wehmhöner, D., Jänsch, L. and Braun, H.P. (2004) Proteomic approach to characterize the supramolecular organization of photosystems in higher plants. *Phytochemistry*, **65**, 1683–1692.

Heinemeyer, J., Lewejohann, D. and Braun, H.P. (2005) Blue-native gel electrophoresis for the characterization of protein complexes in plants. In: Thiellement, H. (ed) *Plant Proteomics*. Methods in Molecular Biology Series, Humana Press, Totowa, USA (in press).

Ho, Y. et al. (2002) Systematic identification of protein complexes in *Saccharomyces cerevisiae* by mass spectrometry. *Nature*, **415**, 180–182.

Hu, C.D., Chinenov, Y. and Kerppola, T.K. (2002) Visualization of interactions among bZIP and Rel family proteins in living cells using bimolecular fluorescence complementation. *Mol. Cell*, **9**, 789–798.

Huang, Z., Andrianov, V.M., Han, Y. and Howell, S.H. (2001) Identification of *Arabidopsis* proteins that interact with the cauliflower mosaic virus (CaMV) movement protein. *Plant Mol. Biol.*, **47**, 663–675.

Huges et al. (2000) Functional discovery via a compendium of expression profiles. *Cell*, **102**, 109–126.

Immink, R.G. and Angenent, G.C. (2002) Transcription factors do it together: the hows and whys of studying protein–protein interactions. *Trend. Plant Sci.*, **7**, 531–534.

Innes, R.W. (1998) Genetic dissection of R gene signal transduction pathways. *Curr. Opin. Plant Biol.*, **1**, 299–304.

Jänsch, L., Kruft, V., Schmitz, U.K. and Braun, H.P. (1996) New insights into the composition, molecular mass and stoichiometry of the protein complexes of plant mitochondria. *Plant J.*, **9**, 357–368.

Kaufmann, K., Anfang, N., Saedler, H. and Theissen, G. (2005) Mutant analysis, protein–protein interactions and subcellular localization of the *Arabidopsis* B_{sister} (ABS) protein. *Mol. Gen. Genom.*, **274**, 103–118.

Kjell, J., Rasmusson, A.G., Larsson, H. and Widell, S. (2004) Protein complexes of the plant plasma membrane resolved by Blue native PAGE. *Physiol. Plant.*, **121**, 546–555.

Krause, F., Reifschneider, N.H., Vocke, D., Seelert, H., Rexroth, S. and Dencher, N.A. (2004) 'Respirasome'-like supercomplexes in green leaf mitochondria of spinach. *J. Biol. Chem.*, **279**, 48369–48375.

Kügler, M., Jänsch, L., Kruft, V., Schmitz, U.K. and Braun, H.P. (1997) Analysis of the chloroplast protein complexes by blue-native polyacrylamide gel electrophoresis. *Photosynth. Res.*, **53**, 35–44.

Latijnhouwers, M., Hawes, C., Carvalho, C., Oparka, K., Gillingham, A.K. and Boevink, P. (2005) An *Arabidopsis* GRIP domain protein locates to the trans-Golgi and binds the small GTPase ARL1. *Plant J.*, **44**, 459–470.

Lehner, B. and Fraser, A.G. (2004) A first-draft human protein-interaction map. *Genome Biol.*, **5**, 63.

Leterme, S. and Boutry, M. (1993) Purification and preliminary characterization of mitochondrial complex I (NADH:ubiquinone reductase) from broad bean (*Vicea faba* L.). *Plant Physiol.*, **102**, 435–443.

Li, S. et al. (2004) A map of the interactome network of the metazoan *C. elegans*. *Science*, **303**, 540–543.

Maple, J., Aldridge, C. and Moller, S.G. (2005) Plastid division is mediated by combinatorial assembly of plastid division proteins. *Plant J.*, **43**, 811–823.

Meyer, E.H., Giege, P., Gelhaye, E., Rayapuram, N., Ahuja, U., Thony-Meyer, L., Grienen-berger, J.M. and Bonnard, G. (2005) AtCCMH, an essential component of the c-type cytochrome maturation pathway in *Arabidopsis* mitochondria, interacts with apocytochrome c. *Proc. Natl Acad. Sci. USA*, **102**, 16113–16118.

Millar, A.H., Eubel, H., Jänsch, L., Kruft, V., Heazlewood, L. and Braun, H.P. (2004) Mitochondrial cytochrome c oxidase and succinate dehydrogenase contain plant-specific subunits. *Plant Mol. Biol.*, **56**, 77–89.

Mizoguchi, T., Ichimura, K., Yoshida, R. and Shinozaki, K. (2000) MAP kinase cascades in *Arabidopsis*: their roles in stress and hormone responses. *Result. Probl. Cell Differ.*, **27**, 29–38.

Molnar, S.A., Anderson, G.P. and Gross, E.L. (1987) The purification of cytochrome f and plastocyanin using affinity chromatography. *Biochim. Biophys. Acta*, **894**, 327–331.

Muller, B. and Eichacker, L.A. (1999) Assembly of the D1 precursor in monomeric photosystem II reaction center precomplexes precedes chlorophyll a-triggered accumulation of reaction center II in barley etioplasts. *Plant Cell*, **11**, 2365–2377.

Obrdlik, P., El-Bakkoury, M., Hamacher, T., Cappellaro, C., Vilarino, C., Fleischer, C., Ellerbrok, H., Kamuzinzi, R., Ledent, V., Blaudez, D., Sanders, D., Revuelta, J.L., Boles, E., Andre, B. and Frommer, W.B. (2004) K^+ channel interactions detected by a genetic system optimized for systematic studies of membrane protein interactions. *Proc. Natl Acad. Sci. USA*, **101**, 12242–12247.

Ossenbühl, F., Gohre, V., Meurer, J., Krieger-Liszkay, A., Rochaix, J.D. and Eichacker, L.A. (2004) Efficient assembly of photosystem II in *Chlamydomonas reinhardtii* requires Alb3.1p, a homolog of *Arabidopsis* ALBINO3. *Plant Cell*, **16**, 1790–1800.

Pasch, J.C., Nickelsen, J. and Schunemann, D. (2005) The yeast split-ubiquitin system to study chloroplast membrane protein interactions. *Appl. Microbiol. Biotechnol.*, **30**, 1–8.

Peltier, J.B., Ripoll, D.R., Friso, G., Rudella, A., Cai, Y., Ytterberg, J., Giacomelli, L., Pillardy, J. and von Wijk, K.J. (2004) Clp protease complexes from photosynthetic and non-photosynthetic plastids and mitochondria of plants, their predicted three-dimensional structures, and functional implications. *J. Biol. Chem.*, **279**, 4768–4781.

Perales, M., Eubel, H., Heinemeyer, J., Colaneri, A., Zabaleta, E. and Braun, H.P. (2005) Disruption of a nuclear gene encoding a mitochondrial gamma carbonic anhydrase reduces complex I and supercomplex I + III_2 levels and alters mitochondrial physiology in *Arabidopsis*. *J. Mol. Biol.*, **350**, 263–277.

Perron, K., Goldschmidt-Clermont, M. and Rochaix, J.D. (2004) A multiprotein complex involved in chloroplast group II intron splicing. *RNA*, **10**, 704–711.

Peter, G.F. and Thornber, J.P. (1991) Electrophoretic procedures for fractionation of photosystems I and II pigment-proteins of higher plants and for determination of their subunit composition. *Method. Plant Biochem.*, **5**, 195–210.

Piotrowski, M., Janowitz, T. and Kneifel, H. (2003) Plant C–N hydrolases and the identification of a plant *N*-carbamoylputrescine amidohydrolase involved in polyamine biosynthesis. *J. Biol. Chem.*, **278**, 1708–1712.

Quail, P.H. (2000) Phytochrome-interacting factors. *Semin. Cell Dev. Biol.*, **11**, 457–466.

Rexroth, S., Meyer zu Tittingdorf, J.M., Krause, F., Dencher, N.A. and Seelert, H. (2003) Thylakoid membrane at altered metabolic state: challenging the forgotten realms of the proteome. *Electrophoresis*, **24**, 2814–2823.

Rigaut, G., Shevchenko, A., Rutz, B., Wilm, M., Mann, M. and Seraphin, B. (1999) A generic protein purification method for protein complex characterization and proteome exploration. *Nat. Biotechnol.*, **17**, 1030–1032.

Rivas, S., Romeis, T. and Jones, J.D. (2002) The Cf-9 disease resistance protein is present in an approximately 420-kilodalton heteromultimeric membrane-associated complex at one molecule per complex. *Plant Cell*, **14**, 689–702.

Rual, J.F. et al. (2005) Towards a proteome-scale map of the human protein–protein interaction network. *Nature*, **437**, 1173–1178.

Rubio, V., Shen, Y., Saijo, Y., Liu, Y., Gusmaroli, G., Dinesh-Kumar, S.P. and Deng, X.W. (2005) An alternative tandem affinity purification strategy applied to *Arabidopsis* protein complex isolation. *Plant. J.*, **41**, 767–778.

Sabar, M., Gagliardi, D., Balk, J. and Leaver, C.J. (2003) ORFB is a subunit of F(1)F(O)-ATP synthase: insight into the basis of cytoplasmic male sterility in sunflower. *EMBO Rep.*, **4**, 1–6.

Sakihama, N., Nagai, K., Ohmori, H., Tomizawa, H., Tsujita, M. and Shin, M. (1992) Immobilized ferredoxins for affinity chromatography of ferredoxin-dependant enzymes. *J. Chromatogr.*, **597**, 147–153.

Salomon, M., Lempert, U. and Rüdiger, W. (2004) Dimerization of the plant photoreceptor phototropin is probably mediated by the LOV1 domain. *FEBS Lett.*, **572**, 8–10.

Sang, Y., Li, Q.H., Rubio, V., Zhang, Y.C., Mao, J., Deng, X.W. and Yang, H.Q. (2005) N-terminal domain-mediated homodimerization is required for photoreceptor activity of *Arabidopsis*. *Plant Cell*, **17**, 1569–1584.

Schägger, H. (2003) Blue native electrophoresis. In: Hunte, C., von Jagow, G. and Schägger, H. (eds) *Membrane Protein Purification and Crystallization: A Practical Guide*. Academic Press, London, UK, pp. 105–130.

Schägger, H. and von Jagow, G. (1991) Blue native electrophoresis for isolation of membrane protein complexes in enzymatically active form. *Anal. Biochem.*, **199**, 223–231.

Schägger, H. and Pfeiffer, K. (2000) Supercomplexes in the respiratory chains of yeast and mammalian mitochondria. *EMBO J.*, **19**, 1777–1783.

Schulze, W.X., Reinders, A., Ward, J., Lalonde, S. and Frommer, W.B. (2003) Interactions between co-expressed *Arabidopsis* sucrose transporters in the split-ubiquitin system. *BMC Biochem.*, **3**, 1471–2091.

Seidel, T., Golldack, D. and Dietz, K.J. (2005) Mapping of C-termini of V-ATPase subunits by *in vivo*-FRET measurements. *FEBS Lett.*, **579**, 4374–4382.

Stelzl, U. et al. (2005) A human protein–protein interaction network: a resource for annotating the proteome. *Cell*, **122**, 957–968.

Sunderhaus, S., Lewejohann, D. and Braun, H.P. (2006a) Two-dimensional blue native/blue native polyacrylamide gel electrophoresis for the characterization of mitochondrial protein complexes and supercomplexes. In: Leister, D. and Herrmann, J.H. (eds) *Mitochondrial Genomics and Proteomics Protocols*. Methods in Molecular Biology Series, Humana Press, Totowa, USA (in press).

Sunderhaus, S., Dudkina, N., Jänsch, L., Klodmann, J., Heinemeyer, J., Perales, M., Zabaleta, E., Boekema, E. and Braun, H.P. (2006) Carbonic anhydrase subunits form a matrix-exposed domain attached to the membrane arm of mitochondrial complex I in plants. *J. Biol. Chem.*, **281**, 6482–6488.

Tetlow, I.J., Wait, R., Lu, Z., Akkasaeng, R., Bowsher, C.G., Esposito, S., Kosar-Hashemi, B., Morell, M.K. and Emes, M.J. (2004) Protein phosphorylation in amyloplasts regulates starch branching enzyme activity and protein–protein interactions. *Plant Cell*, **16**, 694–708.

Tran, L.S., Nakashima, K., Sakuma, Y., Simpson, S.D., Fujita, Y., Maruyama, K., Fujita, M., Seki, M., Shinozaki, K. and Yamaguchi-Shinozaki, K. (2004) Isolation and functional analysis of *Arabidopsis* stress-inducible cis-element in the early responsive to dehydration stress 1 promoter. *Plant Cell*, **16**, 2481–2498.

Tsuchisaka, A. and Theologis, A. (2004) Heterodimeric interactions among the 1-amino-cyclopropane-1-carboxylate synthase polypeptides encoded by the *Arabidopsis* gene family. *Proc. Natl Acad. Sci. USA*, **101**, 2275–2280.

Uetz, P. et al. (2000) A comprehensive analysis of protein–protein interactions in *Saccharomyces cerevisiae*. *Nature*, **403**, 623–627.

van Wijk, K.J., Bingsmark, S., Aro, E.M. and Andersson, B. (1995) *In vitro* synthesis and assembly of photosystem II core proteins. The D1 protein can be incorporated into photosystem II in isolated chloroplasts and thylakoids. *J. Biol. Chem.*, **270**, 25685–25695.

von Mehring, C., Krause, R., Snel, B., Cornell, M., Oliver, S., Fields, S. and Bork, P. (2002) Comparative assessment of large-scale data sets of protein–protein interactions. *Nature*, **417**, 399–403.

Walter, M., Chaban, C., Schutze, K., Batistic, O., Weckermann, K., Nake, C., Blazevic, D., Grefen, C., Schumacher, K., Oecking, C., Harter, K. and Kudla, J. (2004) Visualization of protein interactions in living plant cells using bimolecular fluorescence complementation. *Plant. J.*, **40**, 428–438.

Waugh, D.S. (2005) Making the most of affinity tags. *Trend. Biotechnol.*, **23**, 316–320.

Werhahn, W. and Braun, H.P. (2002) Biochemical dissection of the mitochondrial proteome from *Arabidopsis thaliana* by three-dimensional gel electrophoresis. *Electrophoresis*, **23**, 640–646.

Werhahn, W., Niemeyer, A., Jänsch, L., Kruft, V., Schmitz, U.K. and Braun, H.P. (2001) Purification and characterization of the preprotein translocase of the outer mitochondrial membrane from *Arabidopsis thaliana*: identification of multiple forms of TOM20. *Plant Physiol.*, **125**, 943–954.

Witte, C.P., Noel, L.D., Gielbert, J., Parker, J.E. and Romeis, T. (2004) Rapid one-step protein purification from plant material using the eight-amino acid StrepII epitope. *Plant Mol. Biol.*, **55**, 135–147.

Zerbetto, E., Vergani, L. and Dabbeni-Sala, F. (1997) Quantitation of muscle mitochondrial oxidative phosphorylation enzymes via histochemical staining of blue native polyacrylamide gels. *Electrophoresis*, **18**, 2059–2064.

4.2 Control of cellular redox status

Reactive oxygen/nitrogen species (ROS/RNS) such as hydrogen peroxide (H_2O_2), superoxide, singlet oxygen, NO and the hydroxyl radical can cause damage to lipids, nucleic acids and proteins and are very reactive against protein thiols. In plant tissues, ROS are produced primarily by chloroplasts and peroxisomes. However, some ROS production also takes place in the electron transport chain of the respiratory pathway in the mitochondrion (Bailly, 2004). ROS/RNS moreover have signalling roles (Foyer and Noctor, 2005), and in plants, the production is induced during developmental processes (e.g. seed germination) and by abiotic (e.g. desiccation) or biotic (e.g. pathogen attack) stresses.

A coordinated system for detoxification of ROS is required to remove the different toxic by-products of metabolism and to regulate oxidative signalling. The major control of cellular redox status is the ascorbate–glutathione cycle (Potters et al., 2002; Figure 4.1A). The enzymes of the ascorbate–glutathione cycle in plants occur in the chloroplast, mitochondrion and cytosol. Glutathione, the tripeptide γ–glutamyl–cysteinyl–glycine, is the dominant low-molecular-weight thiol in the cell, present at millimolar concentrations. The ratio of reduced (GSH) to oxidized (GSSG) glutathione is the main control of cellular redox balance including the thiol-disulphide redox status. This ratio is around 100:1 in the cytoplasm of eukaryotic cells (Hwang et al., 1992) and the maintenance of the GSH/GSSG ratio is coupled by glutathione reductase to the availability of NADPH. The reducing environment of the cytosol does not favour disulphide formation. Disulphides formed in cytosolic proteins are therefore likely to be transient and highly sensitive to changes in the redox status of the cell. In contrast to the cytoplasm the GSH/GSSG ratio is around 3:1 in the endoplasmic reticulum (ER). This more oxidizing environment permits stable disulphide bond formation. PDI catalyses disulphide formation and isomerization in the ER, and is responsible for the correct disulphide pairing in nascent secretory proteins. Although PDI has also been identified in chloroplasts of the unicellular green alga *Chlamydomonas reinhardtti* (Trebitsh et al., 2001), disulphide bonds are typically found in proteins that are processed via the secretory pathway and are generally considered to have a structural and stabilizing role. Disulphide bond reduction can be used to regulate these proteins by destabilization and increasing their susceptibility to proteolytic degradation. This mechanism is suggested for the solubilization and breakdown of cereal seed storage proteins during seed germination (Kobrehel et al., 1992). In addition to the role of glutathione in maintenance and regulation of the cellular redox state, glutathionylation of thiol groups in proteins protects them from irreversible oxidation by ROS. The glutaredoxin (Grx) system (Figure 4.1B) is responsible for the subsequent dethiolation, which also requires reducing power from NADPH, supplied via glutathione reductase. Another major system for regulation of protein thiol redox status is the thioredoxin (Trx) system that also uses reducing power from NADPH, via Trx reductases, to reduce disulphide bonds in target proteins (Figure 4.1C). It is noteworthy that several of the proteins known to be regulated by Trxs or Grxs are themselves involved in ROS

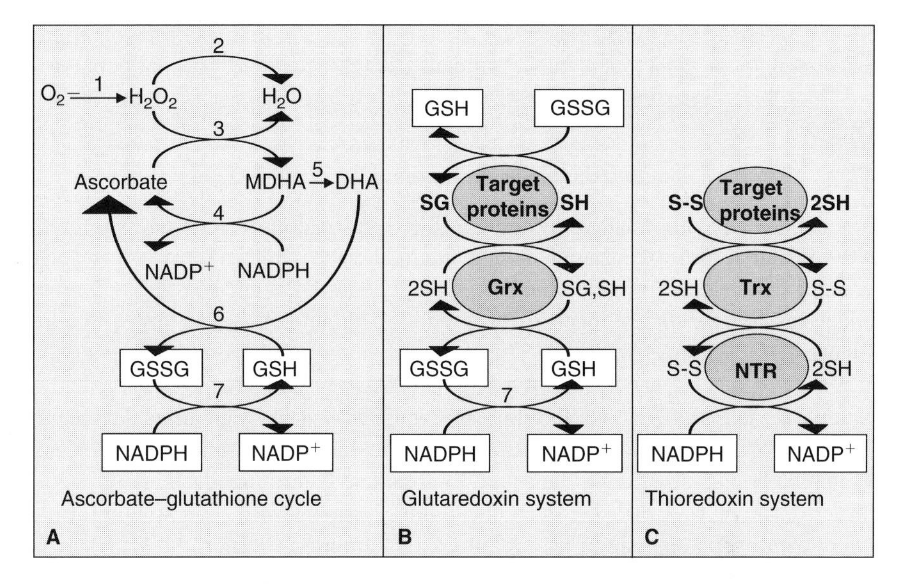

Figure 4.1 Interrelated mechanisms for control of cellular redox status. A: The ascorbate–glutathione cycle for detoxification of ROS. **B:** The glutaredoxin (Grx) system for protein dethiolation. **C:** The thioredoxin (Trx) system for protein disulphide reduction. NTR: NADP-dependent thioredoxin reductase; MDHA: monodehydroascorbate; DHA: dehydroascorbate; SH: free protein thiol group; SG: glutathionylated protein; S—S disulphide bond. Reactions involved: (1) superoxide dismutase; (2) catalase; (3) ascorbate peroxidase; (4) MDHA reductase; (5) non-enzymatic disproportionation of MDHA to DHA (and ascorbate); (6) DHA reductase; (7) glutathione reductase.

metabolism, for example Trx peroxidases are involved in peroxide detoxification (reviewed by Deitz, 2003). Cellular redox control and protein thiol status are intimately linked in a complex regulatory network that is increasingly recognized as central to plant metabolism.

4.2.1 Sequence and structural features of proteins catalysing cysteine redox modifications

PDIs, Trxs and Grxs carry out distinct types of oxidoreductions of cysteine residues. All of these proteins have a conserved active site motif, CXXC/S and share Trx-like folds (Martin, 1995). PDIs typically contain two Trx domains, whereas Trxs and Grxs are single domain proteins. The *Arabidopsis thaliana* genome includes at least 19 and 31 genes coding for Trxs and Grxs, respectively, in contrast to human, yeast and *Escherichia coli* (*E. coli*) which contain only a few Trx and Grx encoding genes (Meyer et al., 2002; Rouhier et al., 2004). The numerous isoforms of plant Trxs are grouped in six families based on their amino acid sequences. The f-, m-, x- and y-type are chloroplastic, the o-type is mitochondrial and the h-type is mainly cytosolic although it has also been identified in the nucleus, mitochondria and phloem sieve tubes (reviewed by Buchanan and Balmer, 2005). Predictions based on amino acid sequences suggest that plant Grxs are also targeted to the cytosol, mitochondria and chloroplast (Rouhier et al., 2004).

4.2.2 Catalytic mechanisms of Trxs and Grxs

Trxs contain two redox active cysteine residues in the conserved active site motif WCG/PPC that form an intramolecular disulphide in the oxidized form. Trxs receive reducing equivalents from electron donors such as NADPH and ferredoxin via Trx reductases, and the dithiol forms of Trxs reduce disulphide bonds in target proteins (Figure 4.1C) (Holmgren, 1985). The surface-exposed thiol group of the most N-terminal cysteine (Cys_1) of the active site of Trx has a relatively low pK_a value, for example 6.7 in *E. coli* Trx (Kallis and Holmgren, 1980). The Cys_1 carries out the initial nucleophilic attack on a target disulphide to form a transient intermolecular disulphide bond, subsequently attacked by the most C-terminal cysteine (Cys_2) of the active site of Trx, releasing the reduced target proteins with concomitant formation of the disulphide bond in Trxs (Kallis and Holmgren, 1980).

Grxs typically contain the active site motif CPYC, although this sequence varies in several Grxs from *Arabidopsis*. Grxs accept reducing equivalents either from NADPH via Grx reductase or from GSH (Holmgren, 1979). Grxs are weaker reductants than Trxs, with redox potentials around -200 mV compared to about -270 mV for Trxs (Åslund et al., 1997). The thiol-pK_a value of the Grx Cys_1 is exceptionally low, for example 3.5 in yeast Grx (Gan et al., 1990), and much lower than in Trxs. Grxs preferentially reduce glutathione mixed disulphides (Gravina and Mieyal, 1993) and account for the majority of deglutathionylation activity in human cells (Chrestensen et al., 2000). Whereas both Cys_1 and Cys_2 are essential in reduction of disulphide bonds by Trxs, Cys_1 is sufficient for protein deglutathionylation by Grxs as it can receive the glutathione group from *S*-glutathionylated proteins by a single nucleophilic attack (Yang et al., 1998). In fact Cys_2 in a number of Grxs from *Arabidopsis thaliana* is replaced by serine (Rouhier et al., 2004). Grxs and Grx mutants that lack Cys_2 retain reductase activity towards *S*-glutathionylated proteins (Yang et al., 1998), dehydroascorbate (DHA) and cysteine SOH in peroxiredoxin (Rouhier et al., 2002a, b). Grxs have also been shown to catalyse protein *S*-glutathionylation (Beer et al., 2004).

While it is often possible to predict a role in redox regulation for proteins containing a CXXC/S motif, the target proteins for these reactions have proved to be highly diverse in sequence, structure and function. Currently, it is still a matter of debate to which extent the various enzyme-catalysed thiol-disulphide exchange reactions are governed by specific protein–protein interactions. Moreover, it is not yet possible to predict whether a protein is likely to be a target for one of the redox-regulatory proteins discussed here. However, some of the coexisting Trx isoforms in plants have been demonstrated to have distinct physiological functions. For instance, the two chloroplastic Trxs, Trx f and Trx m, preferentially reduce fructose-1,6-bisphosphatase and NADP-malate dehydrogenase, respectively, *in vitro* (Wolosiuk et al., 1979). *Arabidopsis thaliana* Trx h isoforms differ in their ability to substitute for endogenous

Trxs in *Saccharomyces cerevisiae* (Mouaheb et al., 1998). In addition, a similar study demonstrated functional diversity of endogenous Trx h isoforms *in vivo* in *Chlamydomonas reinhardtti* (Sarkar et al., 2005). Thus in plants, the presence of numerous protein cysteine oxidoreductases in combination with other redox active components constitute a complex system for protein redox regulation.

4.3 Proteomics techniques for analysis of cysteine modifications

In classical proteome analysis, involving two-dimensional gel electrophoresis (2-DE) and/or mass spectrometry (MS), the general approach has been to treat protein extracts with reducing and alkylating reagents for complete reduction of disulphide bonds and prevention of side reactions of cysteines. In order to study PTMs involving disulphide bonds and cysteine residues, the proteomics techniques have been creatively modified to allow specific labelling of cysteine side chains for determination of the PTMs they undergo in response to different experimental conditions.

4.3.1 Reagents for cysteine labelling

Reduced cysteines are very reactive and can be covalently labelled with a series of thiol-specific reagents (Table 4.1) many of which are invaluable tools for investigation of cysteine modifications in proteomes. Most of these reagents react irreversibly with cysteine residues either through substitution reactions (e.g. alkyl halides such as iodoacetamide (IAM) and iodoacetic acid (IAA)) or double bond addition reactions (e.g. vinylpyridine and *N*-ethylmaleimide (NEM)) (Figure 4.2A,B). Cysteines can also be reversibly labelled by dithiol reagents such as the Ellman's reagent (5,5′-dithio-bis(2-nitrobenzoic acid) (DTNB)) that form a mixed disulphide and can be released upon reduction (Figure 4.2C). In general, cysteine-reactive reagents act on the thiolate ion form of cysteinyl residues, and thus do not react with oxidized cysteine species such as disulphides, mixed disulphides, SOH, SO_2H and SO_3H. It should be noted that some reagents, especially outside the recommended pH-ranges, at a large excess or during prolonged incubation may also react with histidine, methionine, tyrosine and N-terminal amino groups (Bailey, 1967; Lapko et al., 2000; Boja and Fales, 2001).

A wide array of derivatives of cysteine-reactive reagents have been developed that contain chromophores (e.g. 4-dimethylaminophenylazaphenyl-4′-maleimide (BABMI)), fluorophores (e.g. iodoacetamidofluorescein) and radiolabels (e.g. [^{14}C]IAM) for visualization of proteins after sodium dodecyl sulphate-polyacrylamide gel electrophoresis (SDS-PAGE) or 2-DE or for detection after chromatographic separation (Table 4.1). In addition, high-molecular-weight reagents such as maleimide-conjugated polyethylene glycol polymer (maleimide-PEG), shift protein migration patterns in SDS-PAGE (Xiao et al., 2004). Other reagents are designed for selective enrichment or isolation of proteins and peptides. For example, biotin-conjugated reagents are used for avidin affinity chromatography and some reagents with positively

Table 4.1 Thiol-reactive reagents applied in studies of protein disulphides and cysteine oxidoreduction

Reagent	Description[a]	References
Acrylamide	Regularly used for cysteine alkylation in protocols for 2-DE and MS peptide analysis.	Brune (1992), Sechi and Chait (1998)
[2,3,3′-D_3]acrylamide	Deuterium-labelled reagent used to determine cysteine content of peptides based on the isotopic distributions in MS.	Sechi and Chait (1998)
AMS (4-acetamido-4′-maleimidylstilbene-2,2′-disulphonic acid)	Water-soluble fluorophore (322/411 nm) used in combination with SDS-PAGE and MS peptide analysis.	Kobayashi et al. 1997, Vestweber and Schatz (1988), Tie et al. (2004)
APTA (3-acrylamidopropyl)-trimethylammonium chloride	Acrylamide derivative with quaternary amine tag used for enrichment of cysteine-containing peptides by cation exchange chromatography. Induces characteristic fragmentation patterns in tandem MS.	Ren et al. (2004)
Biotin-HPDP (*N*-[6-(biotinamido)hexyl]-3′-(2′-pyridyldithio)propionamide)	Reagent for reversible labelling. Used for isolation of proteins and peptides by avidin affinity chromatography. The chromophore pyridine-2-thione (343 nm) is released upon reaction with thiol.	Jaffrey et al. (2001), Kuncewicz et al. (2003), Lindermayr et al. (2005)
Biotin-maleimide	Used for isolation of proteins and peptides by avidin affinity chromatography.	Lind et al. (2002), Hamnell-Pamment et al. (2005)
Bromoethylamine	Trypsin cleaves at cysteines aminoethylated with this reagent.	Thevis et al. (2003)
DABMI (4-dimethylaminophenylazaphenyl-4′-maleimide)	Chromophore (515 nm) that provides characteristic fragmentation patterns in MALDI-TOF MS.	Borges and Watson (2003)
DTNB	Reagent for reversible labelling. Regularly used for thiol quantification. Stoichiometrically yields the chromophore, 5-mercapto-2-nitrobenzoic acid (412 nm) upon reaction with thiols.	Laragione et al. (2003), Gevaert et al. (2004)
DTPD (4,4′-dithiopyridine)	Reagent for reversible labelling. Smaller molecule than DTNB and mostly uncharged at neutral pH. Stoichiometrically yields the chromophore, 4-thiopyridone (324 nm) upon reaction with thiols	Riener et al. (2002)
Ethylenimine	Trypsin cleaves at cysteines aminoethylated with this reagent.	Plapp et al. (1967)
IAA (Iodoacetic acid)	Negatively charged reagent regularly used for cysteine alkylation.	Galvani et al. (2001)
[^{14}C]IAA	Used for autoradiography in combination with electrophoretic protein separation.	Lee et al. (1998)
IAM (Iodoacetamide)	Regularly used for cysteine alkylation in protocols for 2-DE and MS peptide analysis.	Shevchenko et al. (1996)
[^{14}C]IAM	Used for autoradiography in combination with 2-DE.	Marchand et al. (2004)
IBTP ((4-iodobutyl)triphenyl-phosphonium)	Hydrophobic and positively charged reagent targeted into mitochondria. Used for *in vivo* labelling of mitochondrial proteins in combination with 2-DE.	Lin et al. (2002)

$[D_{15}]$IBTP	Stable isotope-labelled IBTP. Potentially useful for quantitative peptide analysis.	Marley et al. (2005)
ICAT reagents	Pairs of stable isotope labelled and unlabelled reagents with IAM group used for quantitative proteomics. Contain a biotin affinity tag for peptide isolation by avidin affinity chromatography.	Gygi et al. (1999), Sethuraman et al. (2004)
5-iodoacetamidofluorescein	Extensively used water-soluble fluorophore (488/530 nm).	Wu et al. (1998), Baty et al. (2002)
Iodoacetyl-biotin	Used for isolation of proteins and peptides by avidin affinity chromatography.	Kim et al. (2000), Laragione et al. (2003)
Iodoacetyl-PEO-biotin	Alternative to iodoacetyl-biotin with water-soluble PEO spacer arm.	Conrads et al. (2001), Borisov et al. (2002)
Iodoacetyl-cyanine dyes	Spectrally distinct fluorophores, Cy3 (532/580 nm) and Cy5 (633/670 nm), used for quantitative proteomics in 2-D DIGE technique.	Chan et al. (2005)
1,5-I-AEDANS (5–{2((iodoacetyl)amino(ethyl)}(amino)naphthalene-1-sulphonic acid)	Potentially useful fluorophore (336/490 nm) for quantitative proteomics based on peptide enrichment by immobilized metal affinity chromatography followed by LC-MS.	Clements et al. (2005)
Maleimide-cyanine dyes	Spectrally distinct fluorophores, Cy3 (532/580 nm) and Cy5 (633/670 nm), used for quantitative proteomics in 2-D DIGE technique.	Shaw et al. (2003), Maeda et al. (2004)
Maleimide-PEG	High-molecular-weight reagent used to shift protein migration patterns in SDS-PAGE	Xiao et al. (2004)
Methyl methanethiosulphonate	Used for reversible methylthiolation.	Jaffrey et al. 2001; Kuncewicz et al. (2003), Lindermayr et al. (2005)
Monobromobimane	Fluorescent (380/490 nm) when conjugated to thiol. Extensively used in combination with SDS-PAGE and 2-DE.	Yano et al. (2001)
NEM	Regularly used for cysteine alkylation.	Yen et al. (2002)
SBD-F (Ammonium 7-fluoro-2,1,3-benzoxadiazole-4-sulphonate)	Water-soluble fluorophore (380/505 nm) used for quantitative proteomics in combination with LC.	Toriumi and Imai (2003)
2-vinylpyridine	Can be used for cysteine alkylation below neutral pH.	Lindorff-Larsen and Winther (2000)
$[D_4]$2-vinylpyridine	Deuterium-labelled reagent potentially useful for quantitative proteomics.	Sebastiano et al. (2003)
4-vinylpyridine	Regularly used for cysteine alkylation in protocols for 2-DE and MS peptide analysis.	Friedman et al. (1970), Sechi and Chait (1998)

[a]In brackets, the appropriate wavelengths are indicated for chromophores (absorption wavelength) and fluorophores (extinction wavelength/emission wavelength).

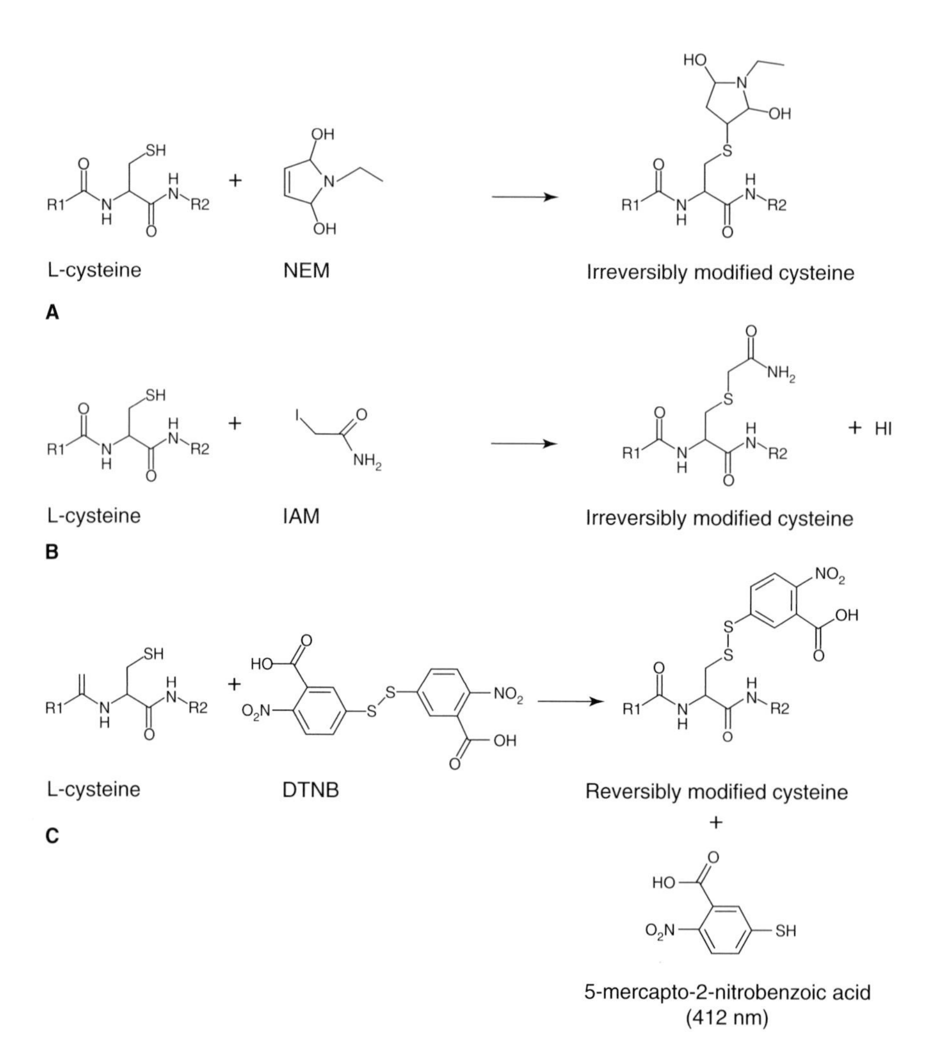

Figure 4.2 Reactions of L-cysteinyl residues with thiol-specific reagents. Examples of reactions between L-cysteinyl residues and thiol-specific reagents. **A:** NEM and other thiol-specific reagents containing reactive double bonds form irreversible conjugates with L-cysteinyl residues through addition reactions. **B:** IAM and other alkyl halides react irreversibly with L-cysteinyl residues through substitution reactions. **C:** Dithiol reagents such as DTNB react reversibly with L-cysteinyl residues to form a mixed disulphide. The released chromophore 5-mercapto-2-nitrobenzoic acid can be monitored spectrophotometrically at 412 nm. R1: N-terminal continuation of the peptide chain; R2: C-terminal continuation of the peptide chain.

charged groups can be used for immobilized metal affinity chromatography and cation exchange chromatography (Laragione et al., 2003; Ren et al., 2004; Clements et al., 2005). Modification of cysteines can also alter protein digestion patterns. Aminoethylation with bromoethylamine converts cysteine into a target for trypsin digestion, and the N-terminal peptide bond of cysteine residues cyanylated with 1-cyano-4-dimethylamino-pyridinium tetrafluoroborate is cleaved under alkaline conditions (Wu and Watson, 1997; Thevis et al., 2003). Other promising reagents for proteome analysis of cysteine oxidoreduction are isotope-coded thiol-specific reagents such as isotope-coded affinity tags (ICAT) reagents regularly used for quantitative proteomics (Gygi et al., 1999). Sethuraman et al. (2004) introduced a technique to quantify oxidation of individual cysteine residues in complex mixtures of proteins using ICAT reagents. Briefly, a sample of a protein mixture exposed to H_2O_2 was treated with the ^{13}C-isotope-coded 'heavy' ICAT reagent and the control sample treated with the ^{12}C-isotope-coded 'light' ICAT reagent. The two samples were combined and then digested with trypsin. Labelled peptides were isolated by avidin affinity chromatography and analysed by liquid chromatography (LC) coupled to tandem MS (LC-MS/MS). Since only reduced thiol groups react with the ICAT reagents, a peptide containing a cysteine oxidized by H_2O_2 will have a light/heavy ratio higher than one. Deuterium-labelled thiol-specific reagent such as [2,3,3′-D_3]acrylamide, [D_4]2-vinylpyridine and [D_{15}](4-iodobutyl)triphenyl-phosphonium ([D_{15}]IBTP), are also potentially useful for quantification of cysteine oxidation (Sechi and Chait, 1998; Sebastiano et al., 2003; Marley et al., 2005). An alternative approach that can be used for quantitative analysis of cysteine oxidation is two-dimensional difference gel electrophoresis (2-D DIGE). In this method, which is conceptually similar to the ICAT approach, the samples to be compared are each labelled with a spectrally distinct thiol-specific fluorescent reagent, combined and separated by 2-DE. Chan et al. (2005) used iodoacetylated cyanine dyes in a 2-D DIGE approach to analyse the redox status of proteins under conditions of oxidative stress.

The chemical structure of cysteine-reactive compounds can influence their applicability for proteome research. Some reagents have been shown to provide increased ionization efficiency in MS analysis (Ren et al., 2004). Furthermore, it has been shown that certain reagents elicit specific fragmentation patterns that are useful for identification of modified peptides by MS/MS (Borisov et al., 2002; Borges and Watson, 2003). When applied to 2-DE, the choice of cysteine reagent can influence streaking (Luche et al., 2004) and charged reagents may cause migration shifts in isoelectric focusing (Shaw et al., 2003).

4.3.2 Disulphide mapping

In spite of the importance of disulphide bonds in structure and regulation, it is still not possible to predict the disulphide bond patterns of proteins from the amino acid sequence, or even to predict whether or not disulphide bonds are likely to be formed. Analysis of protein disulphide bond patterns is often very challenging. For example, the complex inter- and intramolecular disulphide bonding in seed storage proteins is still not well characterized despite considerable effort (Cunsolo et al., 2004). For purified proteins, an established and widely used approach for determination of intramolecular disulphide pairing consists in proteolytic digestion of non-reduced proteins followed by chromatographic peptide separation in reducing and non-reducing conditions. Disulphide paired peptides, which display changed retention times after reduction, are selected for identification by MS (Gorman et al., 2002). Mildly acidic digestion conditions are preferable, since thiol-disulphide exchange reactions are known to occur at neutral, alkaline and strongly acidic pH (Bailey, 1967). Pepsin has often been used for this purpose (Gorman et al., 2002) and its broad specificity is an advantage in disulphide mapping, since peptide fragments containing a single cysteine residue are often generated. However, protein digestion with pepsin often results in a complex mixture of overlapping peptides which may complicate identification by peptide mass fingerprinting. To overcome these problems, Wallis et al. (2001) developed a method adding ^{18}O-labelled water to the digestion mixtures. Briefly, ^{18}O is incorporated in the released C-termini of the produced peptides, and disulphide-bound peptic fragments will thus contain two ^{18}O provided that neither of the fragments contain the C-terminus of the intact protein. In contrast, single chain fragments, regardless of whether they contain a disulphide bond, incorporate only one ^{18}O. MS analysis of the isotope distribution of ^{18}O-labelled peptic fragments can thus distinguish disulphide-paired peptides from other peptides. If proteolytic digestions are carried out at non-acidic pH, the free cysteines are blocked first, to avoid thiol-disulphide exchange (Yen et al., 2000).

Disulphide bonds in digested peptides are commonly validated by comparing mass spectra acquired before and after *in vitro* chemical reduction (e.g. using dithiothreitol (DTT)). Under some circumstances this reduction can be carried out directly in a matrix spot on a matrix-assisted laser desorption/ionization (MALDI) MS target (Fischer et al., 1993). Information about disulphide bond linkages may also be obtained *in situ* in the mass spectrometer. For example, fragmentation of disulphide bonds by in-source or post-source decay has been demonstrated in MALDI time-of-flight (TOF) MS and MALDI ion trap MS (Patterson and Katta, 1994; Qin and Chait, 1997; Jones et al., 1998). Furthermore, cleavage of disulphide bonds is particularly pronounced in the electron capture dissociation MS/MS fragmentation technique that is used in conjunction with Fourier transform ion cyclotron MS (Zubarev et al., 1999).

In the complex protein mixtures that are subject to proteome analysis, it is very difficult to determine the complete disulphide bond connectivity. A simpler objective may be to identify proteins linked by intermolecular disulphide bonds. Such proteins can be identified using 2-D diagonal electrophoresis, a concept originally developed by Sommer and Traut (1974). In the first dimension, proteins are separated by non-reducing SDS-PAGE and in the second dimension – at right angles to the initial direction – by reducing SDS-PAGE. Thus, proteins not linked by intermolecular disulphide bonds will line up on the diagonal, whereas protein partners in intermolecular disulphide bonds will dissociate from each other in the second dimension and migrate below the diagonal. This method combined with MS has been used to identify proteins forming intermolecular disulphide bonds in human myocytes during conditions of oxidative stress (Brennan et al., 2004).

In extracts from stem and leaf of *Arabidopsis thaliana*, Lee et al. (2004) used a novel proteomics approach to identify 65 proteins forming disulphide bonds *in vivo*.

In the first step of this procedure free thiol groups in the extracted proteins were blocked by a combination of IAM and NEM under denaturating conditions. The proteins were treated with tributylphosphine to reduce disulphides and then subjected to a thiol-affinity chromatography with sepharose-bound glutathione 2-pyridyl disulphide as the active group. In the control experiment, treatment with tributylphosphine was omitted. Using this procedure, the proteins containing disulphides *in vivo* were trapped on the affinity resin, subsequently eluted by DTT, separated by SDS-PAGE and identified by MALDI-TOF MS after tryptic digestion. The identified proteins included cytosolic proteins, proteins processed via the secretory pathway, membrane proteins and chloroplastic proteins including four known Trx targets. Also identified was ADP-glucose pyrophosphorylase (AGPase) involved in starch synthesis. In potato, AGPase forms an inactive dimer that is activated by *in vitro* treatment with DTT (Tiessen et al., 2002). Taken together, these studies support a role for redox regulation in the control of starch synthesis.

4.3.3 S-glutathionylation

In studies of plants and plant cells with suppressed glutathione synthesis, a requirement for GSH has been suggested for many developmental processes, for example flowering (Ogawa et al., 2001) and transition from the vegetative to the reproductive growth stage (Yanagida et al., 2004). This emphasizes the vital importance of glutathione for maintenance of the cellular redox state and protection against irreversible oxidation of cysteines.

One of the best characterized examples of the involvement of *S*-glutathionylation in regulatory signalling is the redox regulation of mammalian protein tyrosine phosphatases (PTPs). Growth factors and cytokines that stimulate cell differentiation and proliferation induce a slight increase in intracellular ROS concentration (Meier et al., 1989; Bae et al., 1997). When treated with ROS *in vitro*, PTPs are inactivated by oxidation and *S*-glutathionylation of the active site cysteine (Denu and Tanner, 1998; Lee et al., 1998; Barrett et al., 1999a). Moreover, in a cell culture treated with epidermal growth factor (EGF), the amount of the reduced form of PTP decreased (Lee et al., 1998), and the glutathionylated active site cysteine in PTP was detected by immunoprecipitation and MS (Barrett et al., 1999b). Tyrosine phosphorylation is involved in various processes in plants such as pollen development (Gupta et al., 2002) and stomatal closure (MacRobbie, 2002). Redox regulation of PTPs is likely also to occur in plants, as inhibition of two PTPs from *Arabidopsis thaliana* by H_2O_2 (Meinhard and Grill, 2001; Gupta and Luan, 2003), and *S*-glutathionylation of a soybean PTP (Dixon et al., 2005a) have been demonstrated *in vitro*. Some transcription factors are also regulated by reversible oxidations and *S*-glutathionylation of cysteine residues. Reversible *S*-glutathionylation of the transcription factor, nuclear factor I, has been suggested *in vitro* and *in vivo* to result in loss of DNA-binding activity (Bandyopadhyay et al., 1998).

Immobilized glutathione or nitrosoglutathione have been used for isolation of proteins susceptible to *in vitro* *S*-glutathionylation from a mixture of proteins (Klatt et al., 2000; Eaton and Shattock 2002). Proteins that are *in vivo* *S*-glutathionylated under conditions of oxidative stress have been identified from human cells using a strategy for radiolabelling of cellular glutathione (Fratelli et al., 2002; Ghezzi and Bonetto, 2003). Protein synthesis was blocked by cycloheximide and [^{35}S]-labelled L-cysteine was added to cell cultures to allow selective incorporation of ^{35}S into glutathione. The cells were then exposed to oxidants, diamide (*N,N,N′,N′*-tetramethylazodicarboxamide) and H_2O_2, to induce *S*-glutathionylation and the *S*-glutathionylated proteins were detected by 2-DE and autoradiography. Alternatively, *in vivo S*-glutathionylated proteins can be detected by western blotting using anti-GSH antibodies (Wang et al., 2001; McDonagh et al., 2005).

Another approach used Grx for identification of *in vivo S*-glutathionylated proteins (Lind et al., 2002; Hamnell-Pamment et al., 2005). Here, accessible protein thiols were blocked *in vivo* with NEM and glutathione mixed disulphides were specifically reduced by Grx system after protein extraction. The newly liberated thiol groups were labelled with biotin-maleimide, and the biotionylated proteins were isolated by avidin affinity chromatography and separated by 2-DE. In a gel-free approach, proteins treated likewise with NEM, Grx system and biotin-maleimide were digested with trypsin and the biotionylated peptides were isolated by avidin affinity chromatography for analysis by LC-MS/MS and identification of the *S*-glutathionylated cysteine residues (Hamnell-Pamment et al., 2005).

A cell permeable ethyl ester derivative of GSH conjugated to biotin via a primary amine (Sullivan et al., 2000) has been used to identify proteins that are *S*-glutathionylated *in vivo* in *Arabidopsis thaliana* cell cultures (Ito et al., 2003). Briefly, the extracted proteins were fractionated by gel filtration, ion exchange and hydrophobic interaction chromatography and the biotinylated proteins were detected by using streptavidin-conjugated horseradish peroxidase. A cytosolic triose phosphate isomerase and a putative plastidic aldolase involved in the pathways of glycolysis and gluconeogenesis, respectively, were identified as *S*-glutathionylated proteins. *Arabidopsis* triose phosphate isomerase was produced in recombinant form and shown to be inhibited by *S*-glutathionylation. Inhibition of sugar metabolizing enzymes is suggested to occur under conditions of increased cellular ROS concentrations and negatively regulate the corresponding metabolic pathways to prevent further production of ROS in mitochondria and chloroplasts.

In a similar approach reported recently, *Arabidopsis* cell culture was treated with GSSH conjugated to biotin via a primary amine (GSSG-Biotin). *In vivo S*-glutathionylated proteins were affinity isolated on avidin columns and separated by 2-DE (Dixon et al., 2005b). A total of 79 proteins comprising *S*-glutathionylated proteins and proteins in complex with *S*-glutathionylated proteins were identified by MS (Dixon et al., 2005b).

4.3.4 *Cysteine SOH, SO_2H and SO_3H*

SOH represents a reversible modification of cysteine residues that is mainly formed after exposure to diverse oxidative conditions. In specific cases, SOH is also part of the catalytic cycle of enzymes, for example peroxiredoxins involved in ROS removal (Claiborne et al., 2001). A proteomic method for identification of proteins susceptible to oxidation to SOH was described by Saurin et al. (2004). This procedure was based on selective reduction of SOH in protein extracts by arsenite and subsequent labelling of the reduced thiols by biotin-maleimide. By completely alkylating protein thiols prior to treatment with arsenite, proteins containing SOH were selectively biotinylated, detected by western blotting with streptavidin-conjugated horseradish peroxidase and isolated by avidin affinity chromatography. The method was applied to protein extracts from rat cardiac tissue.

SOH is a reactive species that can be reduced, form a disulphide, or be further oxidized to SO_2H and SO_3H. In contrast to SOH, SO_2H and SO_3H are more stable and are not reduced by common reductants such as DTT. Thus, oxidation to SO_2H and SO_3H can be observed as shifts in the migration of proteins in the first dimension of 2-DE due to the introduction of negative charges (Rabilloud et al., 2002; Wagner et al., 2002). These modifications can also be identified after enzymatic digestion using endoproteinase Asp-N that cleaves peptides on the N-terminal side of aspartic acid, SO_2H and SO_3H, but not cysteine (Drapeau, 1980). SO_2H and SO_3H modifications can thus be identified by peptide mass fingerprinting (Wagner et al., 2002). Furthermore, the mass increase of 32 Da (SO_2H) and 48 Da (SO_3H) relative to unmodified cysteine allows site-specific assignment of modified residues using MS/MS (Rabilloud et al., 2002).

4.3.5 Trxs and disulphide reduction

Trx was originally identified in *E. coli* as an electron donor for ribonucleotide reductase (Laurent et al., 1964). In addition to transferring reducing equivalents, Trxs specifically regulate the activity of various target proteins by disulphide bond reduction. Trx reduced and regulated the activity of the transcription factors NFκB and Ap-1 when transiently over-expressed in human cell cultures, suggesting regulatory roles of Trx in transcriptional processes (Schenk et al., 1994).

In plants, numerous Trx isoforms show diverse sub-cellular localizations (Section 4.2.1). Chloroplastic Trxs were originally identified as factors essential for photoactivation of the Calvin cycle enzyme fructose-1,6-bisphosphatase by reduction of its regulatory disulphide bond (Buchanan et al., 1971; Wolosiuk and Buchanan, 1977). In this light-dependent regulatory pathway, reducing equivalents are transferred from photosystem I to ferredoxin, an iron–sulphur protein and subsequently to Trxs via ferredoxin-dependent Trx reductase (Schürmann and Jacquot, 2000). The mainly cytosolic Trx h is abundant in seeds of different plants and is thought to be involved in the germination process (Yano et al., 2001; Marx et al., 2003). Although the functions of Trx h in seeds are poorly understood, germinating transgenic barley seeds over-expressing wheat Trx h show higher limit dextrinase activity and earlier increase in α-amylase activity compared to the non-transgenic seeds (Cho et al., 1999; Wong et al., 2002). Storage proteins in cereals are released from a disulphide-bound and poorly soluble form during germination, and Trx h has been shown to enhance this process *in vitro* (Kobrehel et al., 1992).

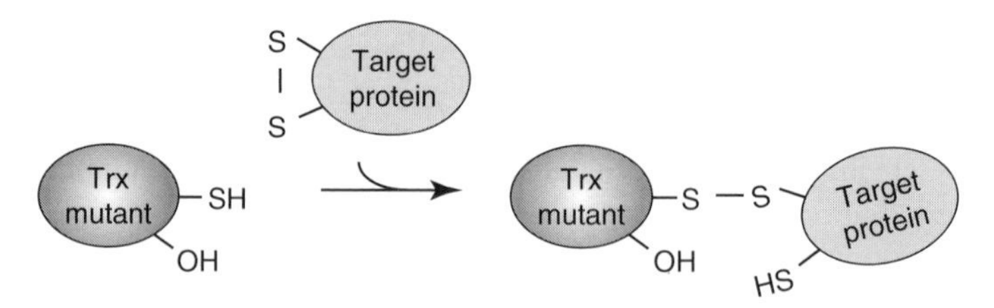

Figure 4.3 Formation of disulphide-bound complexes between thioredoxin Cys–Ser mutants and target proteins. The SH group of the most N-terminal cysteine (Cys_1) in the Trx active site carries out the initial nucleophilic attack on the target disulphide to form a intermolecular disulphide bond, whereas Cys_2 carries out the subsequent release of the reduced target protein. Thus, Trx where Cys_2 is mutated to serine forms disulphide-bound complexes with target proteins.

Recently, proteomics techniques have been applied in several studies for identification of Trx target proteins in plants. The approaches used can be categorized into three groups based on the techniques used to detect the target proteins: (1) isolation of target proteins by utilizing the formation of stable intermolecular disulphide bonds with Cys_2–Trx mutants (Verdoucq et al., 1999); (2) *in vitro* target reduction with Trxs followed by labelling of accessible thiol groups and detection of labelled proteins after 2-DE (Yano et al., 2001); (3) affinity isolation of protein–protein interaction partners of Trxs (Kumar et al., 2004).

The first strategy is based on knowledge of the reaction mechanism for the Trx catalysed reduction of disulphides in target proteins (Section 4.2.2). Thus, covalent intermediate Trx complexes with target proteins have been obtained with Trx having Cys_2 mutated to serine or alanine (Figure 4.3). By introducing polyhistidine extensions on such mutants, complexes formed between target proteins and Trx mutants can be isolated by nickel chromatography. This approach was used in yeast transformed with a mutant of *Arabidopsis* Trx h resulting in identification of a novel type of peroxiredoxin as an *in vivo* target (Verdoucq et al., 1999). Similarly, by using an immobilized Trx h mutant, a peroxiredoxin from *Chlamydomonas reinhardtii* was identified and shown to use Trx h as a hydrogen donor (Goyer et al., 2002). Immobilized Trx mutants have also been used in various other *in vitro* isolations of plant target proteins. Potential target proteins of Trx h have been identified from *Chlamydomonas reinhardtii* (Lemaire et al., 2004), from developing wheat seeds (Wong et al., 2004) and from *Arabidopsis* leaves (Yamazaki et al., 2004). Potential target proteins of f- and m-type chloroplastic Trx have also been identified from a stroma lysate of spinach chloroplasts (Motohashi et al., 2001; Balmer et al., 2003).

Fluorescence labelling of thiol groups has been applied in a survey of Trx targets in peanut seed (Yano et al., 2001). In this study, protein extracts were incubated in the presence of the Trx system (NADPH, Trx reductase and *E. coli* Trx) and subsequently labelled with the thiol-specific fluorescent reagent monobromobimane (mBBr). Control extracts were incubated in the absence of the Trx system. After separation by 2-DE, the labelled proteins were detected under ultraviolet light. By comparing

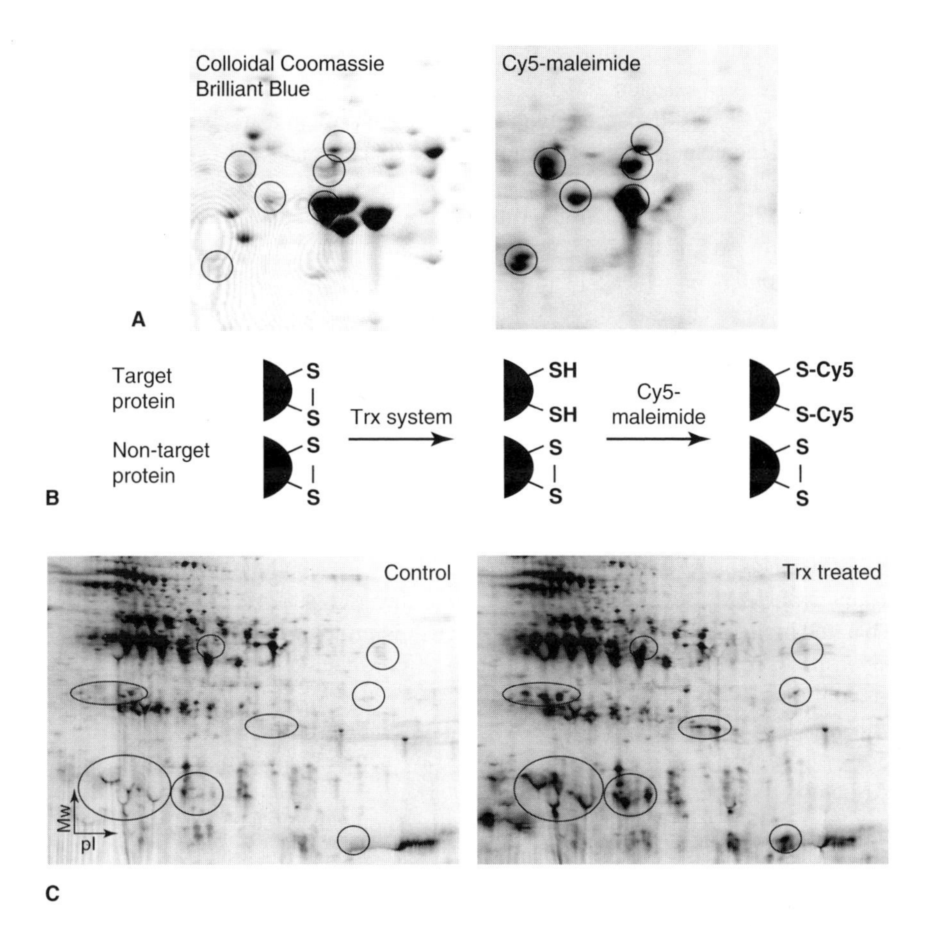

Figure 4.4 A strategy for detection of proteins targeted by Trx based on Cy5 maleimide-labelling of accessible thiol groups and 2-DE. A: Close-up view of a region in a 2D-gel of mature barley seed proteins treated with Cy5 maleimide (without reduction by Trx). Colloidal Coomassie Brilliant Blue-stained proteins (left) and Cy5 maleimide-labelled proteins (right). Proteins labelled with Cy5 maleimide are indicated with circles. **B:** The procedure for target protein identification of Trx based on Cy5 maleimide labelling (see Section 4.3.5). **C:** Cy5 maleimide-labelled proteins in barley seed extracts treated with (right) and without Trx system (left). Additional Cy5 labelled spots in the Trx-treated sample are indicated with circles. Modified from Maeda et al. (2004), with permission.

the resulting fluorescence spot patterns for the control and the Trx-treated sample, a number of proteins reduced *in vitro* by Trx were in-gel digested and identified by MS. A similar approach was used for identification of Trx-target proteins in embryos of germinated barley seeds (Marx et al., 2003). One disadvantage with mBBr is the relatively low sensitivity, thus only abundant proteins can be detected. This motivated the development of a procedure for highly sensitive and specific detection of proteins with accessible thiol groups, using the fluorescent reagent Cy5 maleimide (Maeda et al., 2004; Figure 4.4A). This method was applied on mature and germinated barley seeds, resulting in identification of 16 potential Trx h target proteins by MS analysis of tryptic digests of selected 2D-gel spots (Figure 4.4B,C). In a target protein survey in *Arabidopsis* leaves, the accessible thiol groups in protein extracts were alkylated prior to Trx treatment in order to minimize background signal, and ^{14}C-labelled IAM was used for labelling of protein thiols generated by Trxs (Marchand et al., 2004).

The proteins that have been identified as possible targets of Trxs are diverse in both structure and function, and no common features apart from the necessity for a disulphide bond have yet been identified that would provide a basis for prediction of Trx targets. The techniques discussed above can also result in 'false positive' identifications due to non-specific interactions in affinity chromatography approaches and the presence of multiple proteins in a single 2D-gel spot. For these reasons, an important step in validation of potential target proteins is to specify the disulphide bond that is targeted by Trx. Such information is also valuable for attempts to determine putative specificity determinants for Trxs. To identify Trx-reducible disulphide bonds in a complex mixture of proteins, a differential labelling technique was developed (Maeda et al., 2005; Figure 4.5A). An extract of barley seed proteins was incubated in the presence or absence of barley Trx h followed by labelling of accessible thiol groups with IAM. After desalting to remove the excess IAM, the samples were separated by 2-DE. Prior to the second dimension, proteins in the isoelectric focusing strips were fully reduced and remaining free cysteines were labelled with 4-vinylpyridine. Thus, cysteines involved in Trx-reducible disulphides are modified with 4-vinylpyridine in the control and with IAM in the Trx-treated sample. The two forms of modifications can be distinguished by peptide mass fingerprinting since IAM and 4-vinylpyridine add 57 Da and 105 Da to a labelled cysteine, respectively (resulting in a mass difference for differentially labelled peptides of Δ 48 Da; Figure 4.5B). This procedure was used to analyse proteins in 2D-gel spots that were indicated to contain target proteins on the basis of Cy5 maleimide-labelling patterns (Maeda et al., 2004). By comparing the peptide masses obtained from samples incubated in the presence or absence of Trx, 9 disulphides in 8 proteins were identified as reducible by Trx h. This study provided an indication that some disulphides are more susceptible to Trx h reduction than others. Preferential reduction of one of the two surface-accessible disulphide bonds was observed in the barley α-amylase/subtilisin inhibitor (BASI) (Maeda et al., 2005).

Most of the approaches described so far for identification of Trx target proteins have been applied *in vitro* and therefore only indicate the susceptibility of proteins to undergo reduction. The identifications made using these methods include proteins involved in photosynthesis and carbon metabolism, defence against pathogens, oxidative stress responses and protein folding. Validation of these putative target proteins by other approaches will be required to confirm that they are targeted by Trx.

Finally, proteomics has been used to identify proteins involved in electrostatic protein–protein interactions with Trxs. Although these proteins are not necessarily reduced by Trxs, the interactions may influence the physiological roles of Trxs in redox-regulation mechanisms. In *E. coli*, 80 proteins associated *in vivo* with Trx were identified using a so-called tandem affinity purification tag containing a protein A

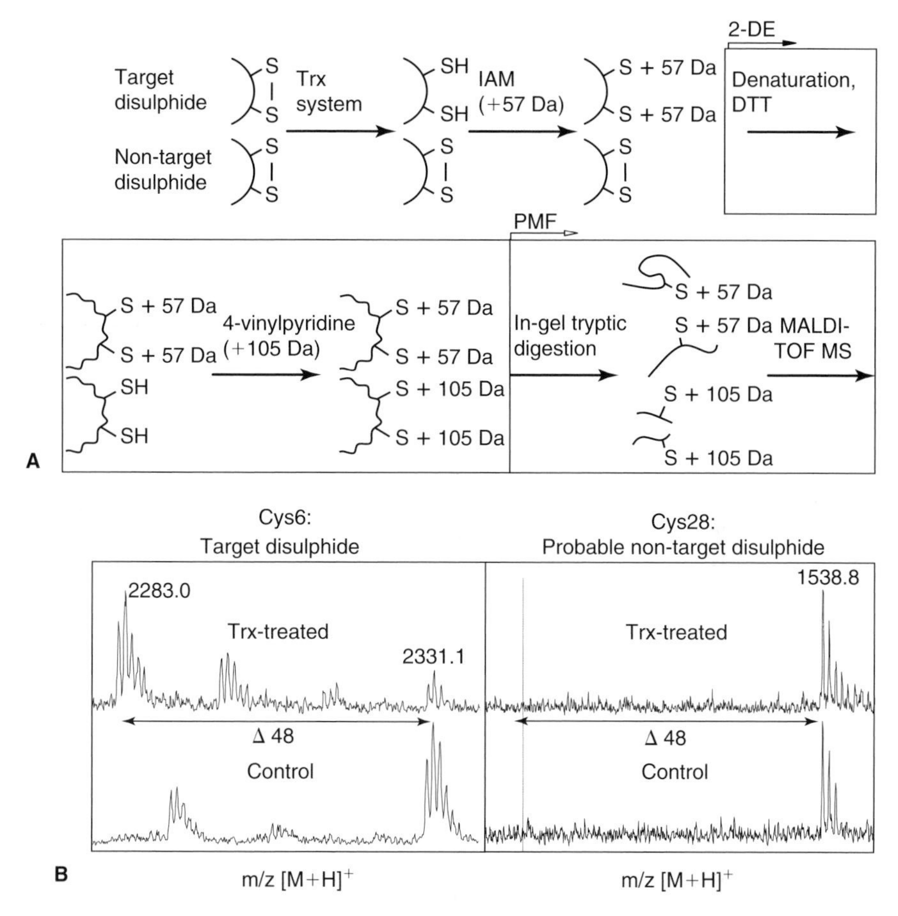

Figure 4.5 Identification of Trx reducible disulphide bonds in barley seed proteins based on differential labelling, 2-DE and MALDI-TOF MS. A: Overview of the differential labelling procedure for identification of Trx h reducible protein disulphides (see Section 4.3.5). **B:** Sections of MALDI-TOF MS spectra for the tryptic peptides of α-amylase inhibitor BMAI-1, showing differential labelling of Cys6 and Cys28, expected to be involved in different disulphide bonds (Maeda et al., 2005). In control (lower spectra), two peaks corresponding to residue 1-19 ($[M + H]^+$ 2331.1 (left)) and residue 26-39 ($[M + H]^+$ 1538.8. (right)) covering Cys6 and Cys28, respectively, in the pyridylethylated form were observed. When treated with Trx h (upper spectra), an additional peak of residue 1-19 at $[M + H]^+$ 2283.0 covering carbamidomethylated Cys6 appeared, showing that Cys6 in BMAI-1 is involved in a Trx h reducible disulphide. A peptide containing the carbamidomethylated form of Cys28 was not observed, indicating that the disulphide bond involving Cys28 is not Trx h reducible. Modified from Maeda et al. (2005), with permission.

moiety and a calmodulin-binding peptide (Kumar et al., 2004). A construct encoding Trx with the tag appended to the C-terminus was expressed in *E. coli*. Complexes associated with the tagged Trx were purified by applying *E. coli* culture extracts sequentially onto immobilized immunoglobulin G and calmodulin. Interaction partners of a Trx from the slime mould *Dictyostelium discoideum* have been identified using yeast two-hybrid analysis (Brodegger et al., 2004). Two of the putative interaction partners identified in this study were also shown to form stable mixed disulphides with a Trx Cys_2 mutant.

To screen for novel interaction partners of Trx f, spinach chloroplast proteins were applied to a column of immobilized Trx f and the retained proteins were separated by 2-DE (Balmer et al., 2004). In this study 27 new Trx f interaction partners were identified, including some proteins involved in translation, protein assembly and adenosine triphosphate (ATP) synthesis.

4.3.6 S-nitrosylation

NO is a short-lived free radical acting as signalling molecule in a variety of processes in plants and other organisms. NO is easily diffusible and reacts with proteins and various other targets. In plants, NO has been shown to be an important messenger in defence against pathogens (Delledonne et al., 1998; Durner et al., 1998). NO also regulates plant developmental processes by promoting germination, leaf senescence and fruit maturation (reviewed in Delledonne, 2005). Furthermore, NO has been shown to delay apoptosis in aleurone cells of barley seeds (Beligni et al., 2002), and induce expression of disease-related genes in tobacco (Durner et al., 1998). This signalling is in part mediated by cyclic guanosine monophosphate (cGMP), a low-molecular-weight second messenger (Durner et al., 1998).

Cysteine nitrosylation is reversible and can modify the properties of proteins. The thiol group in GSH can also react with NO to form *S*-nitrosoglutathione that functions as an NO donor for protein thiols (Kluge et al., 1997; Tsikas et al., 1999). In humans, suppression of apoptosis by NO involves *S*-nitrosylation and inhibition of caspase, a cysteine protease required for apoptosis (Kim et al., 1997, 1998; Li et al., 1999).

Jaffrey et al. (2001) developed the biotin switch method (Figure 4.6) for isolation and identification of *in vitro* and *in vivo* *S*-nitrosylated proteins (Kuncewicz et al., 2003, Martinez-Ruiz et al., 2005). Briefly, free thiol forms of cysteine residues are blocked with the thiol-reactive reagent, methyl methanethiosulphonate. Subsequently, SNO groups are specifically reduced with ascorbate and the resulting free thiol groups are labelled with the thiol-specific biotin reagent, *N*-[6-(biotinamido) hexyl]-3′-(2′-pyridyldithio)propionamide (Biotin-HPDP).

Using the biotin switch method in combination with nano-LC-MS/MS, 63 *in vitro* *S*-nitrosylated proteins were identified in *Arabidopsis* cell cultures (Lindermayr et al., 2005). These included stress-related proteins, signalling and regulating proteins, redox-related proteins, cytoskeleton proteins, metabolic proteins and others. Moreover, 52 proteins from *Arabidopsis* leaves were identified to be *in vivo* *S*-nitrosylated, including several photosynthetic enzymes. These results suggest that *S*-nitrosylation, in addition to Trx-mediated regulation, is involved in the control of photosynthesis mediated through cysteine modifications (Lindermayr et al., 2005).

S-nitrosylation can also be detected using anti-SNO antibodies. These have been used to detect *S*-nitrosylation *in vivo* (Gow et al., 2002) and for immunoprecipitation

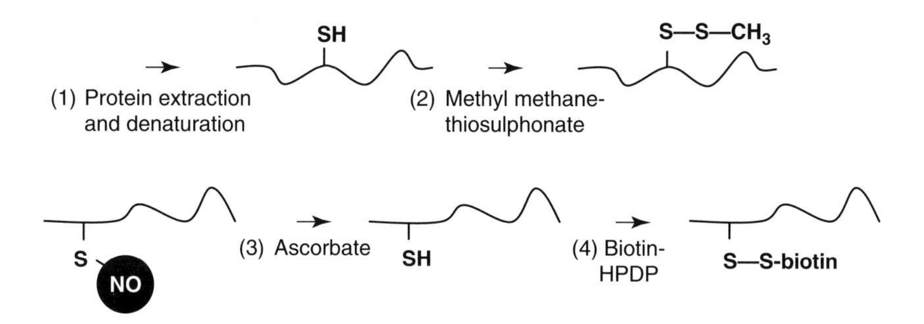

Figure 4.6 The biotin switch method for identification of *S*-nitrosylated proteins. Following protein extraction and denaturation (1), protein thiols are methylthiolated by treatment with methyl methanethiosulphonate (2). Subsequently, nitrosylated cysteine residues are reduced to the thiol forms by treatment with ascorbate (3) and the newly liberated thiol groups are labelled with a biotin-conjugated thiol-specific reagent, biotin-HPDP (4) to specifically purify *S*-nitrosylated proteins by avidin affinity chromatography (Jaffrey et al., 2001).

of *S*-nitrosylated proteins (Martinez-Ruiz et al., 2005). Despite the lability of SNO group, site-specific determination of modified cysteine residues has been successfully demonstrated for *in vitro S*-nitrosylated proteins using MS/MS (Martinez-Ruiz et al., 2005; Taldone et al., 2005).

4.4 Conclusions and perspectives

The importance of cysteine oxidoreduction in control of plant signalling and metabolism is becoming more and more apparent as proteomics techniques have been adapted for detection of cysteine modifications. The unique reactivity of cysteine thiol groups means that they are susceptible to modification by many molecules present in plant cells under different conditions, and it is likely that even more modification mechanisms remain to be identified. For each modification identified there may also be an associated regulatory protein to catalyse its removal, as in the case of the recently identified sulphiredoxin (Biteau et al., 2003).

Proteomics studies have led to the identification of many proteins that undergo cysteine oxidoreduction under various experimental conditions *in vitro* and *in vivo*. However, proteomics approaches require improvements if the whole picture of cysteine oxidoreduction in proteins is to be resolved. Proteome analysis of cysteine oxidoreduction has mostly been focused on abundant water-soluble proteins. Moreover, the identified candidate proteins for redox regulation must be validated by other means before the biological significance of the regulatory mechanism can be evaluated. In order to be biologically relevant, an experimentally observed cysteine oxidoreduction should occur under physiological conditions in response to a signal. The modification must also affect the function of the protein. Thus, although large numbers of proteins susceptible to cysteine modification have been identified by proteomics to date, challenging tasks of characterizing their biological relevance still remain.

The recent development of proteomics tools such as ICAT for quantitative determination of cysteine modifications holds great potential for comparative analysis of cysteine modifications in individual proteins from complex samples. These tools will undoubtedly further our understanding of oxidative modifications and the mechanisms behind redox regulation in plants.

Acknowledgements

The authors acknowledge the Danish Technical Research Council (STVF) for financial support. C.F. is supported by a grant from the Danish Agricultural and Veterinary Research Council (SJVF). K.M. is supported by a Ph.D. scholarship from the Technical University of Denmark.

References

Åslund, F., Berndt, K.D. and Holmgren, A. (1997) Redox potentials of glutaredoxins and other thiol-disulfide oxidoreductases of the thioredoxin superfamily determined by direct protein-protein redox equilibria. *J. Biol. Chem.*, **272**, 30780–30786.

Bae, Y.S., Kang, S.W., Seo, M.S., Baines, I.C., Tekle, E., Chock, P.B. and Rhee, S.G. (1997) Epidermal growth factor (EGF)-induced generation of hydrogen peroxide. Role in EGF receptor-mediated tyrosine phosphorylation. *J. Biol. Chem.*, **272**, 217–221.

Bailey, J.L. (1967) *Techniques in Protein Chemistry*, Elsevier publishing company, Amsterdam.

Bailly, C. (2004) Active oxygen species and antioxidants in seed biology. *Seed Sci. Res.*, **14**, 93–107.

Balmer, Y., Koller, A., del Val, G., Manieri, W., Schurmann, P. and Buchanan, B.B. (2003) Proteomics gives insight into the regulatory function of chloroplast thioredoxins. *Proc. Natl Acad. Sci. USA*, **100**, 370–375.

Balmer, Y., Koller, A., del Val, G., Schürmann, P. and Buchanan, B.B. (2004) Proteomics uncovers proteins interacting electrostatically with thioredoxin in chloroplasts. *Photosynth. Res.*, **79**, 275–280.

Bandyopadhyay, S., Starke, D.W., Mieyal, J.J. and Gronostajski, R.M. (1998) Thioltransferase (glutaredoxin) reactivates the DNA-binding activity of oxidation-inactivated nuclear factor I. *J. Biol. Chem.*, **273**, 392–397.

Barrett, W.C., DeGnore, J.P., Keng, Y.F., Zhang, Z.Y., Yim, M.B. and Chock, P.B. (1999a) Roles of superoxide radical anion in signal transduction mediated by reversible regulation of protein-tyrosine phosphatase 1B. *J. Biol. Chem.*, **274**, 34543–34546.

Barrett, W.C., DeGnore, J.P., Konig, S., Fales, H.M., Keng, Y.F., Zhang, Z.Y., Yim, M.B. and Chock, P.B. (1999b) Regulation of PTP1B via glutathionylation of the active site cysteine 215. *Biochemistry*, **38**, 6699–6705.

Baty, J.W., Hampton, M.B. and Winterbourn, C.C. (2002) Detection of oxidant sensitive thiol proteins by fluorescence labeling and two-dimensional electrophoresis. *Proteomics*, **2**, 1261–1266.

Beer, S.M., Taylor, E.R., Brown, S.E., Dahm, C.C., Costa, N.J., Runswick, M.J. and Murphy, M.P. (2004) Glutaredoxin 2 catalyzes the reversible oxidation and glutathionylation of mitochondrial membrane thiol proteins: implications for mitochondrial redox regulation and antioxidant defense. *J. Biol. Chem.*, **279**, 47939–47951.

Beligni, M.V., Fath, A., Bethke, P.C., Lamattina, L. and Jones, R.L. (2002) Nitric oxide acts as an antioxidant and delays programmed cell death in barley aleurone layers. *Plant Physiol.*, **129**, 1642–1650.

Biteau, B., Labarre, J. and Toledano, M.B. (2003) ATP-dependent reduction of cysteine-sulphinic acid by *S. cerevisiae* sulphiredoxin. *Nature*, **425**, 980–984.

Boja, E.S. and Fales, H.M. (2001) Overalkylation of a protein digest with iodoacetamide. *Anal. Chem.*, **73**, 3576–3582.

Borges, C.R. and Watson, J.T. (2003) Recognition of cysteine-containing peptides through prompt fragmentation of the 4-dimethylaminophenylazophenyl-4′-maleimide derivative during analysis by MALDI-MS. *Protein Sci.*, **12**, 1567–1572.

Borisov, O.V., Goshe, M.B., Conrads, T.P., Rakov, V.S., Veenstra, T.D. and Smith, R.D. (2002) Low-energy collision-induced dissociation fragmentation analysis of cysteinyl-modified peptides. *Anal. Chem.*, **74**, 2284–2292.

Brennan, J.P., Wait, R., Begum, S., Bell, J.R., Dunn, M.J. and Eaton, P. (2004) Detection and mapping of widespread intermolecular protein disulfide formation during cardiac oxidative stress using proteomics with diagonal electrophoresis. *J. Biol. Chem.*, **279**, 41352–41360.

Brodegger, T., Stockmann, A., Oberstrass, J., Nellen, W. and Follmann, H. (2004) Novel thioredoxin targets in *Dictyostelium discoideum* identified by two-hybrid analysis: interactions of thioredoxin with elongation factor 1alpha and yeast alcohol dehydrogenase. *Biol. Chem.*, **385**, 1185–1192.

Brune, D.C. (1992) Alkylation of cysteine with acrylamide for protein sequence analysis. *Anal. Biochem.*, **207**, 285–290.

Buchanan, B.B. and Balmer, Y. (2005) Redox regulation: a broadening horizon. *Annu. Rev. Plant Biol.*, **56**, 187–220.

Buchanan, B.B., Schurmann, P. and Kalberer, P.P. (1971) Ferredoxin-activated fructose diphosphatase of spinach chloroplasts. Resolution of the system, properties of the alkaline fructose diphosphatase component, and physiological significance of the ferredoxin-linked activation. *J. Biol. Chem.*, **246**, 5952–5959.

Chan, H.L., Gharbi, S., Gaffney, P.R., Cramer, R., Waterfield, M.D. and Timms, J.F. (2005) Proteomic analysis of redox- and ErbB2-dependent changes in mammary luminal epithelial cells using cysteine- and lysine-labelling two-dimensional difference gel electrophoresis. *Proteomics*, **5**, 2908–2926.

Cho, M.J., Wong, J.H., Marx, C., Jiang, W., Lemaux, P.G. and Buchanan, B.B. (1999) Overexpression of thioredoxin h leads to enhanced activity of starch debranching enzyme (pullulanase) in barley grain. *Proc. Natl Acad. Sci. USA*, **96**, 14641–14646.

Chrestensen, C.A., Starke, D.W. and Mieyal, J.J. (2000) Acute cadmium exposure inactivates thioltransferase (Glutaredoxin), inhibits intracellular reduction of protein-glutathionyl-mixed disulfides, and initiates apoptosis. *J. Biol. Chem.*, **275**, 26556–26565.

Claiborne, A., Mallett, T.C., Yeh, J.I., Luba, J. and Parsonage, D. (2001) Structural, redox, and mechanistic parameters for cysteine-sulfenic acid function in catalysis and regulation. *Adv. Protein Chem.*, **58**, 215–276.

Clements, A., Johnston, M.V., Larsen, B.S. and McEwen, C.N. (2005) Fluorescence-based peptide labeling and fractionation strategies for analysis of cysteine-containing peptides. *Anal. Chem.*, **77**, 4495–4502.

Conrads, T.P., Alving, K., Veenstra, T.D., Belov, M.E., Anderson, G.A., Anderson, D.J., Lipton, M.S., Pasa-Tolic, L., Udseth, H.R., Chrisler, W.B., Thrall, B.D. and Smith, R.D. (2001) Quantitative analysis of bacterial and mammalian proteomes using a combination of cysteine affinity tags and 15N-metabolic labelling. *Anal. Chem.*, **73**, 2132–2139.

Cunsolo, V., Foti, S. and Saletti, R. (2004) Mass spectrometry in the characterization of cereal seed proteins. *Eur. J. Mass Spectrom.*, **10**, 359–370.

Deitz, K.J. (2003) Plant peroxiredoxins. *Annu. Rev. Plant Biol.*, **54**, 93–107.

Delledonne, M. (2005) NO news is good news for plants. *Curr. Opin. Plant Biol.*, **8**, 390–396.

Delledonne, M., Xia, Y., Dixon, R.A. and Lamb, C. (1998) Nitric oxide functions as a signal in plant disease resistance. *Nature*, **394**, 585–588.

Denu, J.M. and Tanner, K.G. (1998) Specific and reversible inactivation of protein tyrosine phosphatases by hydrogen peroxide: evidence for a sulfenic acid intermediate and implications for redox regulation. *Biochemistry*, **37**, 5633–5642.

Dixon, D.P., Fordham-Skelton, A.P. and Edwards, R. (2005a) Redox regulation of a soybean tyrosine-specific protein phosphatase. *Biochemistry*, **44**, 7696–7703.

Dixon, D.P., Skipsey, M., Grundy, N.M. and Edwards, R. (2005b) Stress-induced protein S-glutathionylation in *Arabidopsis*. *Plant Physiol.*, **138**, 2233–2244.

Drapeau, G.R. (1980) Substrate specificity of a proteolytic enzyme isolated from a mutant of *Pseudomonas fragi*. *J. Biol. Chem.*, **255**, 839–840.

Durner, J., Wendehenne, D. and Klessig, D.F. (1998) Defense gene induction in tobacco by nitric oxide, cyclic GMP, and cyclic ADP-ribose. *Proc. Natl Acad. Sci. USA*, **95**, 10328–10333.

Eaton, P. and Shattock, M.J. (2002) Purification of proteins susceptible to oxidation at cysteine residues: identification of malate dehydrogenase as a target for S-glutathiolation. *Ann. NY Acad. Sci.*, **973**, 529–532.

Fischer, W.H., Rivier, J.E. and Craig, A.G. (1993) *In situ* reduction suitable for matrix-assisted laser desorption/ionization and liquid secondary ionization using tris(2-carboxyethyl)phosphine. *Rapid Commun. Mass Spectrom.*, **7**, 225–228.

Friedman, M., Krull, L.H. and Cavins, J.F. (1970) The chromatographic determination of cystine and cysteine residues in proteins as s-beta-(4-pyridylethyl)cysteine. *J. Biol. Chem.*, **245**, 3868–3871.

Foyer, C.H. and Noctor, G. (2005) Oxidant and antioxidant signalling in plants: a re-evaluation of the concept of oxidative stress in a physiological context. *Plant Cell Environ.*, **28**, 1056–1071.

Fratelli, M., Demol, H., Puype, M., Casagrande, S., Eberini, I., Salmona, M., Bonetto, V., Mengozzi, M., Duffieux, F., Miclet, E., Bachi, A., Vandekerckhove, J., Gianazza, E. and Ghezzi, P. (2002) Identification by redox proteomics of glutathionylated proteins in oxidatively stressed human T lymphocytes. *Proc. Natl Acad. Sci. USA*, **99**, 3505–3510.

Galvani, M., Rovatti, L., Hamdan, M., Herbert, B. and Righetti, P.G. (2001) Protein alkylation in the presence/absence of thiourea in proteome analysis: a matrix assisted laser desorption/ionization-time of flight-mass spectrometry investigation. *Electrophoresis*, **22**, 2066–2074.

Gan, Z.R., Sardana, M.K., Jacobs, J.W. and Polokoff, M.A. (1990) Yeast thioltransferase – the active site cysteines display differential reactivity. *Arch. Biochem. Biophys.*, **282**, 110–115.

Gevaert, K., Ghesquiere, B., Staes, A., Martens, L., Van Damme, J., Thomas, G.R. and Vandekerckhove, J. (2004) Reversible labeling of cysteine-containing peptides allows their specific chromatographic isolation for non-gel proteome studies. *Proteomics*, **4**, 897–908.

Ghezzi, P. and Bonetto, V. (2003) Redox proteomics: identification of oxidatively modified proteins. *Proteomics*, **3**, 1145–1153.

Gorman, J.J., Wallis, T.P. and Pitt, J.J. (2002) Protein disulfide bond determination by mass spectrometry. *Mass Spectrom. Rev.*, **21**, 183–216.

Gow, A.J., Chen, Q.P., Hess, D.T., Day, B.J., Ischiropoulos, H. and Stamler, J.S. (2002) Basal and stimulated protein S-nitrosylation in multiple cell types and tissues. *J. Biol. Chem.*, **277**, 9637–9640.

Goyer, A., Haslekas, C., Miginiac-Maslow, M., Klein, U., Le Marechal, P., Jacquot, J.P. and Decottignies, P. (2002) Isolation and characterization of a thioredoxin-dependent peroxidase from *Chlamydomonas reinhardtii*. *Eur. J. Biochem.*, **269**, 272–282.

Gravina, S.A. and Mieyal, J.J. (1993) Thioltransferase is a specific glutathionyl mixed disulfide oxidoreductase. *Biochemistry*, **32**, 3368–3376.

Gupta, R. and Luan, S. (2003) Redox control of protein tyrosine phosphatases and mitogen-activated protein kinases in plants. *Plant Physiol.*, **132**, 1149–1152.

Gupta, R., Ting, J.T., Sokolov, L.N., Johnson, S.A. and Luan, S. (2002) A tumor suppressor homolog, AtPTEN1, is essential for pollen development in *Arabidopsis*. *Plant Cell*, **14**, 2495–2507.

Gygi, S.P., Rist, B., Gerber, S.A., Turecek, F., Gelb, M.H. and Aebersold, R. (1999) Quantitative analysis of complex protein mixtures using isotope-coded affinity tags. *Nat. Biotechnol.*, **17**, 994–999.

Hamnell-Pamment, Y., Lind, C., Palmberg, C., Bergman, T. and Cotgreave, I.A. (2005) Determination of site-specificity of *S*-glutathionylated cellular proteins. *Biochem. Biophys. Res. Commun.*, **332**, 362–369.

Holmgren, A. (1979) Glutathione-dependent synthesis of deoxyribonucleotides. Characterization of the enzymatic mechanism of *Escherichia coli* glutaredoxin. *J. Biol. Chem.*, **254**, 3672–3678.

Holmgren, A. (1985) Thioredoxin. *Annu. Rev. Biochem.*, **54**, 237–271.

Hwang, C., Sinskey, A.J. and Lodish, H.F. (1992) Oxidized redox state of glutathione in the endoplasmic reticulum. *Science*, **25**, 1496–1502.

Ito, H., Iwabuchi, M. and Ogawa, K. (2003) The sugar-metabolic enzymes aldolase and triose-phosphate isomerase are targets of glutathionylation in *Arabidopsis thaliana*: detection using biotinylated glutathione. *Plant Cell Physiol.*, **44**, 655–660.

Jaffrey, S.R., Erdjument-Bromage, H., Ferris, C.D., Tempst, P. and Snyder, S.H. (2001) Protein S-nitrosylation: a physiological signal for neuronal nitric oxide. *Nat. Cell. Biol.*, **3**, 193–197.

Jones, M.D., Patterson, S.D. and Lu, H.S. (1998) Determination of disulfide bonds in highly bridged disulfide-linked peptides by matrix-assisted laser desorption/ionization mass spectrometry with postsource decay. *Anal. Chem.*, **70**, 136–143.

Kallis, G.B. and Holmgren, A. (1980) Differential reactivity of the functional sulfhydryl groups of cysteine-32 and cysteine-35 present in the reduced form of thioredoxin from *Escherichia coli*. *J. Biol. Chem.*, **255**, 10261–10265.

Kim, Y.M., Talanian, R.V. and Billiar, T.R. (1997) Nitric oxide inhibits apoptosis by preventing increases in caspase-3-like activity via two distinct mechanisms. *J. Biol. Chem.*, **272**, 31138–31148.

Kim, Y.M., Talanian, R.V., Li, J. and Billiar, T.R. (1998) Nitric oxide prevents IL-1β and IFN-γ-inducing factor (IL-18) release from macrophages by inhibiting caspase-1 (IL-1β-converting enzyme). *J. Immunol.*, **161**, 4122–4128.

Kim, J.R., Yoon, H.W., Kwon, K.S., Lee, S.R. and Rhee, S.G. (2000) Identification of proteins containing cysteine residues that are sensitive to oxidation by hydrogen peroxide at neutral pH. *Anal. Biochem.*, **283**, 214–221.

Klatt, P., Pineda Molina, E., Perez-Sala, D. and Lamas, S. (2000) Novel application of S-nitrosoglutathione-Sepharose to identify proteins that are potential targets for S-nitrosoglutathione-induced mixed-disulphide formation. *Biochem. J.*, **349**, 567–578.

Kluge, I., Gutteck-Amsler, U., Zollinger, M. and Do, K.Q. (1997) *S*-nitrosoglutathione in rat cerebellum: identification and quantification by liquid chromatography-mass spectrometry. *J. Neurochem.*, **69**, 2599–2607.

Kobayashi, T., Kishigami, S., Sone, M., Inokuchi, H., Mogi, T. and Ito, K. (1997) Respiratory chain is required to maintain oxidized states of the DsbA-DsbB disulfide bond formation system in aerobically growing *Escherichia coli* cells. *Proc. Natl Acad. Sci. USA*, **94**, 11857–11862.

Kobrehel, K., Wong, J.H., Balogh, A., Kiss, F., Yee, B.C. and Buchanan, B.B. (1992) Specific reduction of wheat storage proteins by thioredoxin h. *Plant Physiol.*, **99**, 919–924.

Kumar, J.K., Tabor, S. and Richardson, C.C. (2004) Proteomic analysis of thioredoxin-targeted proteins in *Escherichia coli*. *Proc. Natl Acad. Sci. USA*, **101**, 3759–3764.

Kuncewicz, T., Sheta, E.A., Goldknopf, I.L. and Kone, B.C. (2003) Proteomic analysis of S-nitrosylated proteins in mesangial cells. *Mol. Cell. Proteomics*, **2**, 156–163.

Lapko, V.N., Smith, D.L. and Smith, J.B. (2000) Identification of an artifact in the mass spectrometry of proteins derivatized with iodoacetamide. *J. Mass Spectrom.*, **35**, 572–575.

Laragione, T., Bonetto, V., Casoni, F., Massignan, T., Bianchi, G., Gianazza, E. and Ghezzi, P. (2003) Redox regulation of surface protein thiols: identification of integrin α-4 as a molecular target by using redox proteomics *Proc. Natl Acad. Sci. USA*, **100**, 14737–14741.

Laurent, T.C., Moore, E.C. and Reichard, P. (1964) Enzymatic synthesis of deoxyribonucleotides. IV. Isolation and characterization of thioredoxin, the hydrogen donor from *Escherichia coli* B. *J. Biol. Chem.*, **239**, 3436–3444.

Lee, S.R., Kwon, K.S., Kim, S.R. and Rhee, S.G. (1998) Reversible inactivation of protein-tyrosine phosphatase 1B in A431 cells stimulated with epidermal growth factor. *J. Biol. Chem.*, **273**, 15366–15372.

Lee, K., Lee, J., Kim, Y., Bae, D., Kang, K.Y., Yoon, S.C. and Lim, D. (2004) Defining the plant disulfide proteome. *Electrophoresis*, **25**, 532–541.

Lemaire, S.D., Guillon, B., Le Marechal, P., Keryer, E., Miginiac-Maslow, M. and Decottignies, P. (2004) New thioredoxin targets in the unicellular photosynthetic eukaryote *Chlamydomonas reinhardtii*. *Proc. Natl Acad. Sci. USA*, **101**, 7475–7480.

Li, J., Bombeck, C.A., Yang, S., Kim, Y.M. and Billiar, T.R. (1999) Nitric oxide suppresses apoptosis via interrupting caspase activation and mitochondrial dysfunction in cultured hepatocytes. *J. Biol. Chem.*, **274**, 17325–17333.

Lin, T.K., Hughes, G., Muratovska, A., Blaikie, F.H., Brookes, P.S., Darley-Usmar, V., Smith, R.A. and Murphy, M.P. (2002) Specific modification of mitochondrial protein thiols in response to oxidative stress: a proteomics approach. *J. Biol. Chem.*, **277**, 17048–17056.

Lind, C., Gerdes, R., Hamnell, Y., Schuppe-Koistinen, I., von Lowenhielm, H.B., Holmgren, A. and Cotgreave, I.A. (2002) Identification of S-glutathionylated cellular proteins during oxidative stress and constitutive metabolism by affinity purification and proteomic analysis. *Arch. Biochem. Biophys.*, **406**, 229–240.

Lindermayr, C., Saalbach, G. and Durner, J. (2005) Proteomic identification of S-nitrosylated proteins in *Arabidopsis*. *Plant Physiol.*, **137**, 921–930.

Lindorff-Larsen, K. and Winther, J.R. (2000) Thiol alkylation below neutral pH. *Anal. Biochem.*. **286**, 308–310.

Luche, S., Diemer, H., Tastet, C., Chevallet, M., Van Dorsselaer, A., Leize-Wagner, E. and Rabilloud, T. (2004) About thiol derivatization and resolution of basic proteins in two-dimensional electrophoresis. *Proteomics*, **4**, 551–561.

MacRobbie, E.A. (2002) Evidence for a role for protein tyrosine phosphatase in the control of ion release from the guard cell vacuole in stomatal closure. *Proc. Natl Acad. Sci. USA*, **99**, 11963–11968.

Maeda, K., Finnie, C. and Svensson, B. (2004) Cy5 maleimide labelling for sensitive detection of free thiols in native protein extracts: identification of seed proteins targeted by barley thioredoxin h isoforms. *Biochem. J.*, **378**, 497–507.

Maeda, K., Finnie, C. and Svensson, B. (2005) Identification of thioredoxin h-reducible disulphides in proteomes by differential labelling of cysteines: insight into recognition and regulation of proteins in barley seeds by thioredoxin h. *Proteomics*, **5**, 1634–1644.

Marchand, C., Le Marechal, P., Meyer, Y., Miginiac-Maslow, M., Issakidis-Bourguet, E. and Decottignies, P. (2004) New targets of *Arabidopsis* thioredoxins revealed by proteomic analysis. *Proteomics*, **4**, 2696–2706.

Marley, K., Mooney, D.T., Clark-Scannell, G., Tong, T.T., Watson, J., Hagen, T.M., Stevens, J.F. and Maier, C.S. (2005) Mass tagging approach for mitochondrial thiol proteins. *J. Proteome Res.*, **4**, 1403–1412.

Martin, J.L. (1995) Thioredoxin – a fold for all reasons. *Structure*, **3**, 245–250.

Martinez-Ruiz, A., Villanueva, L., Gonzalez de Orduna, C., Lopez-Ferrer, D., Higueras, M.A., Tarin, C., Rodriguez-Crespo, I., Vazquez, J. and Lamas, S. (2005) S-nitrosylation of Hsp90 promotes the inhibition of its ATPase and endothelial nitric oxide synthase regulatory activities. *Proc. Natl Acad. Sci. USA*, **102**, 8525–8530.

Marx, C., Wong, J.H. and Buchanan, B.B. (2003) Thioredoxin and germinating barley: targets and protein redox changes. *Planta*, **216**, 454–460.

McDonagh, B., Tyther, R. and Sheehan, D. (2005) Carbonylation and glutathionylation of proteins in the blue mussel *Mytilus edulis* detected by proteomic analysis and Western blotting: Actin as a target for oxidative stress. *Aquat. Toxicol.*, **73**, 315–326.

Meier, B., Radeke, H.H., Selle, S., Younes, M., Sies, H., Resch, K. and Habermehl, G.G. (1989) Human fibroblasts release reactive oxygen species in response to interleukin-1 or tumour necrosis factor-α. *Biochem. J.*, **263**, 539–545.

Meinhard, M. and Grill, E. (2001) Hydrogen peroxide is a regulator of ABI1, a protein phosphatase 2C from *Arabidopsis*. *FEBS Lett.*, **508**, 443–446.

Meyer, Y., Vignols, F. and Reichheld, J.P. (2002) Classification of plant thioredoxins by sequence similarity and intron position; *Method Enzymol.*, **347**, 394–402.

Motohashi, K., Kondoh, A., Stumpp, M.T. and Hisabori, T. (2001) Comprehensive survey of proteins targeted by chloroplast thioredoxin. *Proc. Natl Acad. Sci. USA*, **98**, 11224–11229.

Mouaheb, N., Thomas, D., Verdoucq, L., Monfort, P. and Meyer, Y. (1998) *In vivo* functional discrimination between plant thioredoxins by heterologous expression in the yeast *Saccharomyces cerevisiae*. *Proc. Natl Acad. Sci. USA*, **95**, 3312–3317.

Ogawa, K., Tasaka, Y., Mino, M., Tanaka, Y. and Iwabuchi, M. (2001) Association of glutathione with flowering in *Arabidopsis thaliana*. *Plant Cell Physiol.*, **42**, 524–530.

Patterson, S.D. and Katta, V. (1994) Prompt fragmentation of disulfide-linked peptides during matrix-assisted laser desorption ionization mass spectrometry. *Anal. Chem.*, **66**, 3727–3732.

Plapp, B.V., Raftery, M.A. and Cole, R.D. (1967) The tryptic digestion of S-aminoethylated ribonuclease. *J. Biol. Chem.*, **242**, 265–270.

Potters, G., de Gara, L., Asard, H. and Horemans, N. (2002) Ascorbate and Gutathione: guardians of the cell cycle, partners in crime? *Plant Physiol. Biochem.*, **40**, 537–548.

Qin, J. and Chait, B.T. (1997) Identification and characterization of posttranslational modifications of proteins by MALDI ion trap mass spectrometry. *Anal. Chem.*, **69**, 4002–4009.

Rabilloud, T., Heller, M., Gasnier, F., Luche, S., Rey, C., Aebersold, R., Benahmed, M., Louisot, P. and Lunardi, J. (2002) Proteomics analysis of cellular response to oxidative stress. Evidence for *in vivo* overoxidation of peroxiredoxins at their active site. *J. Biol. Chem.*, **277**, 19396–19401.

Ren, D., Julka, S., Inerowicz, H.D. and Regnier, F.E. (2004) Enrichment of cysteine-containing peptides from tryptic digests using a quaternary amine tag. *Anal. Chem.*, **76**, 4522–4530.

Riener, C.K., Kada, G., and Gruber, H.J. (2002) Quick measurement of protein sulfhydryls with Ellman's reagent and with 4,4′-dithiodipyridine. *Anal. Bioanal. Chem.*, **373**, 266–276.

Rouhier, N., Gelhaye, E. and Jacquot, J.P. (2002a) Exploring the active site of plant glutaredoxin by site-directed mutagenesis. *FEBS Lett.*, **511**, 145–149.

Rouhier, N., Gelhaye, E. and Jacquot, J.P. (2002b) Glutaredoxin-dependent peroxiredoxin from poplar: protein-protein interaction and catalytic mechanism. *J. Biol. Chem.*, **277**, 13609–13614.

Rouhier, N., Gelhaye, E. and Jacquot, J.P. (2004) Plant glutaredoxins: still mysterious reducing systems. *Cell Mol. Life Sci.*, **61**, 1266–1277.

Sarkar, N., Lemaire, S., Wu-Scharf, D., Issakidis-Bourguet, E. and Cerutti, H. (2005) Functional specialization of *Chlamydomonas reinhardtii* cytosolic thioredoxin h1 in the response to alkylation-induced DNA damage. *Eukaryot. Cell.*, **4**, 262–273.

Saurin, A.T., Neubert, H., Brennan, J.P. and Eaton, P. (2004) Widespread sulfenic acid formation in tissues in response to hydrogen peroxide. *Proc. Natl Acad. Sci. USA*, **101**, 17982–17987.

Schenk, H., Klein, M., Erdbrugger, W., Droge, W. and Schulze-Osthoff, K. (1994) Distinct effects of thioredoxin and antioxidants on the activation of transcription factors NF-κB and AP-1. *Proc. Natl Acad. Sci. USA*, **91**, 1672–1676.

Schürmann, P. and Jacquot, J.P. (2000) Plant thioredoxin systems revisited. *Annu. Rev. Plant Physiol. Plant Mol. Biol.*, **51**, 371–400.

Sebastiano, R., Citterio, A., Lapadula, M. and Righetti, P.G. (2003) A new deuterated alkylating agent for quantitative proteomics. *Rapid Commun. Mass Spectrom.*, **17**, 2380–2386.

Sechi, S. and Chait, B.T. (1998) Modification of cysteine residues by alkylation. A tool in peptide mapping and protein identification. *Anal. Chem.*, **70**, 5150–5158.

Sethuraman, M., McComb, M.E., Huang, H., Huang, S., Heibeck, T., Costello, C.E. and Cohen, R.A. (2004) Isotope-coded affinity tag (ICAT) approach to redox proteomics: identification and quantitation of oxidant-sensitive cysteine thiols in complex protein mixtures. *J. Proteome Res.*, **3**, 1228–1233.

Shevchenko, A., Wilm, M., Vorm, O. and Mann, M. (1996) Mass spectrometric sequencing of proteins silver-stained polyacrylamide gels. *Anal. Chem.*, **68**, 850–858.

Shaw, J., Rowlinson, R., Nickson, J., Stone, T., Sweet, A., Williams, K. and Tonge, R. (2003) Evaluation of saturation labelling two-dimensional difference gel electrophoresis fluorescent dyes. *Proteomics*, **3**, 1181–1195.

Sommer, A. and Traut, R.R. (1974) Diagonal polyacrylamide-dodecyl sulfate gel electrophoresis for the identification of ribosomal proteins crosslinked with methyl-4-mercaptobutyrimidate. *Proc. Natl Acad. Sci. USA*, **71**, 3946–3950.

Taldone, F. S., Tummala, M., Goldstein, E. J., Ryzhov, V., Ravi, K., Black, S. M. (2005) Studying the S-nitrosylation of model peptides and eNOS protein by mass spectrometry. *Nitric Oxide*, **13**, 176–187.

Thevis, M., Ogorzalek Loo, R.R. and Loo, J.A. (2003) In-gel derivatization of proteins for cysteine-specific cleavages and their analysis by mass spectrometry. *J. Proteome Res.*, **2**, 163–172.

Tie, J.K., Jin, D.Y., Loiselle, D.R., Pope, R.M., Straight, D.L. and Stafford, D.W. (2004) Chemical modification of cysteine residues is a misleading indicator of their status as active site residues in the vitamin K-dependent gamma-glutamyl carboxylation reaction. *J. Biol. Chem.*, **279**, 54079–54087.

Tiessen, A., Hendriks, J.H., Stitt, M., Branscheid, A., Gibon, Y., Farre, E.M. and Geigenberger, P. (2002) Starch synthesis in potato tubers is regulated by post-translational redox modification of ADP-glucose pyrophosphorylase: a novel regulatory mechanism linking starch synthesis to the sucrose supply. *Plant Cell*, **14**, 2191–2213.

Toriumi, C. and Imai, K. (2003) An identification method for altered proteins in tissues utilizing fluorescence derivatization, liquid chromatography, tandem mass spectrometry, and a database-searching algorithm. *Anal. Chem.*, **75**, 3725–3730.

Trebitsh, T., Meiri, E., Ostersetzer, O., Adam, Z. and Danon, A. (2001) The protein disulfide isomerase-like RB60 is partitioned between stroma and thylakoids in *Chlamydomonas reinhardtii* chloroplasts. *J. Biol. Chem.*, **276**, 4564–4569.

Tsikas, D., Sandmann, J., Rossa, S., Gutzki, F.M. and Frolich, J.C. (1999) Investigations of S-transnitrosylation reactions between low- and high-molecular-weight S-nitroso compounds and their thiols by high-performance liquid chromatography and gas chromatography-mass spectrometry. *Anal. Biochem.*, **270**, 231–241.

Verdoucq, L., Vignols, F., Jacquot, J.P., Chartier, Y. and Meyer, Y. (1999) *In vivo* characterization of a thioredoxin h target protein defines a new peroxiredoxin family. *J. Biol. Chem.*, **274**, 19714–19722.

Vestweber, D. and Schatz, G. (1988) Mitochondria can import artificial precursor proteins containing a branched polypeptide chain or a carboxy-terminal stilbene disulfonate. *J. Cell Biol.*, **107**, 2045–2049.

Wagner, E., Luche, S., Penna, L., Chevallet, M., Van Dorsselaer, A., Leize-Wagner, E. and Rabilloud, T. (2002) A method for detection of overoxidation of cysteines: peroxiredoxins are oxidized *in vivo* at the active-site cysteine during oxidative stress. *Biochem. J.*, **366**, 777–785.

Wallis, T.P., Pitt, J.J. and Gorman, J.J. (2001) Identification of disulfide-linked peptides by isotope profiles produced by peptic digestion of proteins in 50% ^{18}O water. *Protein Sci.*, **10**, 2251–2271.

Wang, J., Boja, E.S., Tan, W., Tekle, E., Fales, H.M., English, S., Mieyal, J.J. and Chock, P.B. (2001) Reversible glutathionylation regulates actin polymerization in A431 cells. *J. Biol. Chem.*, **276**, 47763–47766.

Wilkinson, B. and Gilbert, H.F. (2004) Protein disulfide isomerase. *Biochim. Biophys. Acta.*, **1699**, 35–44.

Wolosiuk, R.A. and Buchanan, B.B. (1977) Thioredoxin and glutathione regulate photosynthesis in chloroplasts. *Nature*, **266**, 565–567.

Wolosiuk, R.A., Crawford, N.A., Yee, B.C. and Buchanan, B.B. (1979) Isolation of three thioredoxins from spinach leaves. *J. Biol. Chem.*, **254**, 1627–1632.

Wong, J.H., Kim, Y.B., Ren, P.H., Cai, N., Cho, M.J., Hedden, P., Lemaux, P.G. and Buchanan, B.B. (2002) Transgenic barley grain overexpressing thioredoxin shows evidence that the starchy endosperm communicates with the embryo and the aleurone. *Proc. Natl Acad. Sci. USA*, **99**, 16325–16330.

Wong, J.H., Cai, N., Balmer, Y., Tanaka, C.K., Vensel, W.H., Hurkman, W.J. and Buchanan, B.B. (2004) Thioredoxin targets of developing wheat seeds identified by complementary proteomic approaches. *Phytochemistry*, **65**, 1629–1640.

Wu, J. and Watson, J.T. (1997) A novel methodology for assignment of disulfide bond pairings in proteins. *Protein Sci.*, **6**, 391–398.

Wu, Y, Kwon, K.S. and Rhee, S.G. (1998) Probing cellular protein targets of H_2O_2 with fluorescein-conjugated iodoacetamide and antibodies to fluorescein. *FEBS Lett.*, **440**, 111–115.

Xiao, R., Wilkinson, B., Solovyov, A., Winther, J.R., Holmgren, A., Lundstrom-Ljung, J. and Gilbert, H.F. (2004) The contributions of protein disulfide isomerase and its homologues to oxidative protein folding in the yeast endoplasmic reticulum. *J. Biol. Chem.*, **279**, 49780–49786.

Yamazaki, D., Motohashi, K., Kasama, T., Hara, Y. and Hisabori, T. (2004) Target proteins of the cytosolic thioredoxins in *Arabidopsis thaliana. Plant Cell Physiol.*, **45**, 18–27.

Yanagida, M., Mino, M., Iwabuchi, M. and Ogawa, K. (2004) Reduced glutathione is a novel regulator of vernalization-induced bolting in the rosette plant *Eustoma grandiflorum. Plant Cell Physiol.*, **45**, 129–137.

Yang, Y., Jao, S., Nanduri, S., Starke, D.W., Mieyal, J.J. and Qin, J. (1998) Reactivity of the human thioltransferase (glutaredoxin) C7S, C25S, C78S, C82S mutant and NMR solution structure of its glutathionyl mixed disulfide intermediate reflect catalytic specificity. *Biochemistry*, **37**, 17145–17156.

Yano, H., Wong, J.H., Lee, Y.M., Cho, M.J. and Buchanan, B.B. (2001) A strategy for the identification of proteins targeted by thioredoxin. *Proc. Natl Acad. Sci. USA*, **98**, 4794–4799.

Yen, T.Y., Joshi, R.K., Yan, H., Seto, N.O., Palcic, M.M. and Macher, B.A. (2000) Characterization of cysteine residues and disulfide bonds in proteins by liquid chromatography/electrospray ionization tandem mass spectrometry. *J. Mass Spectrom.*, **35**, 990–1002.

Yen, T.Y., Yan, H. and Macher, B.A. (2002) Characterizing closely spaced, complex disulfide bond patterns in peptides and proteins by liquid chromatography/electrospray ionization tandem mass spectrometry. *J. Mass Spectrom.*, **37**, 15–30.

Zubarev, R.A., Kruger, N.A., Fridriksson, E.K., Lewis, M.A., Horn, D.M., Carpenter, B.K. and McLafferty, F.W. (1999) Electron capture dissociation of gaseous multiply-charged proteins is favored at disulfide bonds and other sites of high hydrogen atom affinity. *J. Am. Chem. Soc.*, **121**, 2857–2862.

5 Structural proteomics

Russell L. Wrobel, Craig A. Bingman, Won Bae Jeon, Jikui Song, Dmitriy A. Vinarov, Ronnie O. Frederick, David J. Aceti, Hassan K. Sreenath, Zsolt Zolnai, Frank C. Vojtik, Eduard Bitto, Brian G. Fox, George N. Phillips Jr and John L. Markley

5.1 Introduction

The complete genome sequences of many organisms, including several eukaryotes, have been determined, and the open reading frames (ORFs) predicted from these have revealed a plethora of genes of unknown structure and function. The availability of this sequence data and the growing impact of three-dimensional (3D) structures of proteins on our understanding of sequence to structure relationships, as well as on the elucidation of function, have prompted scientific groups from several countries to undertake high-throughput structural genomics projects (Thornton, 2001). The Center for Eukaryotic Structural Genomics (CESG) at the University of Wisconsin was founded to develop high-throughput and cost-effective methods for determining the structures of eukaryotic proteins (http://www.uwstructuralgenomics.org/). CESG was established in 2001 as one of nine pilot centers funded by the National Institute of General Medical Sciences (NIGMS) of the National Institutes of Health in Phase-1 of the Protein Structure Initiative (PSI-1). The mandate of PSI-1 was to develop high-throughput methods for protein structure determination starting with gene sequences and with the goal of improving our ability to model structures of proteins from their sequences (http://www.nigms.nih.gov/Initiatives/PSI/Background/PilotFacts.htm). CESG decided to focus on developing technology applicable to eukaryotic proteins and began with target selection from *Arabidopsis thaliana*. The reasons were because this was the largest sequenced eukaryotic genome available at the time with a thorough level of annotation and because the National Science Foundation was investing heavily in functional studies of the genome (Initiative, 2000).

Phase-2 of the PSI-2, the production phase, began in 2005 and will continue to 2010 (Service, 2005). This phase includes two types of centers. Four large-scale centers are expected to generate between 3000 and 4000 structures, and six specialized centers will develop novel methods for more quickly determining the structures of proteins that traditionally have been difficult to study (http://www.nigms.nih.gov/Initiatives/PSI/Centers/). CESG was funded as a specialized center and will continue working on plant proteins, but with an expanded scope to include proteins from other eukaryotic organisms, including requests from outside researchers.

Structural genomics demands that gene products be generated given only the putative gene sequence. CESG developed two pathways to obtain proteins (Figure 5.1).

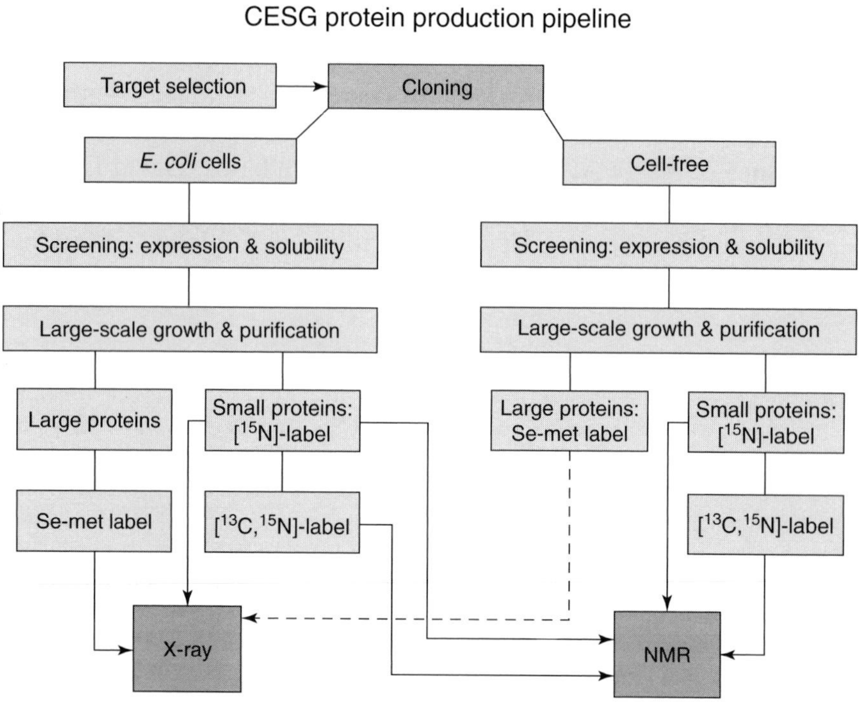

CESG protein production pipeline

Target selection
Cloning
E. coli cells
Cell-free
Screening: expression & solubility
Screening: expression & solubility
Large-scale growth & purification
Large-scale growth & purification
Large proteins
Small proteins: [^{15}N]-label
Large proteins: Se-met label
Small proteins: [^{15}N]-label
Se-met label
[^{13}C,^{15}N]-label
[^{13}C,^{15}N]-label
X-ray
NMR

Figure 5.1 Work flow diagram for protein production pipelines used at the University of Wisconsin Center for Eukaryotic Structural Genomics.

By using semi-automated cloning methods, over 2200 *Arabidopsis* ORFs were cloned into Gateway entry vectors. These have been expressed in *Escherichia coli* using custom-made expression vectors that have amino-terminal tags to facilitate soluble expression, visualization, and purification of recombinant protein. An auto-inducing medium is used to efficiently incorporate either selenomethionine into the polypeptide chain to aid in X-ray diffraction analysis or ^{15}N followed by ^{15}N+^{13}C for nuclear magnetic resonance (NMR) analyses. In addition to an *E. coli*-based expression pipeline, CESG developed a parallel pipeline for expression of ORFs in a wheat germ cell-free system. Robotic systems have been installed that automate the screening and protein production steps in this pipeline. In the 4 years between 10/2001 and 9/2005, over 300 different *Arabidopsis* targets were purified by semi-automated methods at average yields of more than 10 mg per target. Proteins destined for X-ray crystallography were subjected to automated screening against 192 crystallization conditions. Those targets showing promising crystallization results were then rescreened on a second robotic system to optimize the crystallization conditions. Suitable crystals resulting from the second screen were examined for diffraction quality on a home source, and those passing this screen were taken to beam lines at the synchrotron at Argonne National Laboratory for data collection. Targets destined for NMR analysis were first prepared as ^{15}N-labeled samples and screened by ^{1}H–^{15}N correlation spectroscopy for suitability of the target and solution conditions. Only targets that passed this screen were prepared as ^{13}C, ^{15}N-labeled samples for structure determination. NMR data were collected at the National Magnetic Resonance Facility at Madison (NMRFAM) and the Medical College of Wisconsin (MCW). This review focuses on the successful methodologies developed, surveys the progress made with *Arabidopsis* targets, and presents representative examples of solved structures to illustrate the range of results obtained to date.

5.2 Project data handling: Sesame

Data management and sharing are important aspects of modern genomic research. The human genome and structural genomics projects may be regarded as harbingers of the methodology of the future, in that they harness high-throughput activities that take advantage of sophisticated instrumentation and robotics, involve far-flung collaborations, make use of large-scale computing, and have a well-developed ethos of data sharing. The advantages of recording in electronic formats that can be shared with and queried by other research scientists include the capture of both positive and negative experimental results, additional details regarding experimental procedures, and information about project resources such as plasmids, clones, etc. The ability to mine these data for novel scientific results, laboratory best practices, potential collaborations, and other outcomes is also important. Ultimately, the goal is to enable more efficient scientific discovery.

Sesame, the laboratory information management system (LIMS) developed in part by CESG funding is a deployed, Internet-accessible, and currently freely accessible software system (Zolnai et al., 2003). It provides support for structural genomics and related proteomics activities, and for the activities of a traditional protein chemistry or molecular biology laboratory. Sesame was designed to (1) provide a flexible resource for storing, recovering, and disseminating data that fits well into a research environment (multiple views of data and parameters); (2) allow the worldwide community of scientists to utilize the system; (3) support remote collaborations at all stages; (4) provide full user data access security and data storage security; (5) permit data mining exercises and analysis of laboratory operations to identify best practices; and (6) simplify the dissemination, installation, maintenance, and interoperability of software. Moreover, Sesame has been designed to capture the information that has been modeled into the most recent data dictionaries used by the structural databases Protein Data Bank and BioMagResBank (PDB and BMRB (www.bmrb.wisc.edu)) and the National Institute of Health (NIH) PSI structural genomics databases (TargetDB and PepcDB). PepcDB is expected to be a major source of information for bioinformaticians interested in understanding protein expression, solubilization, purification, crystallization, and structural stability in aqueous and other solution environments. Capture of these items will greatly expand the information content and usefulness of these resources. Thus, individual laboratories that use Sesame will be well positioned to contribute results to the PepcDB and the expanded structural databases

(PDB and BMRB) and to use information imported from these and other databases. The Sesame system, therefore, represents a tool with the potential of gaining widespread applicability in individual research laboratories as well as centers that carry out high-throughput activities.

Sesame has been implemented as a multi-tier Java/CORBA application. This architecture is scalable: a Sesame installation can be as small as one computer supporting a typical academic laboratory or as large as a multi-server setup supporting one or more proteomics centers, instrumentation centers, and individual investigators. As the users access Sesame through the web, the system is available to collaborators located anywhere in the world.

5.3 ORF cloning

CESG used the sequence-specific recombination-based cloning system from Gateway (Invitrogen, Carlsbad, CA) to clone approximately 2200 *Arabidopsis* ORFs (Thao et al., 2004). Gateway cloning eliminates the need for restriction digestions to produce directional clones. It also allows for the efficient transfer of target ORFs into multiple expression vectors containing different promoters in combination with different affinity, detection, or solubility tags (Walhout et al., 2000; Hartley et al., 2000).

Arabidopsis ORFs were chosen for the likelihood that they would elucidate sequence-structure relationships or novel fold-function relationships (Kifer et al., 2005). Potential targets were scored on the basis of a variety of quantitative and predicted characteristics identified by software packages developed in-house, by collaborators, or licensed from academic sources. Some negatively scored characteristics were: homology to proteins in the PDB, presence of coiled-coil domains, transmembrane segments, signal or target peptides, number of cysteine residues, the percent complexity, and the percent-predicted disorder of the protein (Oldfield et al., 2005). Based on the results of these analyses, each ORF was allocated a 14-digit 'target score,' with low digits indicative of desirable targets.

The first step of the cloning process was to amplify the chosen ORFs from cDNA reverse transcribed from RNA isolated from *Arabidopsis* T87 callus suspension culture (Kang et al., 2001). A two-step polymerase chain reaction (PCR) procedure was then used to create desirable ORF targets sequences with ends suitable for recombinational Gateway cloning (Thao et al., 2004). The first PCR step uses ORF-specific primers to amplify the target while appending portions of the sequence encoding the tobacco etch virus (TEV) protease recognition site to the 5′ end of the target and the attB2 sites 3′ end of the target. The second PCR step uses 'universal' primers to complete the TEV and attB2 sites, and to make the product Gateway cloning competent. The sequences encoding the TEV protease cleavage site were included in the 5′ portion of the ORF-specific primers to allow subsequent cleavage of the amino acids encoded by the 5′ recombination site and any amino-terminal fusion tags from the expressed protein (Kapust et al., 2002). Amplified products were analyzed by agarose gel electrophoresis, then scored and entered into the database as either

Table 5.1 Summary of results from the *Arabidopsis* protein production pipeline

Activity	Number of instances
Selected targets	4128
PCR+	3012
Entry clone +	2526
Sequence +	2434
Destination clone +	2229
Screening expression +	1477
Screening soluble +	1308
Large-scale cell growth +	1227
Production scale expression +	1132
Soluble production +	969
Tag cleaved +	673
Successfully purified	309
To crystallization screening	237
Crystallized +	97
Diffraction quality crystal +	42
Crystal structure +	33
X-ray PDB deposited	33
HSQC trials	105
HSQC+	34
NMR assigned	16
NMR structure	16
NMR PDB deposited	16
BMRB deposited	16

PCR+ (successful) or PCR− (unsuccessful). The level of attempted targets that were PCR+ (about 60%) was in agreement with results of microarray experiments that indicated a high percentage of gene transcripts in the cDNA (Stolc et al., 2005).

Some selected ORFs were purchased as full-length cDNA clones from the *Arabidopsis* Biological Resource Center (ABRC). These cDNA clones were also used as templates in the two-step PCR method. Gene cloning from these purchased clones was completed with greater than 95% efficiency, which brought the overall PCR+ success for all genes to over 70% (see Table 5.1).

Amplified targets were transferred via the Gateway base pair (BP) reaction to pDONR221 to produce entry clones (Thao et al., 2004). Entry clones were analyzed in two ways. First, the plasmid DNA from the clones was isolated with the aid of a Qiagen BioRobot 8000; then PCR was used to determine if the plasmid contained an inserted ORF. Plasmids containing inserts were entered into the database as Entry Clone +. About 83% of the amplification products made it into entry vectors. In the second round of analysis, the DNA sequence of these entry clones was determined, and any deviation from the expected sequence was added to the database. About 96% of the entry clones had an acceptable DNA sequence. Most of the early clones that failed sequencing had deletions in the primer regions that led to reading frame shifts. This number dropped when higher-quality (i.e. gel-purified) gene-specific primers were used in the initial PCR reaction. Another interesting observation was that the sequences of about 20% of the ORF clones differed from those annotated in The *Arabidopsis* Information Resource (TAIR) database. Most of these were miscalled introns or exons, or splice-site differences. CESG submits the sequences of these ORFs to GenBank, so that they can be used to improve subsequent annotated versions of the *Arabidopsis* genome sequence.

Entry clones with an acceptable DNA sequence were used to prepare *E. coli* expression constructs by the Gateway LR reaction. Again, PCR was used as a screen to determine if the clones contained an inserted ORF. On the basis of this screen, the products were annotated in the database as either Destination Clone + or Destination Clone −. The LR reaction proved to be fairly robust, and over 91% of the sequence acceptable entry clone ORFs were transferred to expression vectors.

The CESG expression vectors were based on pQE80 vectors that utilize the T5 promoter (Qiagen, Valencia, CA). Our modifications were to add the Gateway cassette, including the attB recombination sites, and to add, as alternative amino-terminal tags, an S-tag or tetracysteine motif for visualization, His_6 or His_8 for affinity purification, and maltose-binding protein (MBP) to promote solubility (Kapust and Waugh, 1999; Adams et al., 2002b). Vectors were also created that replaced MBP by other solubility tags such as glutathione-*S*-transferase (GST), thioredoxin (TRX), and NusA (Hammarstrom et al., 2002). A PET version containing the MBP tag that utilized the T7 promoter for expression was also made (Studier et al., 1990).

Some structural genomics groups have used the Gateway system for their cloning needs, but additional cloning methods have been employed. Notably, the Midwest Center for Structural Genomics automated the ligation-independent PCR cloning method (LIC-PCR) for the generation of expression clones (Dieckman et al., 2002). At CESG we are assessing the possibility of using the Flexi®Vector system (Promega, Madison, WI) to meet our cloning needs (Blommel, 2005), because it is less expensive overall than the Gateway system.

5.4 *E. coli* cell-based protein production pipeline

The *E. coli*-based protein production pipeline consists of small-scale expression screening, large-scale production of bacterial cells, and a semi-automated purification method. Small-scale high-throughput screening (HTS) procedure was used to evaluate the expression levels of eukaryotic proteins expressed in *E. coli*. The HTS procedure was based on growing suitable *E. coli* host strains with an expression plasmid of a target protein in small culture volumes arranged in 96-well format. Gene expression was induced by using either isopropyl-beta-D-thiogalactopyranoside (IPTG) (Aceti et al., 2003) or an auto-induction medium (Studier, 2005), and the protein production levels were analyzed by sodium dodecyl sulphate-polyacrylamide gel electrophoresis (SDS-PAGE). Expression screens using two different induction methods were needed to build a database of gene expression profiles and to decide which induction system produces the maximum number of expressed proteins. CESG's system has similarities to other screening methods recently described in the literature (Albala et al., 2000; Knaust and Nordlund, 2001; Scheich et al., 2003, 2004; Luan et al., 2004; Page et al., 2004; Segelke et al., 2004).

Several challenges are encountered when expressing eukaryotic proteins in *E. coli*-based expression systems. Some of these include low levels of gene expression, toxicity of the target protein to the host strain, and improper protein folding that can lead to aggregation and insolubility (Sorensen and Mortensen, 2005b; Swartz, 2001). As a first step in tackling these problems, gene expression screens were set up to identify successfully expressed target proteins. The *E. coli* cultures for the target proteins were grown on a small scale of 400 μL in 96-well growth blocks. This approach enabled convenient optimization of both protein expression and solubility, by testing, in parallel, a variety of culture conditions such as medium composition, growth temperatures, and induction methods and duration of expression. The best combination of culture growth conditions could then be scaled up and tested on a larger scale (liters of cell culture) to produce sufficient protein for purification (Sreenath et al., 2005).

The *E. coli* methionine auxotroph, B834-pRARE2, was used for the small-scale HTS. B834 produces β-galactosidase (encoded by lacZ) and lactose permease (encoded by lacY). These are vital for expression of the target protein, because induction by lactose (the key component of the auto-induction system) requires functional lacY and lacZ genes (Studier, 2005; Hoffman et al., 1995). The role of the pRARE2 plasmid (Novagen, Madison, WI) is to produce seven rare tRNA genes that are deficient in *E. coli*. Eukaryotic genes sometimes fail to express in *E. coli* due to codon bias; the additional rare tRNA genes carried by pRARE2 compensate for the low abundance (Jonasson et al., 2002; Sorensen and Mortensen, 2005a). Finally, the strain, B834-pRARE2 is deficient in the ompT protease, which helps to reduce proteolysis of the target protein during protein purification (Grodberg and Dunn, 1988).

Chemically competent B834-pRARE2 cells were prepared using a one-step transformation and storage solution (Chung et al., 1989). This method provides long-lasting competent cells with high transformation efficiency suitable for HTS of protein production. The method was quick and simple to perform, and produced cells with a transformation rate of about 10^5 cfu/μg of plasmid DNA. Most importantly, it produced competent cells with a long shelf life of up to 3 months at −80°C and worked well with a variety of different host strains.

The next step in expression screening was the manual transformation of the competent B834-pRARE2 cells with the plasmid for the target proteins. The expression plasmids and competent cells were prepared in a 96-well microplate, and transformed cells were spread onto 96 chemically defined PA0.5G agar plates (Studier, 2005). The plates contained the appropriate antibiotics used to maintain selection pressure for the target protein and pRARE2 plasmids. A 96-well growth block containing PA0.5G medium was inoculated with single colonies from the PA0.5G plates. The PA0.5G medium ensured that the cells grow to saturation without basal induction of the target protein. A second growth block containing a chemically defined auto-induction medium (Studier, 2005) was inoculated with cells from the initial PA0.5G growth block. These cells were grown for 1 day, and gene expression was automatically induced by lactose in the medium. Next, the cells were harvested by centrifugation and lysed by sonication. The cell lysates were analyzed by SDS-PAGE to determine the total protein production levels. This approach is similar to other reported small-scale HTS methods that use auto-induction (Luan et al., 2004; Segelke et al., 2004). By comparing the soluble cell lysate fractions isolated by centrifugation with the total expression using SDS-PAGE analysis, target proteins that were expressed, soluble, and proteolyzed from the fusion protein were identified and recorded in our LIMS (Zolnai et al., 2003). Overall, about 70% of the cloned target proteins were expressed in *E. coli* as a fusion to MBP and ~55% these were soluble and suitable for large-scale cell growths.

5.4.1 Large-scale protein production and labeling

X-ray crystallographic studies typically require multi-milligram quantities of highly purified proteins. In addition, a high level of incorporation of selenomethionine is required for the multiwavelength anomalous diffraction (MAD) phasing approach in X-ray crystallography (Hendrickson et al., 1990). NMR studies typically require similar amounts of protein samples with high-level ^{15}N incorporation in order to assess suitability for structure determination as well as a second sample with incorporation of both ^{13}C and ^{15}N required for chemical shift assignments and collection of nuclear overhauser enhancement (NOE) constraints (Palmer et al., 1992).

CESG's high-throughput, cell-based, large-scale protein production employs auto-induction medium for producing both of these types of labeled proteins in *E. coli* (Sreenath et al., 2005; Tyler et al., 2005b). Auto-induction provides many advantages over traditional induction methods for HTP operations, including rapid and reproducible growth of starting inoculum, and elimination of time-consuming and imprecise determinations of the appropriate time for culture induction.

Protocols have been developed and applied in the high-throughput production of selenomethionine – labeled fusion proteins using the conditional methionine auxotroph *E. coli* B834 containing pRARE2 plasmid (Sreenath et al., 2005). The scale-up takes advantage of the conditional auxotrophy to provide rapid and reproducible growth of the starting inoculum. The large-scale growth and expression uses a chemically defined auto-induction medium containing selenomethionine, salts, trace metals, glucose, glycerol, α-lactose, amino acids, and vitamins. Methionine and vitamin B_{12} were intentionally omitted from the medium used for large-scale cell growth in order to promote selenomethionine incorporation. The cell growth was done in 2-L polyethylene terephthalate (PET) beverage bottles in a standard laboratory refrigerated floor shaker for *E. coli* growth (Sreenath et al., 2005). The PET bottles provide sterility, efficient aeration, and achievement of sufficient cell densities required for the expression of recombinant proteins (Millard et al., 2003). The growth cycle from inoculation of the culture bottle through the growth, induction, and expression was time programmed to take ~24 h. From more than 150 expression trials, a 2-L culture of the auto-induction medium produced an average final OD_{600} of ~6 and an average wet cell mass yield of ~14 g/L. For favorably scored expression trials, the average yield of purified, selenomethionine-labeled target protein obtained after proteolysis of the fusion protein was ~30 mg (Sreenath et al., 2005).

The conditional methionine auxotroph *E. coli* B834 was also used for the high-throughput production of [U–^{15}N]- or [U–^{13}C-,U–^{15}N]-labeled proteins (Tyler et al., 2005b). The results from several expression trials in 2 L of the auto-induction medium (500 mL in each of four beverage bottles) gave an average final OD_{600} of ~5 and an average wet cell mass yield of ~9.5 g/L, and an average yield of ~20 mg of the labeled protein of interest.

A scoring method based on denaturing electrophoresis gels was developed to support the screening of large-scale recombinant protein expression in *E. coli* B834 (Sreenath et al., 2005). The visual scoring scheme of protein expression, solubility, and protease reactivity using SDS-PAGE was based on the fractionation pattern and the intensity of Coomassie Blue-stained recombinant proteins relative to molecular weight standards and host protein markers. Each of these properties was scored as high (H), medium (M), or weak (W). These scores provided systematic information on total recombinant protein expression, the solubility of the protein, and fusion protein proteolysis with TEV protease. This protocol also defines a standardized format for sample preparation, gel loading, and digital documentation. The scoring scheme was used to focus purification efforts on proteins with the highest likelihood of success in purification.

5.4.2 Protein purification

One of the essential requirements for successful structural proteomics is a purification pipeline capable of producing large numbers of recombinant proteins in sufficient quantity and purity for structural studies. As the biochemical properties of eukaryotic proteins are highly diverse, it is difficult to develop a generic method for protein purification. The use of affinity tags such as poly-histidine $(His)_n$ where (n = 6, 7, or 8), GST, and MBP are typically used for purifying proteins in a high-throughput mode (Hammarstrom et al., 2002; Jeon et al., 2005). CESG established an *E. coli* cell-based protein production pipeline utilizing a $(His)_{6-8}$-MBP fusion tag system. This combination overcame the low solubility of some recombinant proteins and allowed a generic purification strategy based on Ni-IMAC (Amersham Biosciences, Piscataway, NJ). Protein purification has been largely automated at CESG through the use of ÄKTA purifier platforms that are controlled by Unicon 4.12 software (Amersham Biosciences). By use of this system, up to six proteins are purified per chromatography run.

The protein purification protocol includes sonication to lyse the *E. coli* cells (Step 1), preparation of a soluble fraction by centrifugation (Step 2), sample loading onto the Ni-affinity column (Step 3), initial purification of fusion protein using an imidazole gradient (Step 4), desalting of fusion protein into TEV proteolysis buffer by using a HiPrep 26/10 desalting column (Step 5), cleavage of the fusion tag from the target protein by TEV proteolysis (Step 6), purification of target protein from the fusion tags by a subtractive Ni-affinity column (Step 7), polishing of the target protein with a Superdex 200 26/60 gel filtration column (Step 8), desalting into a buffer specified according to the target pI using a HiPrep 26/10 desalting column (Step 9), concentration of the desalted target to 10 mg/mL (Step 10), and drop freezing the target into liquid nitrogen (Step 11) with storage of the frozen target at −80°C (Step 12). All 12 steps are completed within a 5-day cycle. A 300 μg aliquot of each purified protein is stored in acid washed metal-free Eppendorf tubes for subsequent quality control studies.

The LIMS (Zolnai et al., 2003) was used to capture and store data generated during the purification, including protein yield, images of SDS-polyacrylamide gels, UV-visible spectra, and quality control results. The purity of most target proteins obtained was greater than 95% with an average yield of 32 mg from a 2-L culture volume. When the purity of target protein was less than desired, polishing steps using either ion exchange and/or size exclusion were used to improve the purity to greater than 95%.

In order to improve the efficiency of the protein purification pipeline, production data have been analyzed by various criteria (Jeon et al., 2005). The purification success rate of targets with molecular weight (MW) values in the 10–50 kDa was 39–42%, whereas those with a MW greater than 50 kDa had a 23% success rate (Figure 5.2A). Purifications of targets with MW values less than 10 kDa were the least successful,

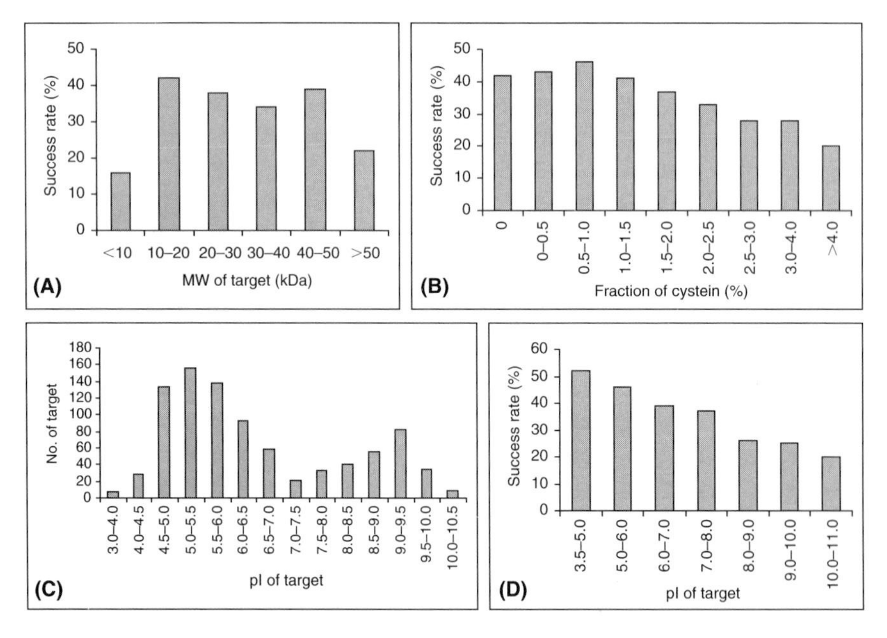

Figure 5.2 Analysis of factors influencing the success of semi-automated purification of protein targets. A: Purification success rate as a function of the molecular weight of the target protein. **B:** Purification success rate as a function of percentage of cysteine residues in the sequence. **C:** Distribution of the pI values in targets for which purification was attempted. **D:** Purification success rate as a function of the pI of the target.

with a rate of only 16%. Thus, our current pipeline is best suited for proteins between 10 and 50 kDa in size, and other methods may be needed for either smaller (less than 10 kDa) or larger (greater than 50 kDa) proteins. The distribution of the proportion of cysteine in targets investigated was unimodal, with an average value of 1.60%. For targets with cysteine fraction values greater than 0.5%, the purification success rate decreased linearly with an increasing percentage of cysteine in a target (Figure 5.2B). Our target selection process discriminates against proteins with large numbers of cysteine residues, and the purification success data generally support this strategy. The distribution of the calculated pI values of selected targets was also bimodal, with both acidic and basic proteins studied (Figure 5.2C). The purification success rate decreased linearly as the pI value of the target increased (Figure 5.2D).

5.5 Wheat germ cell-free protein production

Cell-free methods for protein synthesis with extracts from prokaryotic (Kramer et al., 1999) or eukaryotic (Clemens and Prujin, 1999) sources offer an alternative to the *E. coli* cell-based platforms that are the mainstay of most structural proteomics centers. In cooperation with the Cell-Free Sciences Co. Ltd. (Yokohama, Japan), CESG has investigated the potential of wheat germ cell-free protein production as an enabling technology for NMR-based structural proteomics (Endo, 2001; Madin et al., 2000; Sawasaki et al., 2002). This protein production platform has potential advantages over *E. coli* cell-based and *E. coli* cell-free approaches. A larger proportion of smaller protein targets are produced in folded, soluble form by the wheat germ cell-free approach than by the *E. coli* cell approach. In a comparison of 96 randomly chosen *A. thaliana* targets carried through CESG's wheat germ cell-free and *E. coli* cell-based pipelines to ^{15}N-labeled proteins, eight proteins from the cell-free pipeline, versus five proteins from the cell-based pipeline, were found suitable for NMR structural analysis on the basis of ^{1}H–^{15}N correlation NMR spectra (Tyler et al., 2005a). The platform utilizes a single construct for all targets without any redesign of the DNA or RNA; thus it offers advantages over commercial cell-free methods utilizing *E. coli* extracts (Betton, 2003) that require multiple constructs or redesign of the ORF. The wheat germ cell-free protocol employs no additives such as polyethylene glycol, which has been found to improve yields in *E. coli* S30 cell-free synthesis (Kim, 2000). Due to the small volume requirements for screening (25–50 μL) and protein production for structural studies (4–12 mL), the method is amenable to automation. This successful implementation of cell-free protein expression avoids specific problems associated with cell harvesting, cell lysis, and precolumn manipulations. Moreover, cell-free systems permit labeling strategies that cannot be achieved in whole cell systems because of biological reactions that scramble labels. One example is stereo-array isotope labeling (SAIL) through the incorporation of specially labeled amino acids developed by Professor Kainosho at the Tokyo Metropolitan University (Torizawa et al., 2005). The incorporation of SAIL amino acids leads to reduced spectral congestion, unambiguous chiral assignments, improved resolution of scalar and dipolar couplings, narrower line widths, and simplified relaxation analysis. These benefits should increase the molecular weight limit for routine NMR structure determinations of proteins.

The wheat germ cell-free pipeline protocol developed at the CESG for NMR structure determinations consists of four steps: (1) creation of a plasmid used for *in vitro* transcription; (2) automated small-scale (25 μL) screening to check the level of protein production and solubility; (3) automated larger scale (4–12 mL) production of [U–^{15}N] protein used to evaluate whether solution conditions can be found that render the target suitable for NMR structure determination (soluble, monodisperse, folded, and stable); and (4) automated production of sufficient [U–^{13}C,^{15}N]-labeled protein for multidimensional, multi-NMR data collection. The detailed cell-free His$_6$-tag protocol including cloning, small-scale screening, large-scale protein production, protein purification, and NMR sample preparation has been described previously (Vinarov et al., 2004). Briefly, a well-defined series of cloning steps is used to create a DNA plasmid containing the target gene and 5′- and 3′-extensions that promote efficient transcription and translation. Small-scale protein expression and purification trials are carried out in 96-well format on a GeneDecoder1000™ (CFS, Japan) small-scale automated protein synthesizer. Successful candidates from these screens (those estimated to yield under larger-scale production greater than or equal to 0.5 mg/mL target protein with solubility greater than 75%) are then selected for larger scale protein expression on the Protemist10™ (CFS, Japan) large-scale automated protein synthesizer on ^{15}N-labeled amino acids. Purified [U–^{15}N] proteins are then assayed by ^{1}H–^{15}N correlation spectroscopy for their suitability as structural candidates (they must be folded, monodisperse, and stable at room temperature for at least 14 days). Targets that pass these tests are then prepared as [U–^{13}C,^{15}N] protein samples on Protemist10™.

The wheat germ cell-free platform has been used to screen ~1000 eukaryotic ORFs, with more than 400 of these from the *A. thaliana* genome. Of the *Arabidopsis* ORFs, 125 produced protein that was more than 75% soluble, 76 that had high expression and solubility, and more than 75% went forward into the 4-mL production phase with uniform ^{15}N-labeling. Of these, 17 yielded ^{1}H–^{15}N heteronuclear single-quantum correlation (HSQC) spectra indicative of folded, non-aggregating protein. To date, 12 targets have passed the HSQC and protein stability screen: 10 of these targets have been produced with incorporation of [U–^{13}C,^{15}N] amino acids and one has been prepared with incorporation of SAIL amino acids. Six structures have been determined thus far via this platform.

5.6 Mass spectrometry of purified proteins for quality assurance and analysis

Mass spectrometry (MS) has become an extremely useful technique in protein research (Loo, 1997; Mann and Wilm, 1995; Cohen and Chait, 2001). CESG uses MS as a high-throughput step for determining protein molecular weight, protein identity,

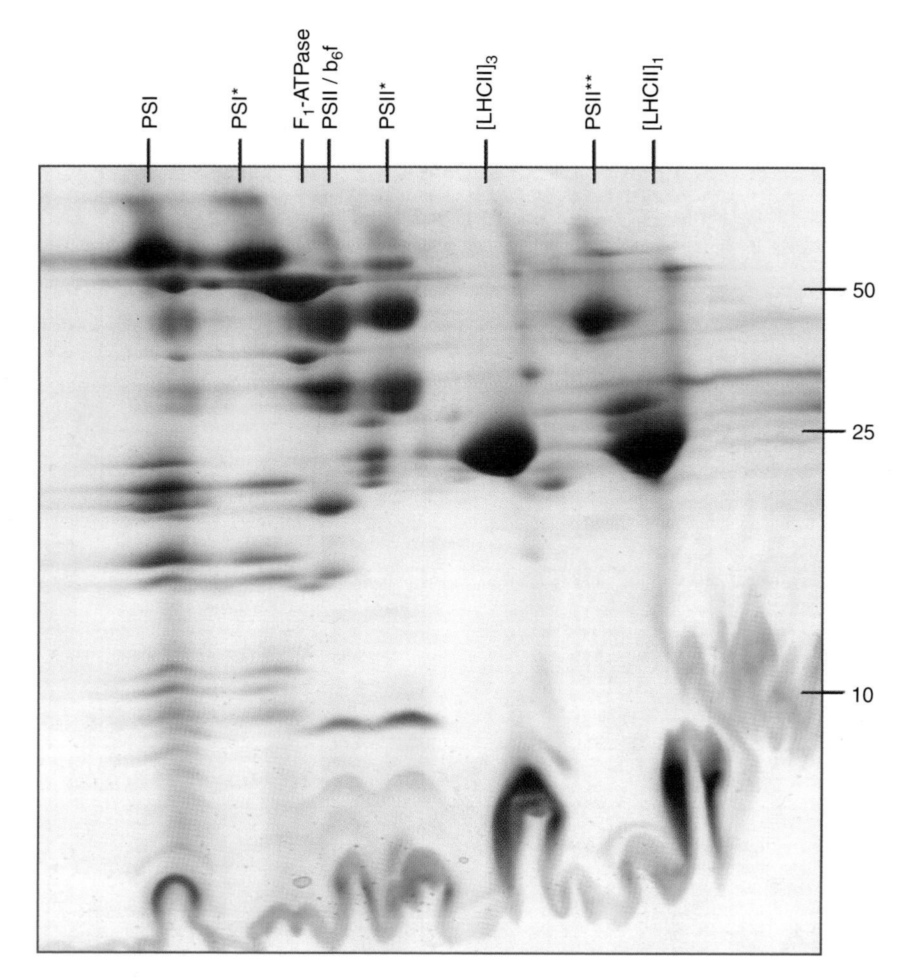

Plate 3.1 Resolution of chloroplast protein complexes from *Arabidopsis* by 2D Blue-native/SDS PAGE. Protein complexes of purified chloroplast membranes were solubilized by a buffer containing 1% dodecylmaltoside and subsequently separated by gel electrophoresis as outlined in Heinemeyer et al. (2005). The 2D gel was Coomassie-stained. Molecular masses of standard proteins are given to the right and the identities of resolved protein complexes above the gel. PSI: photosystem I; PSI*: subcomplex of photosystem I; F_1-ATPase: F1 part of the chloroplast ATP synthase complex; PSII: photosystem II; b_6f: cytochrome b_6f complex; PSII* and PSII**: subcomplexes of photosystem II; $[LHCII]_3$: trimeric light harvesting complex II; $[LHCII]_1$: monomeric light harvesting protein II. The green dye front at the bottom of the gel represents free chlorophyll liberated from chlorophyll-protein complexes due to the presence of SDS during the second gel dimension.

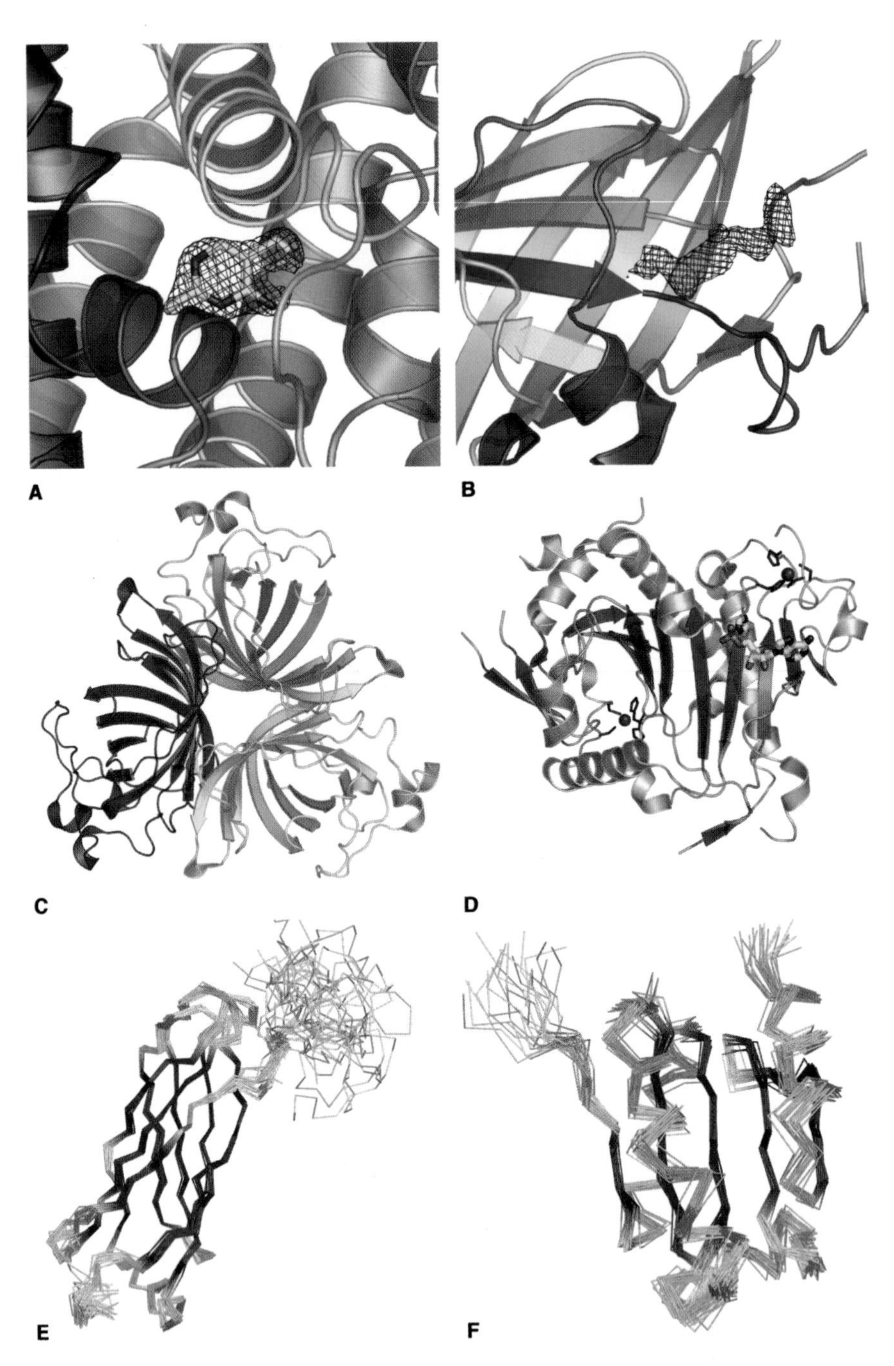

Plate 5.1 Illustrative structures. A: Active site of At3g16990.1 (PDB ID 1Q4M) revealed bound substrate, 4-amino-5-hydroxymethyl-2-methylpyrimidine. **B:** Density of an unknown ligand found in the structure of allene oxide cyclase At3g25760.1 (PDB ID 1ZVC). **C:** Allene oxide cyclase At3g25770.1 (PDB ID 1Z8K) forms a biological trimer. **D:** A covalent adenylyl-enzyme intermediate of At5g18200. His186 is covalently modified by AMP. The enzyme contains two zinc atoms (red spheres) coordinated by two histidine and two cysteine residues (blue), respectively. **E:** Ensemble of 20 solution structures of At1g01470.1 (PDB ID XO8). **F:** Ensemble of 20 solution structures of At3g51030.1 (PDB ID 1XFL).

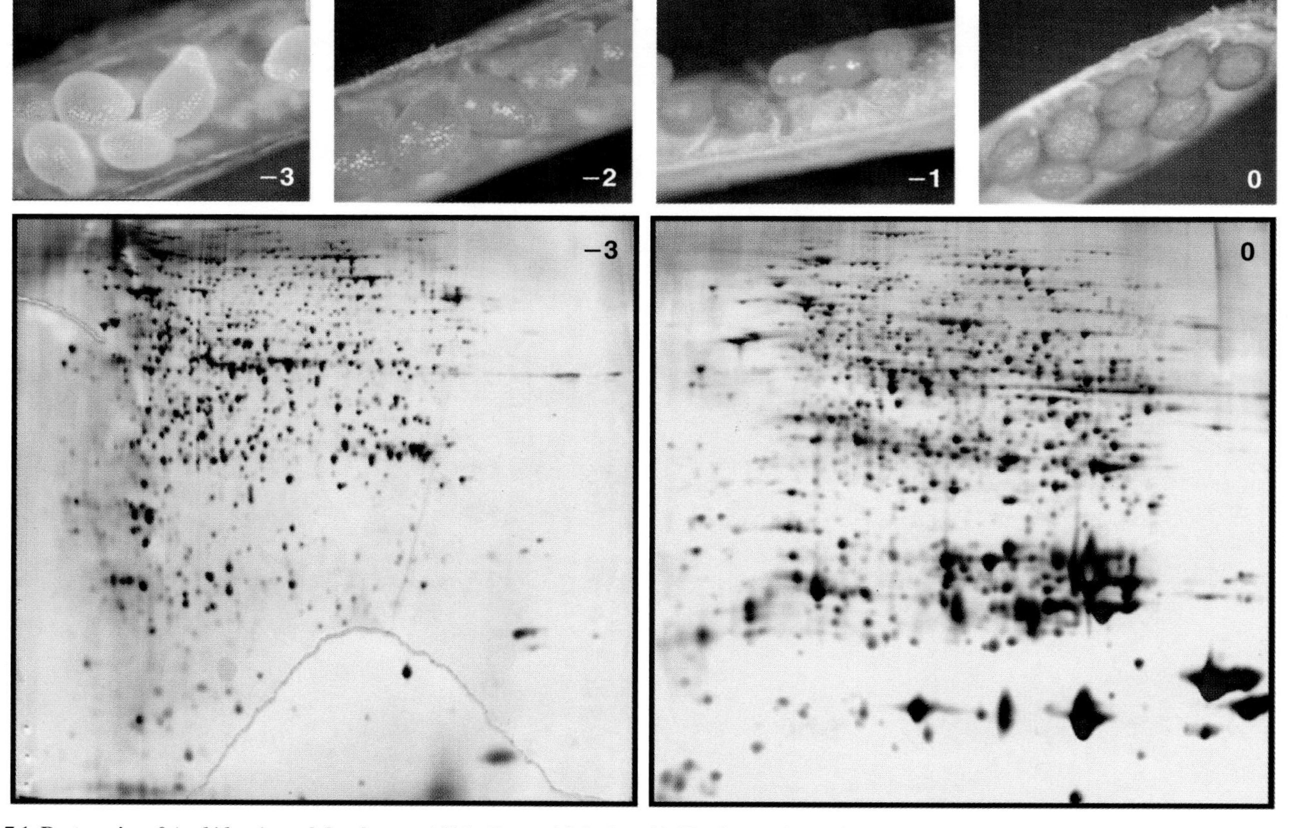

Plate 7.1 Proteomics of *Arabidopsis* seed development (Job, C. unpublished work). The figure aims at illustrating the approach of functional proteomics based on differential 2-DE. Developmental stages are shown on the top (from −3 to 0, immature embryos to mature dry embryos). Total soluble proteins were extracted and analyzed by 2-DE (bottom) from isolated immature (stage −3) and dry mature (stage 0) embryos as described in Job et al. (2005). About 100 μg protein was loaded onto the gels.

and degree of incorporation of selenomethionine, ^{15}N, and ^{13}C labels (Sreenath et al., 2005; Tyler et al., 2005b). All purified proteins are characterized for approximate molecular weight and purity by matrix-assisted laser desorption/ionization time-of-flight (MALDI-TOF) MS, electrospray ionization MS (ESI-MS), and SDS-PAGE before delivery to X-ray and NMR structure determination groups. MALDI-TOF MS is useful for scanning higher molecular weight molecules and is relatively forgiving of salts and other contaminants (Mann and Wilm, 1995). ESI-MS is more useful for lower molecular weight molecules. The mass accuracy of ESI-MS is greater than that of MALDI-TOF by an order of magnitude; ESI-MS supports protein molecular weight determination to within several Daltons and can reveal the extent of label incorporation (Cohen and Chait, 2001; Mann and Wilm, 1995). ESI-MS also has been used to identify the multimeric nature of a large number of proteins, both those covalently bound by disulfide bonds and those non-covalently bound (Loo, 1997; Smith et al., 2004). Amino acid analysis is used as a supplementary technique in the fairly rare occasions when MS fails to reveal label incorporation. When the identity of a protein must be confirmed, proteolytic digestion combined with liquid chromatography/tandem mass spectrometry (LC-MS/MS) is used to determine the amino acid sequence of the peptides (Mann and Wilm, 1995). In addition, inductively coupled plasma mass spectrometry (ICP-MS) and ESI-MS are used to detect and study metal and non-metal ligands (Loo, 1997; Smith et al., 2004; Di Tullio et al., 2005).

A total of 577 unique protein samples were processed to conclusion by MS and denaturing gel electrophoresis. Those judged to be degraded, insufficiently pure, or poorly incorporated with label were not passed on to the structure determination groups. Proteins of incorrect identification were renamed and reinserted into the pipeline whenever possible. The great majority of incorrectly identified proteins were the result of mishandling errors in the protein production pipeline. Proteins that were not identified by initial MS analysis were further analyzed by LC-MS/MS to confirm or refute the identity of the protein. The final acceptance rate of target proteins was 83%. Throughput was approximately nine MALDI-TOF and ESI, and four LC-MS/MS experiments per week. The degree of incorporation of selenomethionine, ^{15}N, and ^{13}C was determined for all labeled samples. Mean selenomethionine incorporation as determined by ESI-MS was 90% (Sreenath et al., 2005). ^{15}N and ^{13}C incorporation were each consistently greater than 95% (Tyler et al., 2005b). MS was also used to determine the extent of reductive methylation when this procedure (Rypniewski et al., 1993) was used to potentially improve crystallization; mean reductive methylation was roughly 90% (data not shown).

Protein oligomers were identified by MALDI-TOF MS. Since non-covalent bonds often dissociate during preparation of samples for MS, it was hypothesized that the subunits were most likely joined by disulfide bonds; this was later confirmed by comparison of reduced versus non-reduced samples on SDS-PAGE (Smith et al., 2004).

MS has also aided in the identification of proteins containing metal ions and other non-metal protein bound ligands. Thus ICP-MS showed calcium in At3g03401.1, iron in At5g51720.1 and At1g79260.1, and nickel in At5g11950.1 and At3g17210.1.

A putative serine protease inhibitor with structure solved by CESG (PDB ID 2APJ) was found to have additional mass corresponding to a derivative of phenylmethyl sulfonyl fluoride added to the purification buffers as a protease inhibitor. A second protein, predicted by sequence homology to be a flavoprotein, exhibited an ESI-MS peak corresponding to the predicted combined mass of the protein and flavin adenine dinucleotide (At5g21482.1).

5.7 Crystallomics and X-ray structure determination

The major bottlenecks in producing high-quality 3D models within a high-throughput structural genomics project are essentially identical to the most significant landmarks in a traditional research lab: obtaining lead crystals, optimizing crystals to a size and quality suitable for diffraction experiments, screening individual crystals for cryoprotection and diffraction quality, determining phases, obtaining an initial trace of the molecule, building and refining models. A major emphasis of the protein structure initiative has been to increase the yield of each step of high-throughput structure determination not only to facilitate structural genomics, but also to develop technology for accelerating the path from gene to structure in traditional research labs. In the following, the current practices in X-ray crystallography at CESG are described, along with a chronicle of recent successes with high-throughput structure determination of plant proteins.

5.7.1 Initial screening

Drop freezing protein samples provides time to conduct extensive QA critical for high-throughput operations. Purified proteins that pass quality assurance, certifying their identity and purity, come to the crystallography unit in the form of frozen droplets stored at −80°C. Aliquots are removed for initial screening. The overall movement of samples and the composition of crystallization screens are tracked with Sesame (Zolnai et al., 2003).

For the last two years, initial crystallization screening has been conducted either with multi-channel pipettes or with a Tecan Genesis™ liquid handling robot using Corning 192 condition vapor-diffusion plates and using a non-commercial screen. Crystals were grown at both ambient temperature and at 4°C. Images were collected over a 4-week period using a CrystalScore™ (Diversified Scientific, Birmingham, AL) or CrystalFarm™ (Discovery Partners International, San Diego, CA) imaging systems. A ‘0’ to ‘10’ scoring scheme was developed to rapidly score each well image. Targets with either crystals or promising paracrystalline material with a score of ‘5’ or better were then passed to individual crystallographers for optimization.

Our project has also conducted preliminary investigations on the use of microfluidic methods for crystallization screening using the Fluidigm Topaz™ (Fluidigm, South San Francisco, CA) system. In this system, solutions of precipitant and macromolecule are placed together by pressure in microfluidic channels in a ‘chip’ constructed of photo-etched layers of silicone Room Temperature Vulcanizing (RTV) laid down over an optically transparent plastic substrate (Hansen and Quake, 2003; van der Woerd et al., 2003) Topaz™ technology requires only 10–20 nL protein sample per crystallization trial. Preliminary experiments have indicated a high degree of global correlation between sitting drop vapor-diffusion experiments at 1-μL scale and microfluidics at 10-nL scale: protein targets that crystallize in vapor diffusion experiments also crystallize in nanoliter microfluidics experiments. However, results are not always well correlated at a finer grain: the precise crystallization solutions yielding crystals in microfluidics experiments may or may not yield crystals in vapor diffusion experiments. Crystallization experiments in the Topaz system are complete within 1 week, as opposed to 1 month by conventional methods, and the optical properties of the chips and image recognition software allow rapid and efficient discovery of positive crystallization results.

Our experiences with the Topaz™ system have led us to propose a two-tier approach for integrating microfluidics into our pipeline. A microscale protein sample of 10–20 μL could be prepared by either cell-free or cell-based expression technology, and screened for crystallizability. Since a successful structure determination often requires around 1 mL of sample at 10 mg/mL, a large-scale sample would then be prepared only for proteins shown to crystallize in microfluidic experiments. Other microfluidics systems for protein crystallization are under development (Mao et al., 2002; Zheng et al., 2004; Hansen et al., 2004), but none appear to have attained the maturity and flexibility of the Topaz™ system.

The cost of protein production and other supplies has thus far largely restricted examination of samples by microfluidics to the realm of 'rare and fine' protein samples obtained from tissue culture expression. However, if the cost of generating a small amount of protein sample with adequate purity for crystallization trials can be reduced, microfluidics technology may find a comfortable place in high-throughput structural genomics.

5.7.2 Optimizations and salvage

Optimizations of promising crystal leads were described using the Well module of Sesame and using either grid or stochastic (Segelke, 2001) methods. The Well module constructs worklists describing the optimization that were then executed on the Tecan Genesis™ (Tecan US, Durham, NC). In the event that no workable crystals were obtained, protein modification by reductive methylation (Rypniewski et al., 1993) has been a productive strategy for CESG, netting three structures that would have been otherwise lost. Several other structural genomics projects have adopted reductive methylation as a salvage strategy for recovering poorly behaved target proteins (Liu et al., 2005). HTS of additives has also yielded several structures that were initially refractory to traditional optimization approaches.

Once crystals larger than 20 μm (Sliz et al., 2003) are obtained, they are screened for diffraction on the home source Bruker AXS (www.bruker-axs.de/) (Madison, WI) Microstar™ generator and Proteum™ charge-coupled device (CCD) detector. Our experience has shown that it is important to screen as many crystal leads for a given protein as early as possible, because visual quality of crystals may not be a reliable indicator of diffracting power. MicroRT capillary replacements allow crystals to be extracted from a high-density crystallization plate and screened for diffraction in a few minutes (Kanalin et al., 2005). Unless the crystal was grown from a solution that already has a cryoprotectant, many crystals may be required to screen additives that allow the crystal to be flash cooled to liquid nitrogen temperature while maintaining internal order (Rodgers, 1994). This is a labor-intensive process that has been facilitated by making putative cryoprotectants using the Tecan Genesis™ and will be facilitated in the future by the Bruker Bruno™ crystal handling robot. Interestingly, the first cryoprotectant used for flash-cooling proteins, Paratone-N (Hope, 1988), has returned to favor in a high-throughput context, because it does not require passing the crystal through a series of cryoprotectant solutions.

Technology for incorporating selenomethionine into recombinant proteins (Hendrickson et al., 1990), advances in phase determination by MAD (Hendrickson, 1991), development of flash-cooling techniques for extending the lifetime of protein crystals in X-ray beams (Hope, 1988) and completion of the first beamlines at third-generation synchrotron sources (Walsh et al., 1999) heralded the structural genomics revolution. The potent combination of these technologies virtually eliminated the need to make heavy atom derivatives of proteins, preserved crystals long enough to complete multiple data sets at different wavelengths, and delivered the intense beams of tunable X-rays required for production-level MAD experiments.

After a last-chance bioinformatics survey (Jaroszewski et al., 2000) to find potential atomic models for molecular replacement (Vagin and Teplyakov, 1997), the vast majority of CESG crystal structures have been determined by multi- or single-wavelength anomalous diffraction data acquired on selenomethionyl proteins at the Advanced Photon Source. The selenium substructure is typically determined by direct methods (Smith et al., 1998; Adams et al., 2002a). Modern phasing programs such as *autoSHARP* (Bricogne et al., 2003) followed by robust density modification and phase improvement programs like *ARP/WARP* (Perrakis et al., 1997) or *SOLVE/RESOLVE* (Terwilliger and Berendzen, 1999; Terwilliger, 2000) can produce nearly complete starting models of the protein structure for crystals that diffract to beyond 2.5 Å resolution. Substantial manual intervention in the tracing process is still required for poorly diffracting crystals (Jones et al., 1991) where even contemporary density modification and auto-tracing programs lose traction. The atomic model is completed by iterative map fitting (McRee, 1999; Emsley and Cowtan, 2004) and refinement (Bailey, 1994; Murshudov et al., 1997; Brunger et al., 1998). The final model is validated for stereochemical and conformational reasonableness (Laskowski et al., 1993; Davis et al., 2004) and deposited in the PDB (Berman et al., 2002).

Plant proteins seem to behave in a similar fashion to our overall pan-eukaryotic target set. Not surprisingly, our efficacy in working with proteins from mesophilic, complex organisms was lower than that of projects focusing on thermophilic prokaryotes (Page and Stevens, 2004). However, our crystallization success rate for eukaryotic proteins is comparable to, or exceeds, the published efficacy of other structural genomics projects where less than or equal to 10% of target proteins screened for crystallization progressed to PDB deposition (Liu et al., 2005).

5.7.3 Future directions

Over the last year and a half, the fraction of PDB depositions from proteins screened by crystallography at CESG has risen from 5% to over 10%. Significant bottlenecks still exist. Robust and reliable algorithms for classifying images of crystallization experiments would reduce the labor required to identify crystals from initial screens. Uniform application of salvage techniques pioneered in PSI-1 should recover more diffracting crystals from targets that have already been cloned and purified. Incremental improvements in screening cryoprotected crystals for diffraction quality would assist in finding rare, well-diffracting crystals from targets with problematic crystal forms. Microfluidic experiments hold the promise of generating rapid assessment of the crystallizability of a given protein, or multiple truncated forms of one protein. Improvements in these areas would allow more labor to be devoted to the step unlikely to be automated: application of the final craftsmanship of a crystallographer polishing and rationalizing a high-quality 3D model of a protein new to science.

5.8 NMR screening and structure determination

CESG has established an efficient protocol for high-throughput protein structural analysis by NMR spectroscopy. The whole process includes the following major steps: sample screening, NMR data collection, chemical shift assignment and structural calculation, and structure-based functional annotation.

5.8.1 Sample screening

The objective of the NMR sample-screening procedure is to determine the gross state (folded, partially folded, unfolded, aggregated) and stability of each protein under a given set of solution conditions. This information is used to identify protein target/solution condition combinations that are likely to yield structures, targets that show promise and may become suitable through redesign of the peptide chain and/or modification of the solution conditions and targets that are unsuitable for NMR structural analysis. Screening can consist of as many as four stages carried out with a [U–^{15}N]-protein (Figure 5.3): (1) prescreening; (2) screening and stability; (3) NMR suitability test; and (4) buffer/target modification. Two-dimensional (2D) [^{1}H,^{15}N]-HSQC was routinely used for NMR sample screening. Prescreening, in which one-dimensional (1D) [^{1}H,^{15}N]-HSQC and 1D ^{1}H spectrum were collected, is required only when the protein sample is too dilute for 2D HSQC. Criteria used in categorizing protein samples are: (a) the chemical shift dispersion of the peaks, particularly in the ^{1}H dimension (a ^{1}H^{N} dispersion of greater than 3 ppm is considered good); (b) the peak count (the result is considered good if the number of ^{1}H–^{15}N peaks resolved is between

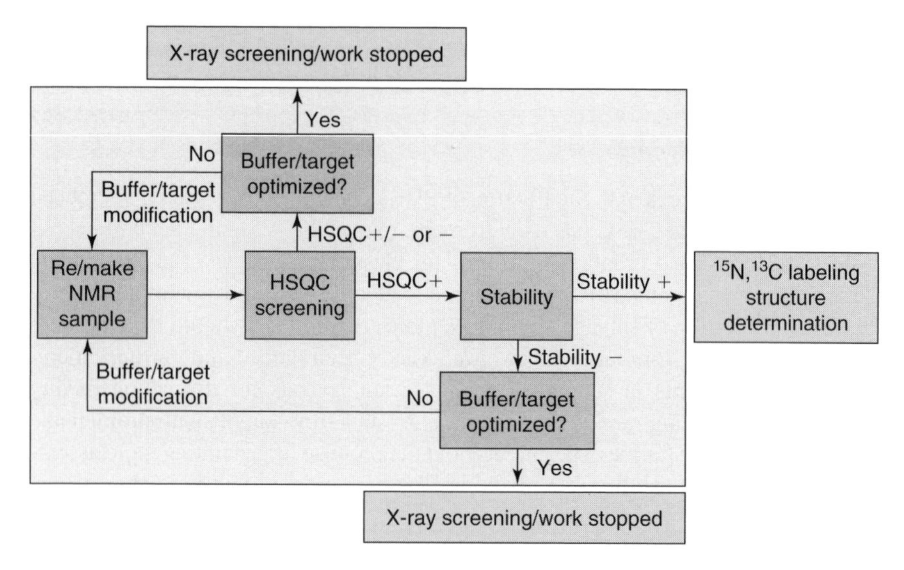

Figure 5.3 Work flow diagram for NMR sample screening.

80% and 120% of the expected number); and (c) the uniformity of peak intensities. On the basis of these spectral results, the protein sample was classified as HSQC+, HSQC+/−, or HSQC− and entered into our LIMS (Zolnai et al., 2003). The sample was marked as HSQC+ if all of the three criteria (a–c) were satisfied; it was marked as HSQC+/− if criterion (a) was met but either (b) or (c) were not; otherwise it was marked as HSQC−.

Proteins in the HSQC+ category then proceed to the stability test, in which the 2D [^{1}H,^{15}N]-HSQC spectrum was recollected after incubation for 7 days at room temperature. This spectrum was then compared with the original spectrum to monitor long-term stability, monodispersity, and solubility. If a sample passes the stability test, additional NMR data sets are collected (backbone ^{15}N relaxation and translational diffusion measurements) to assess the overall tertiary structure and presence of disordered stretches. After a suitable NMR target was identified, a [U–^{15}N,U–^{13}C]-labeled sample of the target was prepared and used in collecting NMR data for the structure determination. If the sample falls into the HSQC+/− or HSQC− category, domain optimization and screening of buffer condition will be considered. The microdrop analysis approach (Lepre and Moore, 1998) enables the screening of a large set of solution conditions with little protein and has been adopted by CESG. Proteins that fail the screening test were pulled off the target queue for NMR structure determination.

5.8.2 NMR data collection

Within 2 to 3 weeks, a defined set of heteronuclear NMR spectra were recorded on an NMR spectrometer (generally 500 or 600 MHz for ^{1}H) equipped with a triple-resonance (^{1}H, ^{13}C, ^{15}N, with ^{2}H lock) cryogenic probe. The temperature of the sample was normally held at 25°C. Data sets that were collected for chemical shift assignment and structure determination include: 2D [^{1}H,^{15}N]-HSQC, 2D [^{1}H,^{13}C]-HSQC, HNCACB, CBCA(CO)NH, HBA(CO)NH, H(CCO)NH, C(CO)NH, HCCH-TOCSY, HNCO, ^{15}N edited [^{1}H,^{1}H] NOESY, and ^{13}C edited [^{1}H,^{1}H] NOESY. The spectra were processed with the NMRPipe software package (Delaglio et al., 1995).

5.8.3 Chemical shift assignment and structural calculation

Sequential backbone resonance assignment and side chain assignment of the target were achieved by the automated assignment program combined with subsequent spectral analysis. The GARANT (General Algorithm for Resonance AssignmeNT, www.mol.biol.ethz.ch/groups/wuthrich_group/software/garant) (Bartels et al., 1996) software program or the recently developed in-house package probabilistic identification of spin systems and their assignments including coil-helix inference as output (PISTACHIO) (Eghbalnia et al., 2005) were used for automated backbone and side-chain assignments. The TALOS (Torsion Angle Likelihood Obtained from Shift and Sequence Similarity, spin.niddk.nih.gov/bax/software/TALOS/index.html) (Cornilescu et al., 1999) software package was used to predict ϕ and y torsion angles from the assigned chemical shifts as restraints for structure calculations. The automatic NOE assignment module in CYANA (Combined assignment and dYnamics Nlgorithm for NMR Applications, www.las.jp/prod/cyana/eg/Main.html) (Herrmann et al., 2002) or ARIA (Ambiguous Restraints for Iterative Assisgnment, www.pasteur.fr/recherche/unites/Binfs/aria/) (Linge et al., 2001) was used to generate initial NOE assignment and structural ensemble. Additional NOE assignments were then added and erroneous ones corrected through examination of NMR spectra prior to recalculation of structures. In the final round of structural calculation, on the basis of their lowest target function or energy, 20 structures were chosen for refinement using the water refinement protocol in Crystallography and NMR System (CNS, http://cns.csb-yale.edu/vl.o/). The final 20 NMR conformers representing the protein structure were validated by Procheck-NMR (Laskowski et al., 1996).

5.8.4 Structure-based functional annotation

After a protein structure was determined, it was submitted to the VAST (http://www.ncbi.nlm.nih.gov/Structure/VAST/vastsearch.html) or DALI (network service for comparing protein structures in three-dimensions, http://www. ebi.ac.uk/dali/) server to determine proteins with related structures. In addition, a search of the protein sequence database (UniProt, http://www.pir.uniprot.org/) was used to identify possible families to which a protein may belong. The structure modeling program MOLMOL (software for display and analysis of structures, www.mol.biol.ethz.ch/groups/wuthrich-group/software) (Koradi et al., 1996) was used to identify the secondary structural elements. Finally, the protein–protein or protein–ligand interactions inferred for the target may be further investigated by NMR spectroscopy.

5.9 Illustrative structures and functional implications

Structural genomics pilot projects carried out worldwide have shown success in lowering the costs of protein production and structure determination by X-ray crystallography and NMR spectroscopy. The progress of *A. thaliana* targets at each step in the CESG pipelines, as extracted from the Sesame database, is shown in Table 5.1. Although considerable attrition occurred at each step, the overall success rate from expression clone to deposited structure was similar to those reported by other structural genomics groups.

Critical problems remain to be solved. Most structural genomics centers are reporting overall success rates from target selection through structure determination and refinement in the range between 1% and 10%, with lower success rates for eukaryotic proteins. In the interest of efficiency and cost savings, it is important to analyze where failures occur and to devise strategies to minimize these. The most effective routes for improvement involve a combination of bioinformatics and small-scale screening. Bioinformatics relies on prior information and mathematical models for correlating success rates with gene sequences. Small-scale screening offers the most economical way of testing whether a target will proceed through the critical stages leading to a structure: whether it has been cloned, supports the production of sufficient soluble protein, crystallizes for X-ray analysis, or is suitable for solution-state NMR analysis. CESG will continue employing two production technologies, *E. coli* cell-based and wheat germ cell-free, both for small-scale screening and for protein production for structure determinations. Future efficiencies are anticipated from adopting Flexi®Vector cloning method (Promega, Madison, WI) to allow a seamless switch between these two expression platforms, and by obtaining tighter linkages between the predictive results of small-scale expression testing and large-scale protein production.

CESG has solved the structures of 47 *Arabidopsis* proteins (Table 5.2). Fourteen proteins were solved by NMR spectroscopy, 31 by X-ray crystallography, and two by both methods. At the time they were selected most of these proteins had less than 30% sequence identity over 100 or more residues to proteins with structures deposited in the PDB. Thus, they were selected with the goal of improving the ability to map from sequence to structure and are identified in Table 5.2 as 'sequence/structure space' targets. The functions of four of the targets had been determined experimentally, but their structures were not close to any available from the PDB.

CESG will continue to consider outside requests for structures in the second phase of the Protein Structure Initiative (2005–2010). A number of the structures of *Arabidopsis* proteins resulted from targets chosen in this way, and we encourage investigators to contact us about their interests. Our perception is that the success rate with proteins having a known biological/biochemical function is higher than with sequence/structure space targets. This may be because they are known to be 'real' proteins and already known cofactors, ligands, or substrates can be added to screening buffers to improve the odds of achieving a diffraction quality crystal or acceptable NMR spectrum.

Table 5.2 Structures of *Arabidopsis* proteins solved by the CESG (through 10/01/2005)

AGI code	PDB ID	Putative function	Solved by	Sequence/structure space target?[a]	New functional insights?	Outside request?	Other features?	Reference
At1g01470.1	1XO8	LEA14, unknown	NMR	Y	Y	N	Fibronectin type III fold	Tyler et al. (2005), Singh et al. (2005)
At1g05000.1	1XRI	Putative phosphoprotein phosphatase	X-ray	Y	Y	N		
At1g07440.1	1XQ1	Putative tropinone reductase	X-ray	N	Y	N		
At1g16640.1	1YEL	Nucleic acid binding	NMR	Y	Y	N		Waltner et al. (2005)
At1g23820.1	1XJ5	Spermidine synthase	X-ray	N	Y	Y		
At1g24000.1	1VJH	Allergen, binding pocket	X-ray	Y	Y	N	Dimer	
At1g35720.1	1YCN	Annexin	X-ray	N	Y	Y		
At1g77680.1	1VJI	12-oxophytodienoate reductase isoform 1	X-ray	N	Known	Y	FMN	Fox et al. (2005)
At1g77540.1	1XO4	Acetyltransferase	NMR	Y	Y	N		
At1g77540.1	1XMT	Acetyltransferase	X-ray	Y	Y	N		
At1g79260.1	2A13	Fatty acid binding protein like	X-ray	Y	Y	N		
At2g03760.1	1Q44	Putative sulfotransferase	X-ray	Y	Y	N		Smith et al. (2004)
At2g06050.1	1Q45	12-oxophytodienoate reductase isoform 3	X-ray	N	Known	Y	FMN	Malone et al. (2005)
At2g17340.1	1XFI	Metal binding, putative active site residues identified	X-ray	Y	Y	Y	Metal bound	Bitto et al. (2005)
At2g19940.2	1XYG	Arginine biosynthesis	X-ray	Y	Y	N	Tetramer	
At2g23090.1	1WVK	Unknown	NMR	Y	N	N		
At2g24940.1	1T0G	Steroid binding protein	NMR	Y	Y	N		Song et al. (2004)
At2g31350.1	1XM8	Glyoxalase II	X-ray	N	Y	N	Metal bound	
At2g34160.1	1VM0	DNA binding, alba-like protein	X-ray	Y	Y	N	K, NO_3 ions bound	

(*Continued*)

Table 5.2 (*Continued*)

AGI code	PDB ID	Putative function	Solved by	Sequence/ structure space target?[a]	New functional insights?	Outside request?	Other features?	Reference
At2g37210.1	2A33	Lysine decarboxylase like	X-ray	Y	Y	N		
At2g38220.1	2A3L	Adenosine 5′-monophosphate deaminase	X-ray	Y	Y	N	Inhibitor bound	Han et al. (2005)
At2g43510.1	1JXC	Trypsin inhibitor	NMR	N	N	N		Zhao et al. (2002)
At2g46140.1	1YYC	Unknown	NMR	Y	N	N		
At3g01050.1	1SE9	Unknown	NMR	Y	N	N		Vinarov et al. (2004)
At3g03250.1	1Z90	Putative UDP-glucose pyrophosphorylase	X-ray	Y	Y	N		
At3g03410.1	1TIZ	EF-hand/Ca-binding sensors or signal modulators	NMR	Y	Y	N		Song et al. (2004)
At3g03773.1	1XO9	Unknown	NMR	Y	N	N		
At3g04780.1	1XOY	Unknown	NMR	Y	N	N		Song et al. (2005)
At3g16990.1	1Q4M	Thiamine-biosynthesis pathway	X-ray	Y	N	N	Binds HMP[b], SO_4	Blommel et al. (2004), Benach et al. (2005)
At3g17210.1	1Q53	Unknown	NMR	Y	N	N	Dimer	Lytle et al. (2004)
At3g17210.1	1Q4R	Unknown	X-ray	Y	N	N	Dimer, Mg	Bingman et al. (2004)
At3g21360.1	1Y0Z	Fe(II) αKG dioxygenase	X-ray	Y	Y	N	Fe(II), 2 active site confirmations	Bitto et al. (2005)
At3g22680.1	1VK5	Binds CHAPS detergent in crystal	X-ray	Y	Y	N	CHAPS, Dimer	Allard et al. (2005)
At3g25760.1	1ZVC	Allene oxide cyclase	X-ray	Y	Known	Y	Trimer, unknown density, Mg	

At3g25770.1	1Z8K	Allene oxide cyclase	X-ray	Y	Known	Y	Trimer, unknown density, Mg	
At3g51030.1	1XFL	TRX h1	NMR	N	Y	N	Redox-dependant chemical shift	Peterson et al. (2005)
At4g09670.1	1YDW	Oxidoreductase, AX110p like	X-ray	N	Y	N		
At4g34215.1	2APJ	Unknown, putative esterase	X-ray	Y	Y	N	O-Benzylsulfonyl serine	
At5g01610.1	1YDU	Unknown	NMR	Y	N	N		
At5g01750.1	1ZXU	Tubby-like protein	X-ray	Y	Y	N		
At5g02240.1	1XQ6	Short-chain dehydrogenase	X-ray	Y	Y	N	Alternative space group	
At5g02240.1	1YBM	Short-chain dehydrogenase	X-ray	Y	Y	N	Alternative space group	
At5g06450.1	1VK0	Exonuclease	X-ray	Y	Y	N	Hexamer	
At5g08170.1	1VKP	Deaminase	X-ray	Y	Known	Y	Dimer	
At5g11950.1	1YDH	Lysine-decarboxylase-like protein	X-ray	Y	Y	N		
At5g18200-AMP.1	1Z84	ADP-glucose phosphorylase	X-ray	Y	Y	Y	Dimer, AMP, metals	
At5g18200.1	1ZWJ	Uridyltransferase?	X-ray	Y	Y	N	Dimer, metal	
At5g22580.1	1RJJ	Unknown	NMR	Y	N	N	Dimer	Cornilescu et al. (2004)
At5g48480.1	1XY7	Structural similarity to bleomycin and fosfomycin resistance proteins	X-ray	Y	Y	N		
At5g56660.1	1XMB	IAA amino acid hydrolase	X-ray	Y	Known	N		
At5g66040.1	1TQ1	Unknown	NMR	Y	N	N		

CHAPS: Zwitterionic detergent that combines features of bile salts and *N*-alkyl sulfobetaines; dimethyl-(3-sulfonatopropyl)-[3-[4-[(3*R*,5*R*,7*R*,12*R*)-3,7,12-trihydoxy-10,13-dimethyl-2,3,4,5,6,7,8,9,11,12,14,15,16,17-tetradecahydro-1*H*-cyclopenta[*a*]phenanthren-17-yl]pentanoylamino]propyl]azanium).
[a]At time they were selected, proteins had less than 30% sequence identity over 100 or more residues to proteins with structures in the PDB.
[b]4-amino-5-hydroxymethyl-2-methylpyrimidine.

Several examples of new insights provided by our structural analyses are now provided. One of the first structures CESG solved, At3g16990.1, contained an unidentified electron density (Blommel et al., 2004 #202). The Northeast Structural Genomics Consortium has since solved the structure of a prokaryotic homolog and found the electron density at analogous position to be 4-amino-5-hydroxymethyl-2-methylpyrimidine phosphate, the product of the reaction and an intermediate in the thiamine-biosynthesis pathway (Benach et al., 2005). Reinspection of the electron density maps of At3g16990.1 (see Plate 5.1A in Plate section) in the light of this finding revealed that the ligand bound in this enzyme is its substrate, 4-amino-5-hydroxymethyl-2-methylpyrimidine.

An example of an individual investigator's suggested study is that of At3g25760.1 and At3g25770.1 (see Plate 5.1C in Plate section). These two proteins are allene oxide cyclases, known enzymes in the jasmonic acid synthesis pathway. Although they share extensive sequence similarity, we chose to send both through our pipeline to increase the chances of getting a structure. Both crystallized and yielded X-ray structures. The structures contain an unidentified electron density, most likely of an organic molecule (see Plate 5.1B in Plate section). This density suggests thelocation of the enzyme active site. Investigations are underway to determine the identity of this density and to crystallize the protein in the presence of substrate analogs.

Most of our solved structures have provided new insights into the function of the protein, leading to testable hypotheses that can be addressed by functional analyzes. One case in point, At5g18200.1 (see Plate 5.1D in Plate section) was a sequence/structure space target that had limited amino acid sequence similarity (less than 22%) to a bacterial galactose uridyltransferase (GalT). The crystal revealed a structure similar to that of GalT. However, functional studies carried out in collaboration with Perry Frey's laboratory at the University of Wisconsin, Madison, WI, showed that the protein had no GalT-like activity but instead catalyzed adenylyl transfer in the reaction of ADP-glucose with phosphate to form glucose-1-phosphate (McCoy et al., 2006). The identification of the enzymatic activity led to a crystal structure of the protein with an adenosine monophosphate (AMP) intermediate covalently bonded to His^{186} Protein Data Bank identifier (PDB ID 1Z84). This newly discovered enzymatic reaction could have important implications in starch metabolism.

As another example, the late embryogenesis abundant (LEA) proteins are a heterogeneous group expressed in later stages of embryogenesis in seed embryos. LEA14 (At1g01470.1; see Plate 5.1E in Plate section) was a sequence/structure space target that was solved by NMR spectroscopy (Singh et al., 2005). The closest structural homolog turned out to be the fibronectin type III domain. Fibronectins in animal cells have been shown to be involved in wound healing. Similarly, LEA14 has been implicated in wound and desiccation response in plants. Thus, At1g01470.1 protein may represent a structural motif that plays a role in a cell's response to fluid loss so as to alleviate potential damage.

Another structure solved by NMR was of At3g51030.1 (see Plate 5.1F in Plate Section), the h1 family member of the 19 TRX family members in the *Arabidopsis* genome (Peterson et al., 2005). The NMR data identified a conformational transition dependent on the redox state of a disulfide bridge that affects the active site of this protein.

The power of modern structural genomic approaches has manifested itself in several ways. First, it fosters the discovery of new functions for proteins for which there are no clues. The knowledge of what functions are present is an essential prerequisite to the claim of fuller understanding of a cell or an organism. Second, our understanding of known protein systems is advanced by using the 3D structure as a basis to more elaborate studies of function. This understanding will continue to provide direct benefit to the advancement of basic scientific research, and can also lead to better efforts to develop plant-based biotechnology applications. Indeed, it is our hope that these structures will stimulate further investigations aimed at determining their cellular and mechanistic functions.

Acknowledgements

This work is supported by NIH/NIGMS grant numbers 1 U54 GM074901-01 (JLM; 07/01/05-06/30/10) and P50 GM64598 (JLM; 01/01/02-08/31/05). Mass spectrometry (MALDI-TOF, ESI, and LC-MS/MS) was performed by the University of Wisconsin-Madison Biotechnology Center by Gregorz Sabat, James Brown, and Amy C. Harms. ICP-MS was performed at the University of Wisconsin-Madison Department of Water Science and Engineering by Joel Overdier and Martin Shafer. Amino acid analysis was performed at the W.M. Keck Foundation Biotechnology Resource Laboratory at Yale University by Fernando M. Pineda and J. Myron Crawford. We thank Perry Frey, Jason Mccoy, and Paul Blommel for the use of unpublished results. We further thank all the past and present members of CESG whose hard work made this chapter possible.

References

Aceti, D.J., Blommel, P.G., Endo, Y., Fox, B.G., Frederick, R.O., Hegeman, A.D., Jeon, W.B., Kimball, T.L., Lee, J.M., Newman, C.S., Peterson, F.C., Sawasaki, T., Seder, K.D., Sussman, M.R., Ulrich, E.L., Wrobel, R.L., Thao, S., Vinarov, D.A., Volkman, B.F. and Zhao, Q. (2003) Role of nucleic acid and protein manipulation technologies in high-throughput structural biology efforts. In: Steinbüchel, A. (ed) *Biopolymers*. Wiley-VCH, Weinheim.

Adams, P.D., Grosse-Kunstleve, R.W., Hung, L.W., Ioerger, T.R., Mccoy, A.J., Moriarty, N.W., Read, R.J., Sacchettini, J.C., Sauter, N.K. and Terwilliger, T.C. (2002a) PHENIX: building new software for automated crystallographic structure determination. *Acta Crystallogr. D. Biol. Crystallogr.*, **58**, 1948–1954.

Adams, S.R., Campbell, R.E., Gross, L.A., Martin, B.R., Walkup, G.K., Yao, Y., Llopis, J. and Tsien, R.Y. (2002b) New biarsenical ligands and tetracysteine motifs for protein labeling *in vitro* and *in vivo*: synthesis and biological applications. *J. Am. Chem. Soc.*, **124**, 6063–6076.

Albala, J.S., Franke, K., Mcconnell, I.R., Pak, K.L., Folta, P.A., Rubinfeld, B., Davies, A.H., Lennon, G.G. and Clark, R. (2000) From genes to proteins: high-throughput expression and purification of the human proteome. *J. Cell Biochem.*, **80**, 187–191.

Bailey, S. (1994) The Ccp4 Suite – Programs for protein crystallography. *Acta Crystallogr. D. Biol. Crystallogr.*, **50**, 760–763.

Bartels, C., Billeter, M., Güntert, P. and Wüthrich, K. (1996) Automated sequence-specific NMR assignment of homologous proteins using the program GARANT. *J. Biomol. NMR*, **7**, 207–213.

Benach, J., Edstrom, W.C., Lee, I., Das, K., Cooper, B., Xiao, R., Liu, J., Rost, B., Acton, T.B., Montelione, G.T. and Hunt, J.F. (2005) The 2.35 a structure of the tena homolog from pyrococcus furiosus supports an enzymatic function in thiamine metabolism. *Acta Crystallogr. D. Biol. Crystallogr.*, **61**, 589.

Berman, H.M., Battistuz, T., Bhat, T.N., Bluhm, W.F., Bourne, P.E., Burkhardt, K., Feng, Z., Gilliland, G.L., Iype, L., Jain, S., Fagan, P., Marvin, J., Padilla, D., Ravichandran, V., Schneider, B., Thanki, N., Weissig, H., Westbrook, J.D. and Zardecki, C. (2002) The protein data bank. *Acta Crystallogr. D. Biol. Crystallogr.*, **58**, 899–907.

Betton, J.M. (2003) Rapid translation system (RTS): a promising alternative for recombinant protein production. *Curr. Protein Pept. Sci.*, **4**, 73–80.

Blommel, P.G., Smith, D.W., Bingman, C.A., Dyer, D.H., Rayment, I., Holden, H.M., Fox, B.G. and Phillips, G.N. Jr. (2004) Crystal structure of gene locus At3g16990 from *Arabidopsis thaliana. Proteins*, **57**, 221–222.

Blommel, P.G., Martin, P.A., Wrobel, R.L., Steffen, E., Fox, B.G. (2005) High efficiency single step production of expression plasmids from cDNA clones using the Flori Vector Cloning System. *Protein Expres. Purif.* 2005 Dec 5; (Epub ahead of print].

Bricogne, G., Vonrhein, C., Flensburg, C., Schiltz, M. and Paciorek, W. (2003) Generation, representation and flow of phase information in structure determination: recent developments in and around SHARP 2.0. *Acta Crystallogr. D. Biol. Crystallogr.*, **59**, 2023–2030.

Brunger, A.T., Adams, P.D., Clore, G.M., Delano, W.L., Gros, P., Grosse-Kunstleve, R.W., Jiang, J.S., Kuszewski, J., Nilges, M., Pannu, N.S., Read, R.J., Rice, L.M., Simonson, T. and Warren, G.L. (1998) Crystallography and NMR system: a new software suite for macromolecular structure determination. *Acta Crystallogr. D. Biol. Crystallogr.*, **54**, 905–921.

Chung, C.T., Niemela, S.L. and Miller, R.H. (1989) One-step preparation of competent *Escherichia coli*: transformation and storage of bacterial cells in the same solution. *Proc. Natl Acad. Sci. USA*, **86**, 2172–2175.

Clemens, M.M. and Prujin, G.J. (1999) *Protein Expression. A practical approach.* Oxford, Oxford University Press, UK.

Cohen, S.L. and Chait, B.T. (2001) Mass spectrometry as a tool for protein crystallography. *Annu. Rev. Bioph. Biomol. Struc.*, **30**, 67–85.

Cornilescu, G., Delaglio, F. and Bax, A. (1999) Protein backbone angle restraints from searching a database for chemical shift and sequence homology. *J. Biom. NMR*, **13**, 289–302.

Davis, I.W., Murray, L.W., Richardson, J.S. and Richardson, D.C. (2004) MOLPROBITY: structure validation and all-atom contact analysis for nucleic acids and their complexes. *Nucl. Acids Res.*, **32**, W615–W619.

Delaglio, F., Grzesiek, S., Vuister, G.W., Zhu, G., Pfeifer, J. and Bax, A. (1995) NMRPipe: a multidimensional spectral processing system based on UNIX pipes. *J. Biomol. NMR*, **6**, 277–293.

Di Tullio, A., Reale, S. and De Angelis, F. (2005) Molecular recognition by mass spectrometry. *J. Mass. Spectrom.*, **40**, 845–865.

Dieckman, L., Gu, M., Stols, L., Donnelly, M.I. and Collart, F.R. (2002) High-throughput methods for gene cloning and expression. *Protein Expres. Purif.*, **25**, 1–7.

Eghbalnia, H.R., Bahrami, A., Wang, L., Assadi, A. and Markley, J.L. (2005) Probabilistic identification of spin systems and their assignments including coil-helix inference as output (PISTACHIO). *J. Biomol. NMR*, **32**, 219–233.

Emsley, P. and Cowtan, K. (2004) Coot: model-building tools for molecular graphics. *Acta Crystallogr. D. Biol. Crystallogr.*, **60**, 2126–2132.

Endo, Y. (2001) Genomics to Proteomics: A high-throughput cell-free protein synthesis system for practical use. *The 3rd ORCS International Symposium on Ribosome Engineering.* Tsukuba, Japan.

Grodberg, J. and Dunn, J.J. (1988) ompT encodes the *Escherichia coli* outer membrane protease that cleaves T7 RNA polymerase during purification. *J. Bacteriol.*, **170**, 1245–1253.

Hammarstrom, M., Hellgren, N., Van Den Berg, S., Berglund, H. and Hard, T. (2002) Rapid screening for improved solubility of small human proteins produced as fusion proteins in *Escherichia coli*. *Protein Sci.*, **11**, 313–321.

Hansen, C. and Quake, S.R. (2003) Microfluidics in structural biology: smaller, faster em leader better. *Curr. Opin. Struct. Biol.*, **13**, 538–544.

Hansen, C.L., Sommer, M.O. and Quake, S.R. (2004) Systematic investigation of protein phase behavior with a microfluidic formulator. *Proc. Natl Acad. Sci. USA*, **101**, 14431–14436.

Hartley, J.L., Temple, G.F. and Brasch, M.A. (2000) DNA cloning using *in vitro* site-specific recombination. *Genome Res.*, **10**, 1788–1795.

Hendrickson, W.A. (1991) Determination of macromolecular structures from anomalous diffraction of synchrotron radiation. *Science*, **254**, 51–58.

Hendrickson, W.A., Horton, J.R. and Lemaster, D.M. (1990) Selenomethionyl proteins produced for analysis by multiwavelength anomalous diffraction (MAD): a vehicle for direct determination of three-dimensional structure. *EMBO J.*, **9**, 1665–1672.

Herrmann, T., Guntert, P. and Wuthrich, K. (2002) Protein NMR structure determination with automated NOE assignment using the new software CANDID and the torsion angle dynamics algorithm DYANA. *J. Mol. Biol.*, **319**, 209–227.

Hoffman, B.J., Broadwater, J.A., Johnson, P., Harper, J., Fox, B.G. and Kenealy, W.R. (1995) Lactose fed-batch overexpression of recombinant metalloproteins in *Escherichia coli* BL21 (DE3): process control yielding high levels of metal-incorporated, soluble protein. *Protein Expres. Purif.*, **6**, 646–654.

Hope, H. (1988) Cryocrystallography of biological macromolecules: a generally applicable method. *Acta Crystallogr. B*, **44 (Part 1)**, 22–26.

Initiative, A.G. (2000) Analysis of the genome sequence of the flowering plant *Arabidopsis thaliana*. *Nature*, **408**, 796–815.

Jaroszewski, L., Rychlewski, L. and Godzik, A. (2000) Improving the quality of twilight-zone alignments. *Protein Sci.*, **9**, 1487–1496.

Jeon, W.B., Aceti, D.J., Bingman, C.A., Vojtik, F.C., Olson, A.C., Ellefson, J.M., Mccombs, J.E., Sreenath, H.K., Blommel, P.G., Seder, K.D., Burns, B.T., Geetha, H.V., Harms, A.C., Sabat, G., Sussman, M.R., Fox, B.G. and Phillips, J.G.N. (2005) High-throughput purification and quality assurance of *Arabidopsis thaliana* proteins for eukaryotic structural genomics. *J. Struct. Funct. Genom.*, **31**, 1–5.

Jonasson, P., Liljeqvist, S., Nygren, P.A. and Stahl, S. (2002) Genetic design for facilitated production and recovery of recombinant proteins in *Escherichia coli*. *Biotechnol. Appl. Biochem.*, **35**, 91–105.

Jones, T.A., Zou, J.Y., Cowan, S.W. and Kjeldgaard, M. (1991) Improved methods for building protein models in electron density maps and the location of errors in these models. *Acta Crystallogr. A*, **47 (Part 2)**, 110–119.

Kanalin, Y., Kmetko, J., Bartnik, A., Stewart, A., Gillilan, R., Lobkovsky, E. and Thorne, R.E. (2005) A new sample mounting technique for room temperature macromolecular crystallography. *J. Appl. Crystallogr.*, **38**, 333–339.

Kang, B.H., Busse, J.S., Dickey, C., Rancour, D.M. and Bednarek, S.Y. (2001) The *Arabidopsis* cell plate-associated dynamin-like protein, ADL1Ap, is required for multiple stages of plant growth and development. *Plant Physiol.*, **126**, 47–68.

Kapust, R.B. and Waugh, D.S. (1999) *Escherichia coli* maltose-binding protein is uncommonly effective at promoting the solubility of polypeptides to which it is fused. *Protein Sci.*, **8**, 1668–1674.

Kapust, R.B., Tozser, J., Copeland, T.D. and Waugh, D.S. (2002) The P1′ specificity of tobacco etch virus protease. *Biochem. Bioph. Res. Commun.*, **294**, 949–955.

Kifer, I., Sasson, O. and Linial, M. (2005) Predicting fold novelty based on ProtoNet hierarchical classification. *Bioinformatics*, **21**, 1020–1027.

Kim, D.M.S., Swartz, J.R. (2000) Prolonging cell-free protein synthesis by selective reagent additions. *Biotechnol. Progr.*, **16**, 385–390.

Knaust, R.K. and Nordlund, P. (2001) Screening for soluble expression of recombinant proteins in a 96-well format. *Anal. Biochem.*, **297**, 79–85.

Koradi, R., Billeter, M. and Wuthrich, K. (1996) MOLMOL: a program for display and analysis of macromolecular structures. *J. Mol. Graph.*, **14**, 51–55, 29–32.

Kramer, G., Kudlicki, W. and Hardesty, B. (1999) *Protein Expression. A practical approach*. Oxford University Press, Oxford, UK.

Laskowski, R.A., Macarthur, M.W., Moss, D.S. and Thornton, J.M. (1993) Procheck – a program to check the stereochemical quality of protein structures. *J. Appl. Crystallogr.*, **26**, 283–291.

Laskowski, R.A., Rullmannn, J.A., Macarthur, M.W., Kaptein, R. and Thornton, J.M. (1996) AQUA and PROCHECK-NMR: programs for checking the quality of protein structures solved by NMR. *J. Biomol. NMR*, **8**, 477–486.

Lepre, C.A. and Moore, J.M. (1998) Microdrop screening: a rapid method to optimize solvent conditions for NMR spectroscopy of proteins. *J. Biomol. NMR*, **12**, 493–499.

Linge, J.P., O'donoghue, S.I. and Nilges, M. (2001) Automated assignment of ambiguous nuclear overhauser effects with ARIA. *Method. Enzymol.*, **339**, 71–90.

Liu, Z.J., Tempel, W., Ng, J.D., Lin, D., Shah, A.K., Chen, L., Horanyi, P.S., Habel, J.E., Kataeva, I.A., Xu, H., Yang, H., Chang, J.C., Huang, L., Chang, S.H., Zhou, W., Lee, D., Praissman, J.L., Zhang, H., Newton, M.G., Rose, J.P., Richardson, J.S., Richardson, D.C. and Wang, B.C. (2005) The high-throughput protein-to-structure pipeline at SECSG. *Acta Crystallogr. D. Biol. Crystallogr.*, **61**, 679–684.

Loo, J.A. (1997) Studying noncovalent protein complexes by electrospray ionization mass spectrometry. *Mass Spectrom. Rev.*, **16**, 1–23.

Luan, C.H., Qiu, S., Finley, J.B., Carson, M., Gray, R.J., Huang, W., Johnson, D., Tsao, J., Reboul, J., Vaglio, P., Hill, D.E., Vidal, M., Delucas, L.J. and Luo, M. (2004) High-throughput expression of *C. elegans* proteins. *Genome. Res.*, **14**, 2102–2110.

Madin, K., Sawasaki, T., Ogasawara, T. and Endo, Y. (2000) A highly efficient and robust cell-free protein synthesis system prepared from wheat embryos: plants apparently contain a suicide system directed at ribosomes. *Proc. Natl Acad. Sci. USA*, **97**, 559.

Mann, M. and Wilm, M. (1995) Electrospray mass spectrometry for protein characterization. *Trend. Biochem. Sci.*, **20**, 219–224.

Mao, H., Yang, T. and Cremer, P.S. (2002) A microfluidic device with a linear temperature gradient for parallel and combinatorial measurements. *J. Am. Chem. Soc.*, **124**, 4432–4435.

McCoy, J.G., Arabshahi, A., Bitto, E., Bingman, C.A., Ruzicka, F.J., Frey, P.A. and Phillips Jr., G.N. (2006) Structure and mechanism of an ADP-glucose phosphorylase from *Arabidopsis thaliana*. *Biochemistry*, **10**, 3154–3162.

Mcree, D.E. (1999) XtalView Xfit – A versatile program for manipulating atomic coordinates and electron density. *J. Struct. Biol.*, **125**, 156–165.

Millard, C.S., Stols, L., Quartey, P., Kim, Y., Dementieva, I. and Donnelly, M.I. (2003) A less laborious approach to the high-throughput production of recombinant proteins in *Escherichia coli* using 2-liter plastic bottles. *Protein Expres. Purif.*, **29**, 311–320.

Murshudov, G.N., Vagin, A.A. and Dodson, E.J. (1997) Refinement of macromolecular structures by the maximum-likelihood method. *Acta Crystallogr. D. Biol. Crystallogr.*, **53**, 240–255.

Oldfield, C.J., Ulrich, E.L., Cheng, Y., Dunker, A.K. and Markley, J.L. (2005) Addressing the intrinsic disorder bottleneck in structural proteomics. *Proteins*, **59**, 444–453.

Page, R. and Stevens, R.C. (2004) Crystallization data mining in structural genomics: using positive and negative results to optimize protein crystallization screens. *Methods*, **34**, 373–389.

Page, R., Moy, K., Sims, E.C., Velasquez, J., Mcmanus, B., Grittini, C., Clayton, T.L. and Stevens, R.C. (2004) Scalable high-throughput micro-expression device for recombinant proteins. *Biotechniques*, **37**, 364, 366, 368 passim.

Palmer, A.G., Fairbrother, W.J., Cavanagh, J., Wright, P.E. and Rance, M. (1992) Improved resolution in three-dimensional constant-time triple resonance NMR spectroscopy of proteins. *J. Biomol. NMR*, **2**, 103–108.

Perrakis, A., Sixma, T.K., Wilson, K.S. and Lamzin, V.S. (1997) wARP: improvement and extension of crystallographic phases by weighted averaging of multiple-refined dummy atomic models. *Acta Crystallogr. D. Biol. Crystallogr.*, **53**, 448–455.

Peterson, F.C., Lytle, B.L., Sampath, S., Vinarov, D., Tyler, E., Shahan, M., Markley, J.L. and Volkman, B.F. (2005) Solution structure of thioredoxin h1 from *Arabidopsis thaliana*. *Protein Sci.*, **14**, 2195–2200.

Rodgers, D.W. (1994) Cryocrystallography. *Structure*, **2**, 1135–1140.

Rypniewski, W.R., Holden, H.M. and Rayment, I. (1993) Structural consequences of reductive methylation of lysine residues in hen egg white lysozyme: an X-ray analysis at 1.8-A resolution. *Biochemistry*, **32**, 9851–9858.

Sawasaki, T., Ogasawara, T., Morishita, R. and Endo, Y. (2002) A cell-free protein synthesis system for high-throughput proteomics. *Proc. Natl Acad. Sci. USA*, **99**, 14652–14657.

Scheich, C., Sievert, V. and Bussow, K. (2003) An automated method for high-throughput protein purification applied to a comparison of His-tag and GST-tag affinity chromatography. *BMC Biotechnol.*, **3**, 12.

Scheich, C., Niesen, F.H., Seckler, R. and Bussow, K. (2004) An automated *in vitro* protein folding screen applied to a human dynactin subunit. *Protein Sci.*, **13**, 370–380.

Segelke, B.W. (2001) Efficiency analysis of screening protocols used in protein crystallization. *J. Cryst. Growth*, **232**, 553–562.

Segelke, B.W., Schafer, J., Coleman, M.A., Lekin, T.P., Toppani, D., Skowronek, K.J., Kantardjieff, K.A. and Rupp, B. (2004) Laboratory scale structural genomics. *J. Struct. Funct. Genom.*, **5**, 147–157.

Service, R. (2005) Structural biology. Structural genomics, round 2. *Science*, **307**, 1554–1558.

Singh, S., Cornilescu, C.C., Tyler, R.C., Cornilescu, G., Tonelli, M., Lee, M.S. and Markley, J.L. (2005) Solution structure of a late embryogenesis abundant protein (LEA14) from *Arabidopsis thaliana*, a cellular stress-related protein. *Protein Sci.*, **14**, 2601–2609.

Sliz, P., Harrison, S.C. and Rosenbaum, G. (2003) How does radiation damage in protein crystals depend on X-ray dose? *Structure (Cambridge)*, **11**, 13–19.

Smith, G.D., Nagar, B., Rini, J.M., Hauptman, H.A. and Blessing, R.H. (1998) The use of SnB to determine an anomalous scattering substructure. *Acta Crystallogr. D. Biol. Crystallogr.*, **54 (Part 5)**, 799–804.

Smith, J.C., Siu, K.W. and Rafferty, S.P. (2004) Collisional cooling enhances the ability to observe non-covalent interactions within the inducible nitric oxide synthase oxygenase domain: dimerization, complexation, and dissociation. *J. Am. Soc. Mass Spectrom.*, **15**, 629–638.

Sorensen, H.P. and Mortensen, K.K. (2005a) Advanced genetic strategies for recombinant protein expression in *Escherichia coli*. *J. Biotechnol.*, **115**, 113–128.

Sorensen, H.P. and Mortensen, K.K. (2005b) Soluble expression of recombinant proteins in the cytoplasm of *Escherichia coli*. *Microb. Cell Fact*, **4**, 1.

Sreenath, H.K., Bingman, C.A., Buchan, B.W., Seder, K.D., Burns, B.T., Geetha, H.V., Jeon, W.B., Vojtik, F.C., Aceti, D.J., Frederick, R.O., Phillips Jr., G.N. and Fox, B.G. (2005) Protocols for production of selenomethionine-labeled proteins in 2-L polyethylene terephthalate bottles using auto-induction medium. *Protein Expres. Purif.*, **40**, 256–267.

Stolc, V., Samanta, M.P., Tongprasit, W., Sethi, H., Liang, S., Nelson, D.C., Hegeman, A., Nelson, C., Rancour, D., Bednarek, S., Ulrich, E.L., Zhao, Q., Wrobel, R.L., Newman, C.S., Fox, B.G., Phillips Jr., G.N., Markley, J.L. and Sussman, M.R. (2005) Identification of transcribed sequences in *Arabidopsis thaliana* by using high-resolution genome tiling arrays. *Proc. Natl Acad. Sci. USA*, **102**, 4453–4458.

Studier, F.W. (2005) Protein production by auto-induction in high density shaking cultures. *Protein Expres. Purif.*, **41**, 207–234.

Studier, F.W., Rosenberg, A.H., Dunn, J.J. and Dubendorff, J.W. (1990) Use of T7 RNA polymerase to direct expression of cloned genes. *Method. Enzymol.*, **185**, 60–89.

Swartz, J.R. (2001) Advances in *Escherichia coli* production of therapeutic proteins. *Curr. Opin. Biotechnol.*, **12**, 195–201.

Terwilliger, T.C. (2000) Maximum-likelihood density modification. *Acta Crystallogr. D. Biol. Crystallogr.*, **56**, 965–972.

Terwilliger, T.C. and Berendzen, J. (1999) Automated MAD and MIR structure solution. *Acta Crystallogr. D. Biol. Crystallogr.*, **55**, 849–861.

Thao, S., Zhao, Q., Kimball, T., Steffen, E., Blommel, P.G., Riters, M., Newman, C.S., Fox, B.G. and Wrobel, R.L. (2004) Results from high-throughput DNA cloning of *Arabidopsis thaliana* target genes using site-specific recombination. *J. Struct. Funct. Genom.*, **5**, 267–276.

Thornton, J.M. (2001) From genome to function. *Science*, **292**, 2095–2097.

Torizawa, T., Terauchi, T., Ono, A. and Kainosho, M. (2005) [The SAIL method: a new NMR approach for larger protection]. *Tanpakushitsu Kakusan Koso*, **50**, 1375–1381.

Tyler, R.C., Aceti, D.J., Bingman, C.A., Cornilescu, C.C., Fox, B.G., Frederick, R.O., Jeon, W.B., Lee, M.S., Newman, C.S., Peterson, F.C., Phillips Jr., G.N., Shahan, M.N., Singh, S., Song, J., Sreenath, H.K., Tyler, E.M., Ulrich, E.L., Vinarov, D.A., Vojtik, F.C., Volkman, B.F., Wrobel, R.L., Zhao, Q. and Markley, J.L. (2005a) Comparison of cell-based and cell-free protocols for producing target proteins from the *Arabidopsis thaliana* genome for structural studies. *Proteins*, **59**, 633–643.

Tyler, R.C., Sreenath, H.K., Singh, S., Aceti, D.J., Bingman, C.A., Markley, J.L. and Fox, B.G. (2005b) Auto-induction medium for the production of [U-15N]- and [U-13C, U-15N]-labeled proteins for NMR screening and structure determination. *Protein Expres. Purif.*, **40**, 268–278.

Vagin, A. and Teplyakov, A. (1997) MOLREP: an automated program for molecular replacement. *J. Appl. Crystallogr.*, **30**, 1022–1025.

Van Der Woerd, M., Ferree, D. and Pusey, M. (2003) The promise of macromolecular crystallization in microfluidic chips. *J. Struct. Biol.*, **142**, 180–187.

Vinarov, D.A., Lytle, B.L., Peterson, F.C., Tyler, E.M., Volkman, B.F. and Markley, J.L. (2004) Cell-free protein production and labeling protocol for NMR-based Structural Proteomics. *Nature Method*, **1**, 149–153.

Walhout, A.J., Temple, G.F., Brasch, M.A., Hartley, J.L., Lorson, M.A., Van Den Heuvel, S. and Vidal, M. (2000) GATEWAY recombinational cloning: application to the cloning of large numbers of open reading frames or ORFeomes. *Method. Enzymol.*, **328**, 575–592.

Walsh, M.A., Evans, G., Sanishvili, R., Dementieva, I. and Joachimiak, A. (1999) MAD data collection – current trends. *Acta Crystallogr. D. Biol. Crystallogr.*, **55 (Part 10)**, 1726–1732.

Zheng, B., Tice, J.D., Roach, L.S. and Ismagilov, R.F. (2004) A droplet-based, composite PDMS/glass capillary microfluidic system for evaluating protein crystallization conditions by microbatch and vapor-diffusion methods with on-chip X-ray diffraction. *Angew. Chem. Int. Ed. Engl.*, **43**, 2508–2511.

Zolnai, Z., Lee, P.T., Li, J., Chapman, M.R., Newman, C.S., Phillips Jr., G.N., Rayment, I., Ulrich, E.L., Volkman, B.F. and Markley, J.L. (2003) Project management system for structural and functional proteomics: Sesame. *J. Struc. Funct. Genom.*, **4**, 11–23.

6 Cereal proteomics

Setsuko Komatsu

6.1 Introduction

Cereals are an important source of calories for humans, both by direct intake and as the main feed for livestock. Approximately 50% of the calories consumed by the world population originate from three cereals: rice (23%), wheat (17%), and maize (10%) (Khush, 2003). Rice is one of the world's most important agricultural resources, because it is indisputably the only plant species that feeds almost half of the world's population. Rice is also a model plant for biological research, because its genome is smaller than those of other cereals (Devos and Gale, 2000) and it has an important syntenic relationship with the other cereal species (Gale and Devos, 1998). Publication of draft genome sequences for *Oryza sativa* L. ssp. *indica* (Yu et al., 2002) and for *O. sativa* L. ssp. *japonica* (Goff et al., 2002), and a complete map-based sequence of chromosome 1 (Sasaki et al., 2002) and chromosome 4 (Feng et al., 2002) for *O. sativa* L. cv. Nipponbare provides a rich resource for understanding the biological processes of rice. Recently, the International Rice Genome Sequencing Project (2005) presented a map-based, finished-quality sequence that covers 95% of the 389 Mb genome of rice, including virtually all of the euchromatin and two complete centromeres. Once the rice genome is completely sequenced, the challenge ahead for the plant research community will be to identify the function, regulation, and type of post-translational modification of each encoded protein. Also, whereas the genome is static; the proteome is highly dynamic in its response to external and internal cellular events. The responses of the proteome can include changes not only to the relative abundance but also to the post-translational modification of each protein.

Proteomics is a leading technology for the high throughput analysis of proteins on a genome-wide scale. With the completion of genome sequencing projects and the development of improved analytical methods for protein characterization, proteomics has become a major field of functional genomics. The initial objective of proteomics was the large-scale identification of all protein species in a cell or tissue. During the past couple of years, considerable research effort has been applied to the analysis of the rice proteome, and remarkable progress has been made in the systematic functional characterization of proteins in the various tissues and organelles of rice (Komatsu et al., 2003; Komatsu and Tanaka, 2004; Komatsu, 2005). This approach is currently being extended to analyze various functional aspects of proteins. As part of this research, a system for direct differential display using two-dimensional polyacrylamide gel electrophoresis (2D-PAGE) (O'Farrell, 1975) has been developed for the identification of rice proteins that vary in expression under different physiological conditions and among different tissues. This approach readily visualizes proteins, directly and rapidly selects those with altered expression, and then analyzes their structure by comparison with the Rice Proteome Database (Komatsu et al., 2004; http://gene64.dna.affrc.go.jp/RPD/main.html), or by MS and Edman sequencing. This review, drawing from reports on rice, describes the comprehensive analysis and cataloguing of cereal proteins and the functional analysis of cereal using differential proteomics.

6.2 Comprehensive analysis and cataloguing of cereal proteins

6.2.1 Cataloguing of rice proteins using 2D-PAGE

Significant progress has been made towards identifying and cataloguing the proteins of rice tissues and organelles. The capacity to evaluate the functions of rice proteins has been expanded by proteomics analyses of rice embryo (Fukuda et al., 2003) and endosperm (Komatsu et al., 1993), root (Zhong et al., 1997), green (Islam et al., 2004; Zhao et al., 2005) and etiolated shoot (Komatsu et al., 1999a), suspension-cultured cells (Komatsu et al., 1999b), anther (Imin et al., 2001; Kerim et al., 2003), leaf sheath (Shen et al., 2002), and other organs (Tanaka et al., 2004a). Various organelles, such as Golgi (Mikami et al., 2001) and mitochondria (Heazlewood et al., 2003), and other subcellular compartments (Tanaka et al., 2004b) also have been analyzed. Tsugita et al. (1994) analyzed and identified 4892 proteins from nine tissues and one organelle of rice (leaf, stem, root, germ, dark-germinated seedling, seed, bran, chaff, callus, and chloroplast). A more detailed proteomics analysis of rice (cv. Nipponbare) leaf, root, and seed has also been reported (Koller et al., 2002). However, the above studies were all done using different methods, making comparisons among them difficult. To overcome this problem, a consistent methodology based on 2D-PAGE has been used throughout the rice proteome research described below.

6.2.2 The Rice Proteomics Project

Proteomics analysis using 2D-PAGE has the power to monitor global changes that occur in the protein expression of tissues and subcellular compartments. To apply this technique to rice, the Rice Proteomics Project was initiated to apply a consistent methodology to a wide range of rice protein samples. For this project, proteins were extracted from various tissues and subcellular compartments, separated in the first dimension using either isoelectric focusing (IEF) tube gels for the low pH range (4.0–7.0) or linear immobilized pH gradients (IPG) tube gels for the high pH range (6.0–10.0) (Hirano et al., 2000), and then separated in the second dimension by SDS-PAGE. After detection by Coomassie brilliant blue (CBB) staining, proteins were analyzed with Image-Master 2D Elite software (Amersham Biosciences, Piscataway, NJ, USA). The 2D maps of the low and high pH ranges overlapped at around pH 6.0. To obtain N-terminal amino acid sequence tags by Edman sequencing, the proteins, after separation by 2D-PAGE, were electroblotted onto a polyvinylidene difluoride (PVDF) membrane and detected by CBB staining. Internal amino acid sequences were determined by sequencing the peptides obtained by peptide mapping with *S. aureus* V8 protease (Cleveland et al., 1977). The spots and bands were excised from the membrane and applied to a gas-phase protein sequencer (Procise, Applied Biosystems, Foster City, CA, USA). For MS, individual protein spots were excised from the gel and digested with the site-specific protease trypsin, resulting in a set of tryptic peptides. The peptides were extracted, and their masses were measured by matrix-assisted laser desorption/ionization time-of-flight mass spectrometry (MALDI-TOF MS) (Voyager, Applied Biosystems, Framingham, MA, USA). The list of measured peptide masses was compared with the masses of the predicted tryptic peptides for each entry in the sequence database using Mascot software (Matrix Science Ltd., London, UK), which provides a score indicating the probability of a true positive identification.

Proteins were extracted from a total of 23 rice samples that were either tissue-specific or specific to a subcellular location. From these various tissues and subcellular fractions, 13,129 proteins have been identified and, so far, the amino acid sequences of 5676 separate proteins have been determined and added to the database. Using these data, the Rice Proteome Database was constructed and its website (http://gene64.dna.affrc.go.jp/RPD/main.html) now provides extensive information on the progress of rice proteome research (Komatsu et al., 2004). Proteomics analysis of various tissues and organelles has revealed diverse functional categories of proteins. Although many ubiquitous proteins have been identified that share similar functions in different tissues and organelles, most of the proteins are specific to certain tissues or subcellular compartments. These results highlight the diversity of proteomes within the rice plant, and hence the urgent need to analyze additional tissues and subcellular organelles to gain a comprehensive understanding of the proteins encoded by the rice genome. A major advantage of the Rice Proteome Database, in which known proteins are recorded along with the timing and location of their expression, is the wealth of newly identified proteins on which further experiments can be conducted at the biochemical and molecular levels.

6.2.2.1 Proteomics analysis of rice tissues

The Rice Proteomics Project has analyzed many tissue-specific samples, including suspension-cultured cells, endosperm, embryo, crown, seedling root, seedling leaf sheath, seedling leaf blade, stem, mature plant root, mature plant leaf sheath, mature plant leaf blade, anthers, and panicle before heading, after heading and 1 week after flowering, and some of these data are described here. For example, proteins were extracted from leaf sheath, root, and suspension-cultured cells with lysis buffer (O'Farrell, 1975) and separated by 2D-PAGE. The 2D maps of the proteins were found to be consistently reproducible throughout different experiments. Computer analysis using Image-Master 2D Elite software revealed numerous distinct protein spots on the rice tissue 2D maps: 431 in the leaf sheath, 508 in the root, and 962 in suspension-cultured cells.

After 2D-PAGE, the amino acid sequences of the protein spots were analyzed by MS and Edman sequencing. Initially, 79 leaf sheath proteins were analyzed by Edman sequencing, which yielded N-terminal sequences for 13 of these 79 proteins. Next, 66 distinct visible protein spots on the 2D gels were randomly selected and analyzed systematically by MS. Using these two approaches, the amino acid sequences of a total of 79 leaf sheath proteins were determined successfully. From the root, 94 proteins were analyzed by Edman sequencing, which resulted in N-terminal amino acid sequences for 38 of them. A further 35 proteins were excised from the gels and analyzed by MS. A total of 73 root proteins were successfully sequenced using these two methods. Finally, 140 proteins from suspension-cultured cells were identified, of which 90 were analyzed by MS and 50 by Edman sequencing. The N-terminal amino acid sequences of many proteins could not be determined, and these proteins were inferred to have a blocking group at their N-terminus. The proportion of N-blocked proteins was 46% (6/13) in the leaf sheath, 56% (56/94) in the root, and 46% (38/82) in the suspension-cultured cell samples. This result is consistent with a previous report in which 134 rice proteins were subjected to sequencing and 79 proteins (59%) were found to have blocked N-termini (Tsugita et al., 1994).

To determine the functions of the more abundant proteins in the rice tissue proteomes, the identified proteins were categorized according to the scheme used by Bevan et al. (1998) to analyze the sequence of chromosome 4 of *Arabidopsis thaliana*. In the rice leaf sheath proteome, no clear function could be predicted for 20% of the proteins. This category includes both proteins whose functions have not yet been identified (11%) and those with no detectable homology to other proteins predicted in the data files (9%). Significant fractions of the proteins were involved in metabolism (18%), energy production (12%), and defense (12%). Of the 79 proteins analyzed, the largest subset was involved in primary and secondary metabolism (20%), reflecting the complex photoautotrophic metabolism of the rice leaf sheath. The large number of proteins (12%) involved in energy production might be associated with either photosynthetic electron transport or ATP synthesis reactions. The defense response category accounted for 12% of the proteins and included functions associated with defense against disease and other environmental changes.

In the root proteome, the most abundantly represented category was disease- and defense-related proteins. This functional category included metallothioneine, glutathione *S*-transferase, chitinase, NBS-LRR-type resistance protein, antifungal protein R, thaumatin-like protein, superoxide dismutase (SOD) [Cu–Zn], type-1 pathogenesis-related (PR) protein, and salt-induced protein (SALT). The root proteome also included the root-specific protein RCc3, which was previously reported to be expressed in the elongation and maturation zones of primary and secondary roots, and in the root caps (Xu et al., 1995).

In the proteome of suspension-cultured cells, energy and metabolism were the most abundant functional categories. Glyceraldehyde-3-phosphate dehydrogenase and phosphoenolpyruvate carboxylase kinase, included in the energy category, were the most abundantly expressed proteins in the proteome of the suspension-cultured cells. Proteins in the metabolism category included those involved in the metabolism of starch (isoamylase and α-amylase), amino acids (glutamine synthase and glutamine synthetase), and lipids (oleosin isoform, sterol 14-demethylase, and lipoxygenase L-3). Although suspension-cultured cells are non-photosynthetic, some photosynthesis-related proteins were identified such as chlorophyll a/b binding protein, oxygen-evolving enhancer protein 2 (OEE2), and Rieske Fe–S precursor protein.

Metabolism was a well-represented category in all three of the tissue proteomes. Proteins with housekeeping functions, which include cell growth and division, transcription, protein synthesis, protein targeting and storage, and cell structure, were identified in all three tissues in almost the same proportions. However, proteins in the defense category in the root and the energy category in suspension-cultured cells were more abundant than they were in the other tissues. Furthermore, the 'no assigned function' category, which included proteins without a homolog in published databases and/or with unknown functions, totaled approximately 20% in all three tissues.

6.2.2.2 Proteomics analysis of subcellular compartments

In addition to the tissue-specific analyses, the Rice Proteomics Project has analyzed biological samples that are specific to a subcellular location such as cell wall, plasma membrane, vacuole membrane, Golgi membrane, mitochondrion, chloroplast, nucleus, and cytosol. Some of these results are described below. The plasma membrane, vacuolar membrane, Golgi membrane, mitochondria, and chloroplast fractions, isolated from rice seedlings and suspension-cultured cells, were solubilized in lysis buffer (O'Farrell, 1975), and the proteins were separated by 2D-PAGE and analyzed with Image-Master 2D Elite software as described above for the tissue analysis. The 2D maps of the various subcellular compartments resolved 464 proteins in the plasma membrane, 141 in the vacuolar membrane, 361 in the Golgi membrane, 672 in mitochondria, and 252 in chloroplasts (Tanaka et al., 2004b).

After 2D-PAGE, abundant proteins were analyzed by Edman sequencing and MS. Edman sequencing showed that the number of N-terminally blocked proteins varied widely among the subcellular compartments. In mitochondria and chloroplasts, respectively, 73% and 60% of the proteins were N-terminally blocked. A much larger proportion of the proteins were N-terminally blocked in the plasma membrane (96%), vacuolar membrane (89%), and Golgi membrane (98%). SOSUI system software, developed for the classification and secondary structure prediction of membrane proteins (http://sosui.proteome.bio.tuat.ca.jp/), was used to analyze the proteins identified in the three membrane samples. This software predicted transmembrane helices for 14 of the 58 plasma membrane proteins, 6 of the 43 vacuolar membrane proteins, and 7 of the 46 Golgi membrane proteins sequenced in this study.

To determine the functions of the more abundant proteins in the subcellular proteomes of rice, the identified proteins were categorized as described above for the tissue analysis. In the plasma membrane proteome of rice, proteins with functions in the metabolism, energy, signal transduction, and defense categories were abundant. By contrast, no proteins in the signal transduction and defense categories were identified in the plasma membrane proteome of *A. thaliana*, although proteins involved in metabolism and energy were present (Santoni et al., 1998).

The vacuolar membrane proteome is still poorly understood, but the vacuolar membrane of plant cells is known to contain two electrogenic proton pumps, H^+-ATPase and H^+-translocating inorganic pyrophosphatase (PPase) (Hedrich and Schroeder, 1989). In the Rice Proteomics Project, the β subunit of the ATPase was identified in the vacuolar membrane proteome, but not the PPase (Hedrich and Schroeder, 1989). The α2 subunit of the 20S proteasome was detected in the vacuolar membrane fraction, providing evidence that vacuoles might participate in the degradation of denatured proteins. The vacuole membrane proteome also included a water channel protein, a member of a family of vacuolar and plasma membrane proteins that transport water molecules with high efficiency and selectivity (Maurel, 1997). Signal transduction proteins were abundant in the vacuolar membrane and included a calmodulin-like Ca^{2+}-binding protein. Ca^{2+} pumps are known to be widely distributed on plant membranes, including the vacuolar membrane (Sze et al., 2000).

The Golgi complex is a multifunctional organelle responsible for the biosynthesis of complex cell-surface polysaccharides, the processing and modification of glycoproteins, and the sorting of polysaccharides and proteins destined for various locations (Staehelin and Moore, 1995). To understand these functions in detail, it will be necessary to survey the proteome of the Golgi body. In the Golgi membrane fraction, the functional categories of metabolism, energy, and defense were represented in abundance (Tanaka et al., 2004b). A reversibly glycosylated polypeptide, previously identified as a protein localized to the Golgi membrane and involved in the synthesis of xyloglucan and possibly other hemicelluloses (Dhugga et al., 1997), was detected in the Golgi membrane proteome.

Mitochondria play a pivotal role in the energy metabolism of eukaryotic cells. An analysis of the rice mitochondrial proteome has been recently reported (Heazlewood et al., 2003), however, there is only partial overlap between that analysis and the results of the Rice Proteomics Project. Heazlewood et al. (2003) detected proteins in several functional classes, including ATP synthases, complex I, complex III, complex IV, general metabolism, and transport. In the Rice Proteomics Project, by contrast, no proteins were identified that corresponded to pyruvate and 2-oxoglutarate dehydrogenase complexes, succinyl-CoA ligase, isocitrate dehydrogenase, malate dehydrogenase, citrate synthase, malic enzyme, complex II, aconitase, glycine decarboxylase H protein, import proteins, oxygen stress proteins, or chaperones. In the mitochondrial proteome determined in the Rice Proteomics Project, numerous proteins were present in the signal transduction category, similar to the membrane proteomes. Signal transduction proteins were not previously identified in the rice (Heazlewood et al., 2003) or *A. thaliana* (Kruft et al., 2001; Millar et al., 2001) mitochondrial proteomes, which were established through the analysis of major protein spots on gels. However, the signal transduction proteins identified in the mitochondrial fraction are present at low abundance in the cell and are predicted to be membrane localized. Thus, to determine whether these signal transduction proteins are indeed membrane localized, it would be necessary to analyze the mitochondrial membrane proteome.

As for chloroplasts, lumenal and peripheral thylakoid proteomes have been reported for pea and *A. thaliana* (Peltier et al., 2000, 2002; Schubert et al., 2002). To characterize the proteome of whole chloroplasts in rice, intact chloroplasts were purified from rice seedlings on linear Percoll gradients and used to prepare protein extracts for analysis (Tanaka et al., 2004b). In the chloroplast proteome, the energy category, including photosynthetic proteins, contained the largest number of proteins.

6.3 Functional analysis of cereal using differential proteomics

6.3.1 Stresses

To grow and develop optimally, plants need to perceive and process information from both their biotic and abiotic surroundings. Because plants are not motile, they have to be especially responsive to environmental changes, including stress conditions. Although the responses of cereals to several stresses are well understood at the physiological and transcriptional level, they are not well understood at the biochemical level. Proteomics approaches to identifying proteins that are differentially regulated in response to environmental conditions are becoming commonplace in post-genomic research in cereals. This review describes initial steps toward determining the physiological significance of some proteins identified in cereals exposed to abiotic and biotic stresses.

6.3.1.1 Cold

Crop plants in tropical and subtropical regions are seriously injured by temperatures below 12°C but above the freezing point (Lyons, 1973). A primary, if not exclusive, effect of chilling is considered to be the phase transition of membrane lipids at the critical temperatures (Raison et al., 1971; Lyons, 1973). The way plants acclimate to cold stress is not well understood at the biochemical level, but rice seedlings exposed to low temperatures show various changes in their transcriptome. For example, microarray analysis has shown that 36 rice genes appear to be induced under cold stress, and that the expression level for several genes reached a maximum after 24 h of cold treatment (Rabbani et al., 2003). Although this gene expression profiling has deepened our understanding of the response of rice to cold stress, it is still unknown how the transcriptional changes are reflected at the translational level. Changes in the transcriptome are not always closely correlated with changes in protein species (Gygi et al., 1999). With this limitation in mind, a proteomic study of rice was carried out to gain a better understanding of the molecular mechanisms of acclimation to cold stress.

In another study, rice seedlings were exposed to a progressively lower temperature stress treatment involving successive shifts from the normal growth temperature to 15, 10, and finally 5°C (Cui et al., 2005). From these seedlings, approximately

1700 protein spots were separated and visualized on CBB-stained 2D-PAGE gels. Sixty protein spots were found to be up-regulated in response to the progressively lower temperature treatment and to display various induction patterns. These cold-responsive proteins included four protein biosynthesis factors, four molecular chaperones, two proteases, eight enzymes involved in biosynthesis of cell wall components, seven antioxidative/detoxifying enzymes, two proteins of unknown function, and proteins linked to energy metabolism and signal transduction. A large proportion of the proteins (43.9%) were predicted to be located in the plastids, implying that the plastid proteome is particularly responsive to cold stress.

6.3.1.2 Drought

Drought tolerance is required in plants that experience prolonged deficits in soil water. Tolerant plants can maintain the water content of their tissues, can survive a reduction in tissue water content, and can recover more completely after rewatering. Drought represents one of the most severe limitations to the productivity of rain-fed lowland and upland rice. Drought is also one of the major factors limiting the yield of sugar beet. Hajheidari et al. (2005) reported that the efficiency of breeding for increased drought tolerance could be greatly improved by the identification of candidate genes for marker-assisted selection. One way to identify potentially important drought-tolerance genes is to analyze drought-induced changes in the proteome. To this end, two genotypes of sugar beet differing in genetic background were cultivated in the field. Certain proteins in these plants showed genotype-specific patterns of up- or down-regulation in response to drought; these proteins included ribulose-1, 5-bisphosphate carboxylase/oxygenase (RuBisCO) plus 11 other proteins involved in redox regulation, oxidative stress, signal transduction, and chaperone activities. Some of these proteins could contribute a physiological advantage under drought, making them potential targets for marker-assisted selection for drought tolerance.

In the case of rice, a proteomic analysis of drought-conditioned leaves of 3-week-old plants has been reported (Salekdeh et al., 2002). In that study, protein expression in the drought-tolerant cultivars IR62266 (lowland *indica*) and CT9993 (upland *indica*) was compared. Of more than 1000 protein spots detected in leaf extracts, 42 proteins showed a significant change in abundance under stress, with 27 of them exhibiting a different response pattern between the two cultivars. The expression of chloroplast SOD [Cu–Zn] changed significantly in opposite directions in the two cultivars in response to drought. Ten days after rewatering, the abundance of all the drought-responsive proteins had returned more-or-less completely to that of the well-watered control. In CT9993 and IR62266, the proteins that increased most in response to drought were S-like RNase homolog, actin depolymerizing factor, and RuBisCO activase, whereas the protein that decreased most was isoflavone reductase-like protein.

6.3.1.3 Salinity

Like drought, high salinity also causes a water deficit in plants. Salt stress is a major abiotic stress in agriculture worldwide, with an estimated 20% of the earth's land mass and nearly half of all irrigated land affected by salinity. Increased salinization of arable land is expected to have devastating global effects, with predictions of 30% land loss within the next 25 years, and up to 50% by the year 2050 (Yan et al., 2005).

Response to salinity is a very complex quantitative trait. The apoplast of plant cells is a dynamic compartment involved in many processes, including maintenance of tissue shape, development, nutrition, signaling, detoxification, and defense. Dani et al. (2005) used *Nicotiana tabacum* plants as a model to investigate changes in apoplast soluble protein composition induced in response to salt stress. Apoplastic fluid was extracted from leaves of control plants and plants exposed to salt stress, using a vacuum infiltration procedure. Quantitative evaluation and statistical analyses of the resolved spots in treated and untreated samples revealed 20 polypeptides whose abundance changed in response to salt stress; among these, two chitinases and a germin-like protein increased significantly and two lipid transfer proteins were expressed entirely *de novo*. Some apoplastic polypeptides, involved in cell wall modifications during plant development, remained largely unchanged.

Rice is generally considered to be sensitive to salinity. A proteomics approach was used to identify rice proteins that increase in abundance under this type of stress in leaf sheath, root, and leaf blade (Abbasi and Komatsu, 2004). In rice leaf sheath exposed to 50 mM NaCl for 24 h, eight proteins consistently showed significant changes in abundance. Of these eight proteins, three were unidentified, but the other five were identified as OEE2, two fructose bisphosphate aldolases, SOD, and functional unknown protein. The study also revealed that increased expression of SOD is a common response to cold (5°C), drought, salt, and abscisic acid treatments in leaf sheath. This finding suggests that the accumulation of SOD in response to salt, drought, and cold stress has a generally protective role against stress conditions. Under salt stress (50 mM NaCl for 24 h), enhanced expression of OEE2 and aldolase was detected in leaf sheath and leaf blade. SOD and one of the unidentified proteins were also detected in salt-stressed leaf sheath and root. Another unidentified protein was expressed in leaf sheath but was below the detection limit in leaf blade and root. These results indicate that specific proteins are enhanced in distinct regions of the rice plant and show a coordinated response to salt stress.

Lee et al. (2004) identified *A. thaliana* membrane proteins involved in salt stress using 2D-PAGE and MALDI-TOF MS. Among the salt-responsive microsomal proteins, two spots that increased as a result of salt treatment were identified as annexins. Annexins comprise a multigene family of Ca^{2+}-dependent membrane-binding proteins and have been extensively studied in animal cells. AnnAt1 is strongly expressed in root but rarely in flower tissue. The results of the *A. thaliana* study suggested that salt stress stimulated translocation of annexins from the cytosol to the membrane and possibly increased turnover of existing protein.

Yan et al. (2005) also reported a proteomic analysis of salt stress-responsive proteins in rice. Three-week-old seedlings were treated with 150 mM NaCl for 24, 48, and 72 h. Based on 2D-PAGE patterns, more than 1100 proteins were reproducibly detected, including 34 that were up-regulated and 20 that were down-regulated.

Three spots were identified as the same protein, enolase. While four of the changed proteins were previously identified as salt stress-responsive proteins, six were novel: UDP-glucose pyrophosphorylase, cytochrome *c* oxidase subunit 6b-1, glutamine synthetase root isozyme, putative nascent polypeptide associated complex alpha chain, putative splicing factor-like protein, and putative actin-binding protein.

6.3.1.4 Ozone

Ozone is a destructive gaseous pollutant that seriously affects human and animal respiration, as well as causing extensive damage to both natural and cultivated plant populations (Chameides et al., 1997). The resistance of rice to ozone (O_3) is a quantitative trait controlled by nuclear genes (Kim et al., 2004). The identification of quantitative trait loci and analysis of molecular markers of O_3 resistance is important for increasing the resistance of rice to O_3 stress. Quantitative trait loci associated with the O_3 resistance of rice were mapped on chromosomes using recombinant inbred lines from a cross between 'Milyang 23' and 'Gihobyeo'. The quantitative trait loci were tightly linked to three markers and were detected in each of three replicates. The association between these markers and O_3 resistance in rice cultivars and doubled haploid populations was analyzed. The markers permit the screening of rice germplasm for O_3 resistance and the introduction of resistance into elite lines in breeding programs. This study, by identifying ozone-related quantitative trait loci, provides an increased understanding of ozone responsiveness in rice and may lead to applications in breeding for enhanced ozone tolerance.

Plant responses to ozone have also been analyzed using a proteomics approach. In rice leaves, ozone caused marked visible necrotic damage and increases in ascorbate peroxidase proteins; these changes were accompanied by rapid changes in the 2D-PAGE protein profile (Agrawal et al., 2002). Out of a total of 56 proteins investigated, 52 protein spots were visually identified to be differentially expressed relative to the control. Ozone caused marked reductions in the major photosynthetic proteins in leaf, including RuBisCO, and the induction of various defense- and stress-related proteins. This research provides evidence for the specific and rapid accumulation of certain proteins, such as PR proteins (OsPR5 and OsPR10), ascorbate peroxidase, SOD [Mn], and the ATP-dependent caseinolytic protease, which could serve as sensitive markers to monitor ozone-related damage in rice.

6.3.1.5 Fungus

The ability of plants to defend themselves against most potential pathogens depends on sensitive perception mechanisms that recognize microbial invaders and subsequently activate defense responses. Rice blast disease, caused by *Magnapoethe grisea*, is the most serious disease of cultivated rice in most rice-growing regions of the world (Valent et al., 1991). The *M. grisea*–rice interaction is a model system for understanding plant disease, not only because of its great economic importance, but also because of the genetic and molecular-genetic tractability of the fungus (Valent et al., 1991).

A proteomics approach has been applied to the study of pathogen-infected rice (Kim et al., 2003). Proteins were extracted from suspension-cultured cells after inoculation with the rice blast fungus *M. grisea*, or treatment with an elicitor or other signal molecules such as jasmonic acid (JA), salicylic acid, or H_2O_2. Analysis by 2D-PAGE identified 14 protein spots that showed increased or decreased expression after these treatments. Of these protein spots, 12 proteins from six different genes were identified. For example, OsPR10, isoflavone reductase-like protein, β-glucosidase, and a putative receptor-like protein kinase were induced by rice blast fungus, whereas six isoforms of probenazole-inducible protein and two isoforms of SALT responded to blast fungus, elicitor, and JA. Western blot analysis to quantify the expression levels of probenazole-inducible protein, OsPR10, and SALT revealed that these proteins, which take part in incompatible interactions, were induced earlier and to a greater extent than were proteins involved in compatible reactions.

Konishi et al. (2001) identified proteins that showed expression changes in rice leaf blade infected with the blast fungus *M. grisea*. Using proteome analysis, the same study also showed that quantitative expression changes in these proteins were greatly influenced by the levels of nitrogen fertilizer. Rice plants that have been exposed to excessive nitrogen fertilizer are more susceptible to blast disease than are those exposed to low levels of nitrogen. In contrast to low-nitrogen rice plants, high-nitrogen rice plants show many more lesions and the sizes of these lesions are larger. Twelve proteins that appeared to change according to the level of nitrogen were identified. For example, the amounts of RuBisCO large and small subunits were increased by a nitrogen top dressing, but the RuBisCO small subunit was decreased after nitrogen top dressing combined with blast fungus infection. After blast fungus infection, PR-1 was induced by a nitrogen top dressing. It was proposed that these proteins might be involved in incompatible interactions in rice plants following blast fungus infection.

6.3.1.6 Virus

Rice yellow mottle virus (RYMV), a member of the genus *Sobemovirus*, is endemic to Africa (Pinel et al., 2000; Abubakar et al., 2003), and is considered to be very detrimental to rice production. With only four open reading frames, this virus could be considered as a model for studying the genetics and genomics of virus resistance (Brugidou et al., 2002).

The response to RYMV infection has been analyzed in cells of two cultivars of rice: *indica* rice IR64, which is susceptible to infection; and *japonica* rice Azucena, which is partially resistant to RYMV (Ventelon-Debout et al., 2004). Of the proteins resolved on 2D-PAGE gels, 64 (40 proteins in IR64 and 24 proteins in Azucena) responded to RYMV infection; 19 differentially regulated proteins were identified for the IR64 cultivar, whereas 13 were identified for the Azucena cultivar. These included proteins in three functional categories: metabolism, stress-related proteins, and translation. This study shows that several proteins regulated by abiotic stress response pathways are also activated by RYMV; these include SALT, heat shock proteins, and SOD. On the other hand, other proteins seem to be more specific to RYMV infection, such as dehydrin and proteins involved in glycolysis.

6.3.2 Hormones

Plant hormones play an important role in many aspects of signal transduction in cells, as well as in several growth and development pathways, such as seed dormancy/germination, stem elongation, leaf expansion, and flower and fruit development.

6.3.2.1 Gibberellin

Gibberellins (GAs) are essential regulators that trigger stem or internodal elongation (Hooley, 1994). Proteins that are regulated by the GA response in rice leaf sheath elongation have been analyzed by the differential display proteome method (Shen et al., 2003). Out of 352 protein spots detected on 2D-PAGE gels, 32 proteins showed modulated expression in leaf sheath for 48 h after GA_3 treatment as compared to the control. Of these proteins, two 56-kDa protein spots, both identified as calreticulin, showed different isoelectric points (pI of 4.0 and 4.3) and different expression levels in the GA_3-treated leaf sheath. The expression level of the pI 4.0 spot was down-regulated in response to GA_3, whereas the pI 4.3 spot was up-regulated. In an earlier study, a calreticulin with a pI value of 4.5, which has been subsequently identified as the pI 4.3 spot, was found to be phosphorylated *in vitro* in short-term suspension-cultured cells, whereas no protein with a pI value of 4.0 was phosphorylated (Komatsu et al., 1996). Together these data suggest that the twin 56-kDa spots represent phosphorylated and unphosphorylated forms of calreticulin, and that the phosphorylated form becomes more abundant in response to GA_3. Interestingly, the shift in the relative abundance of the twin spots in the leaf sheath experiment was prevented in the presence of GA_3 synthesis inhibitors, suggesting that the phosphorylation state of calreticulin may be influenced by GA_3 signal transduction during leaf sheath elongation. In the earlier study, the *in vitro* phosphorylation of calreticulin was shown to occur in short term, but not in long-term suspension-cultured cells (Komatsu et al., 1996), suggesting that transient phosphorylation of calreticulin, which could affect Ca^{2+} homeostasis, might play an important role in signal transduction cascades during developmental regulation (Figure 6.1).

The above study of GA-regulated proteins focused on specific proteins involved in leaf sheath elongation in rice (Shen et al., 2003). Another study of GA-regulated proteins in rice tissues, including the leaf sheath, root, and suspension-cultured cells, has also been reported (Tanaka et al., 2004a). Lists of proteins present in these tissues were constructed and used to investigate the effects of GA_3 treatment. GA_3-regulated proteins in rice leaf sheath, root, and suspension-cultured cells were analyzed by 2D-PAGE, and the expression of 8, 21, and 14 proteins, respectively, was changed by the addition of exogenous GA_3. In the leaf sheath, the proteins that responded to GA_3 were involved in transcriptional regulation (the *Osem* gene and replication protein A1), primary metabolism (fructokinase, lactoylglutathione lyase, and OEE2), and signal transduction (putative receptor-like kinase). In the root tissue, proteins affected by GA_3 treatment appeared to be involved in defense reactions (glutathione *S*-transferase, SOD[Cu–Zn], Bowman-Birk protease inhibitor, glutathione *S*-transferase-dependent dehydroascorbate reductase, and PR-1), suggesting that GA has an essential role in

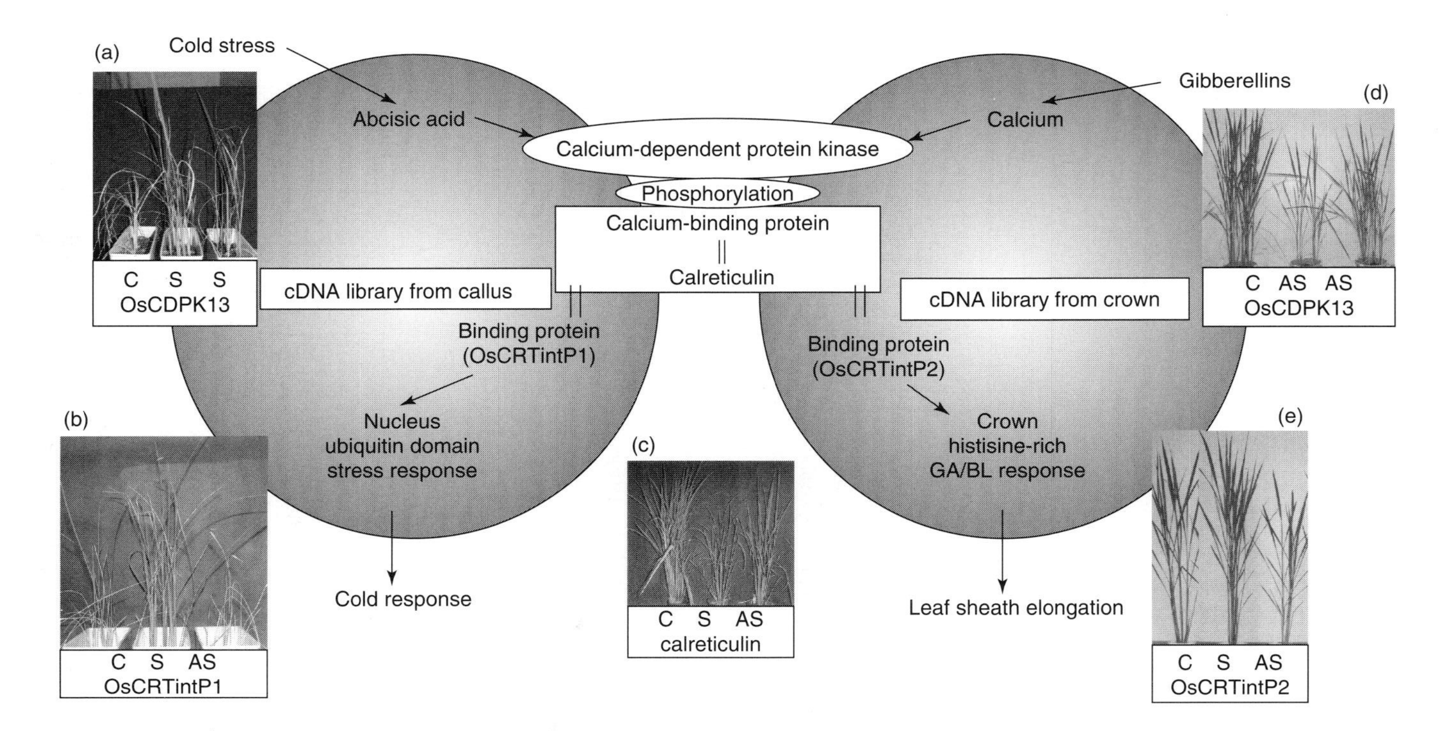

Figure 6.1 Analysis of protein interactions during growth and/or stress conditions (e.g. rice calreticulin). Calreticulin was detected as a GA- and cold-response protein by 2D-PAGE. The immuno-precipitation system with calreticulin was used to identify calcium-dependent protein kinase. The yeast two-hybrid interaction-cloning system was used to identify novel calreticulin interacting proteins. The expression of proteins in transgenic rice was tested using Western blotting. C: vector control; S: sense transgenic rice; AS: antisense transgenic rice. CDPK13: calcium-dependent protein kinase 13; CRT, calreticulin.

defense reactions in rice roots. In suspension-cultured cells, proteins involved in several functions were found to be regulated by GA_3 treatment, such as proteins belonging to the categories of metabolism (formate dehydrogenase and thioredoxin), energy (glyceraldehyde-3-phosphate dehydrogenase), cell growth (growth factor 14-C protein), protein (chaperonin 60), transcription (nucleotide-binding protein 2 and homeobox), defense (phenylalanine ammonia-lyase and glutathione *S*-transferase), signal transduction (small G protein), transport (voltage-dependent anion channel), and hypothetical proteins. The GA-regulated proteins in these tissues may play a significant role in tissue growth stimulated by GA.

6.3.2.2 Brassinosteroid

Brassinosteroids (BRs) are naturally occurring plant steroids with structural similarities to insect and animal steroid hormones. Exogenous application of BRs to plant tissues evokes various growth responses, such as cell elongation, proliferation, differentiation, organ bending, and enhanced stress tolerance (Sasse, 1997). A proteomics approach based on the application of BRs to the lamina joint or root of rice seedlings has been reported (Konishi and Komatsu, 2003). Lamina inclination was markedly stimulated by brassinolide (BL), which is an active BR molecule, whereas the root elongation of rice seedlings was inhibited by BL. On 2D gels, a total of 786 and 508 proteins were detected in extracts from the lamina joint and root of rice seedlings, respectively. BL treatment induced changes in the expression of 9 proteins in the lamina joint and 12 proteins in the root. These proteins were mainly related to photosynthesis in the lamina joint and to stress tolerance in the root. After inclination of the lamina joint caused by BL treatment, degradation of RuBisCO large subunit was observed, suggesting that inclination to receive more light than usual might be associated with degradation of the RuBisCO large subunit.

6.3.2.3 JA

JA is one of the simplest non-traditional plant hormones and has diverse functions, including potential roles in plant defense as part of more complex signaling pathways. JA treatment of the leaf and stem of rice seedlings was found to result in necrosis, accompanied by marked reductions in the abundance of RuBisCO subunits (Rakwal and Komatsu, 2000). JA-treated stem tissues showed especially strong induction of several novel proteins, including a new basic 28-kDa Bowman-Birk proteinase inhibitor and an acidic 17-kDa PR-1 protein. Immunoblot analysis using antibodies generated against these proteins revealed a tissue-specific expression pattern and time-dependent induction after JA treatment. Furthermore, this induction was blocked by a protein synthesis inhibitor, indicating *de novo* protein synthesis. These results indicate that JA affects defense-related gene expression in rice seedlings, as judged by the *de novo* synthesis of novel proteins with potential roles in plant defense.

6.3.2.4 Auxin

Auxin plays a critical role in apical dominance, and in lateral root initiation and emergence (Casimiro et al., 2001). Because auxin coupled with zinc also regulates root formation in rice, a proteomics analysis of seedlings and suspension-cultured cells treated with auxin and zinc was conducted, and seven proteins were found to be up-regulated by this treatment (Oguchi et al., 2004a, b; Yang et al., 2005). Of these proteins, NADPH-dependent oxidoreductase, methylmalonate-semialdehyde dehydrogenase (MMSDH), and elongation factor 1β′ (EF-1β′) were strongly up-regulated, as compared with the untreated control. NADPH oxidoreductase and MMSDH were detected in suspension-cultured cells, root, and leaf sheath, but not in leaf blade. The abundance of MMSDH protein also was increased in GA-treated suspension-cultured cells, as well as in the constitutive GA response mutant *slr1*, indicating that MMSDH is regulated by the GA signal transduction pathway. During root formation stimulated by auxin and zinc, the expression of NADPH oxidoreductase, MMSDH, and EF-1β′ was increased, suggesting that these proteins play an important role in the formation of roots in rice.

6.4 Future prospects of cereal proteomics

Food shortages are one of the most serious global problems in this century. Rice, wheat, and maize provide approximately half of the calories consumed by the world's population. So far, this need has been met by dwarf rice and wheat varieties that were developed by classical plant breeding methods, contributing to the green revolution in the 1960s. However, to meet the expanding food demands of the rapidly growing world population, grain crop production will need to increase by a further 50% by 2025 (Khush, 2003). Furthermore, increased salinization is expected to drastically reduce the amount of arable land available for crop production within the next few decades (Yan et al., 2005). To meet these challenges, genes and proteins that control crop plant architecture and/or stress resistance in a wide range of environments will need to be identified to facilitate the biotechnological improvement of crop productivity. To this end, proteomics is a useful and powerful tool for investigating protein changes induced by various conditions.

In wheat, proteomic methods have been used to identify foam-forming soluble proteins that may play an important role in stabilizing gas bubbles in dough, and thus influence the crumb structure of bread. In this research, many proteins, particularly tritin, were depleted in the dough liquor foam, whereas a number of α-amylase inhibitors were enriched, suggesting that these are among the most strongly surface-active proteins in dough liquor (Salt et al., 2005). The function of thioredoxin in wheat starchy endosperm was also investigated using two proteomic approaches. A comparison of young endosperm to mature endosperm revealed unique sets of proteins functional in processes characteristic of each developmental stage. Flour contained 36 thioredoxin targets, most of which also have been found in isolated developing endosperm (Wong et al., 2004). Vensel et al. (2005) used a combined 2D-PAGE-MS approach to identify over 250 proteins at early and late stages of grain development in wheat. These data provide an insight into the biochemical events taking place during wheat grain development and highlight the value of proteomics in

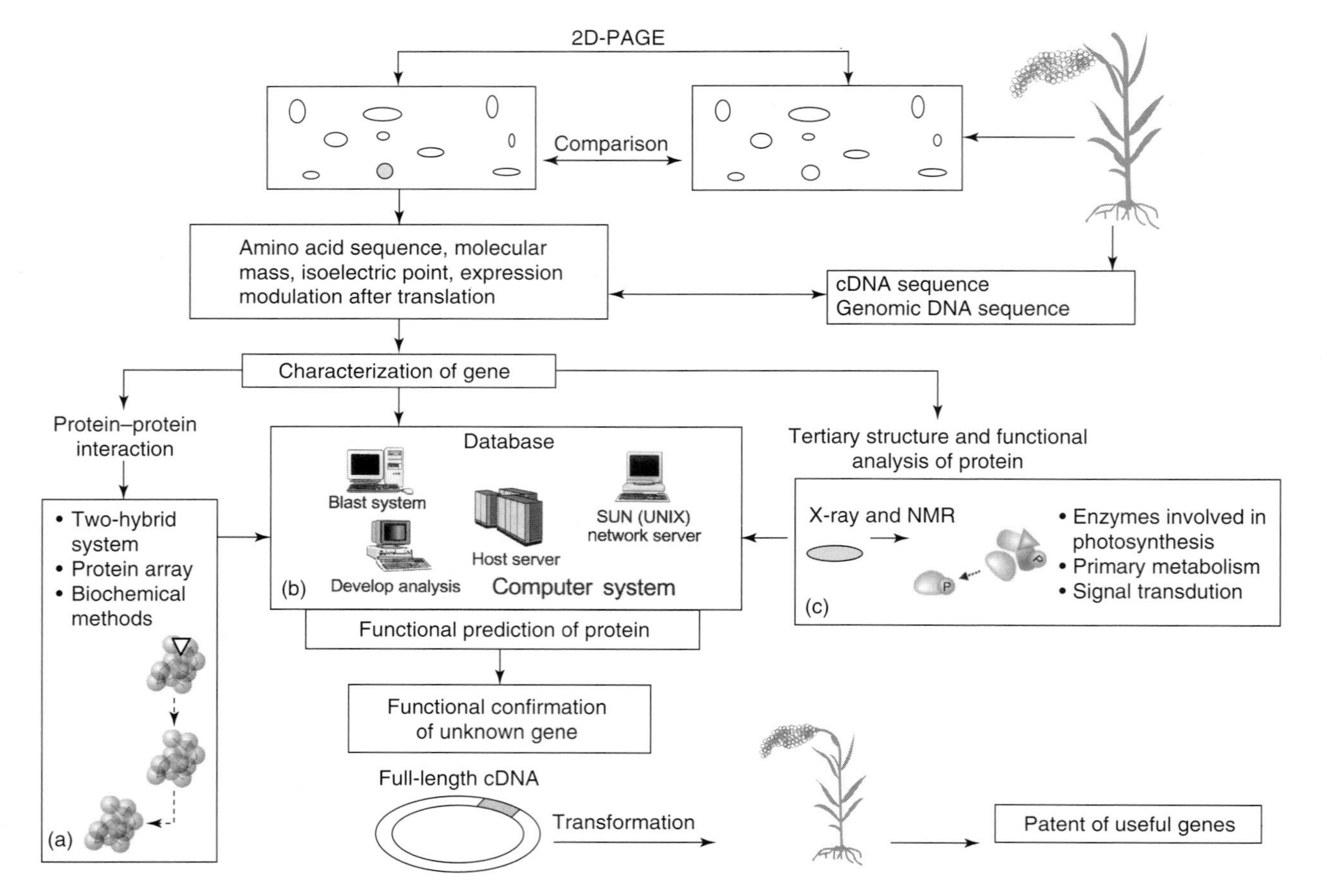

Figure 6.2 Future perspective of cereal proteome research (e.g. rice). The goal is a construction of a database combining data of protein–protein interactions and tertiary structures. After the presumption of function of protein using this database, the useful genes will be isolated.

characterizing complex biochemical processes (Vensel et al., 2005). Furthermore, the proteome maps of wheat will facilitate future studies addressing the effects of genetic and environmental factors on the development and quality of wheat grain.

In maize, drought is a major abiotic stress affecting all levels of plant organization and, in particular, leaf elongation. Several experiments were designed to study the effect of water deficits on leaves at the protein level by taking into account the reduction of leaf elongation (Vencent et al., 2005). The results suggested that the lignin level was lower in leaves of plants subjected to water deficit than in those of well-watered plants. In another study, 2D-PAGE and MS were used to identify proteins that are differentially expressed in response to fungal infection in maize embryos. Among the proteins induced in response to infection are proteins involved in protein synthesis, folding, and stabilization, as well as proteins involved in oxidative stress tolerance (Campo et al., 2004). The proteomes of the maize egg cell and zygote also have been analyzed by 2D-PAGE and MS (Okamoto et al., 2004). This research indicates that the egg cell is especially rich in enzymes of energy metabolism. Furthermore, in the egg cell and zygote, annexin was found to be involved in the exocytosis of cell wall materials, a process induced by a fertilization-triggered increase in cytosolic Ca^{2+} levels. These experiments provide a foundation for the use of proteomics in the design of experiments to address fundamental questions in plant physiology and molecular biology.

Analysis by 2D-PAGE provides a convenient way to study the various proteins that are present in cereals and to identify those that are regulated in response to various growth and/or stress conditions. Knowing where and when individual proteins are synthesized in rice, with respect to the tissue, subcellular compartment, and developmental stage, can also provide clues to their functions. The partial amino acid sequences determined for these proteins will contribute greatly to the field of plant molecular biology, by facilitating the identification of new rice proteins of interest through homology searches. The information thus obtained from the proteomics data will be helpful in predicting the function of the proteins and will aid in their molecular cloning in future experiments. Present-day proteomics research promises to contribute greatly to the development of the high yield, sustainable agriculture of tomorrow (Figure 6.2).

References

Abbasi, F. and Komatsu, S. (2004) A proteomic approach to analyze salt-responsive proteins in rice leaf sheath. *Proteomics*, **4**, 2072–2081.

Abubakar, Z., Ali, F., Pinel, A., Traore, O., N'Guessan, P., Notteghem, J.L., Kimmins, F., Konate, G. and Fargette, D. (2003) Phylogeography of rice yellow mottle virus in Africa. *J. Gen. Virol.*, **84**, 733–743.

Agrawal, G.K., Rakwal, R., Yonekura, M., Kubo, A. and Saji, H. (2002) Proteome analysis of differentially displayed proteins as a tool for investigating ozone stress in rice (*Oryza sativa* L.) seedlings. *Proteomics*, **2**, 947–959.

Bevan, M., Bancroft, I., Bent, E., Love, K., Goodman, H. et al. (1998) Analysis of 1.9 Mb of contagious sequence from chromosome 4 of *Arabidopsis thaliana*. *Nature*, **391**, 485–488.

Brugidou, C., Opalka, N., Yeager, M., Beachy, R.N. and Fauquet, C. (2002) Stability of rice yellow mottle virus and cellular compartmentalization during the infection process in *Oryza sativa* (L.). *Virology*, **297**, 98–108.

Casimiro, I., Marchant, A., Bhalerao, R.P., Beeckman, T., Dhooge, S., Swarup, R., Graham, N., Inze, D., Sandberg, G., Casero, P.J. and Bennett, M. (2001) Auxin transport promotes *Arabidopsis* lateral root initiation. *Plant Cell*, **13**, 843–852.

Chameides, W.L., Saylor, R.D. and Cowling, E.B. (1997) Ozone pollution in the rural United States and the new NAAQS. *Science*, **276**, 916.

Cleveland, D.W., Fisher, S.G., Kirschner, M.W. and Laemmli, U.K. (1977) Peptide mapping by limited proteolysis in sodium dodecyl sulphate and analysis by gel electrophoresis. *J. Biol. Chem.*, **252**, 1102–1106

Campo, P., Carrascal, M., Coca, M., Abian, J. and San Segundo, B. (2004) The defense response of germinating maize embryos against fungal infection: a proteomics approach. *Proteomics*, **4**, 383–396.

Cui, S., Huang, F., Wang, J., Ma, X., Cheng, Y. and Liu, J. (2005) A proteomic analysis of cold stress responses in rice seedlings. *Proteomics*, **5**, 3162–3172.

Dani, V., Simon, W.J., Duranti, M. and Croy, R.R. (2005) Changes in the tobacco leaf apoplast proteome in response to salt stress. *Proteomics*, **5**, 737–745.

Devos, M.K. and Gale, D.M. (2000) Genome relationships: the grass model in current research. *Plant Cell*, **12**, 637–646.

Dhugga, K.S., Tiwari, S.C. and Ray, P.M. (1997) A reversibly glycosylated polypeptide (RGP1) possibly involved in plant cell wall synthesis: purification, gene cloning, and trans-Golgi localization. *Proc. Natl Acad. Sci.*, **94**, 7679–7684.

Feng, Q., Zhang, Y., Hao, P., Wang, S., Fu, G. et al. (2002) Sequence and analysis of rice chromosome 4. *Nature*, **420**, 316–320.

Fukuda, M., Islam, N., Woo, S.H., Yamagishi, A., Takaoka, M. and Hirano, H. (2003) Assessing matrix assisted laser desorption/ionization-time of flight-mass spectrometry as a means of rapid embryo protein identification in rice. *Electrophoresis*, **24**, 1219–1329.

Gale, M.D. and Devos, K.M. (1998) Comparative genetics in the grasses. *Proc. Natl Acad. Sci. USA*, **95**, 1971–1974.

Goff, S.A., Ricke, D., Lan, T.-H., Presting, G., Wang, R. et al. (2002) A draft sequence of rice genome (*Oryza sativa* L. ssp. *japonica*). *Science*, **296**, 92–100.

Gygi, S.P., Rochon, Y., Franza, B.R. and Aebersold, M. (1999) Correlation between protein and mRNA abundance in yeast. *Mol. Cell. Biol.*, **19**, 1720–1730.

Hajheidari, M., Abdollahian-Noghabi, M., Askari, H., Heidari, M., Sadeghian, S.Y., Ober, E.S. and Hosseini Salekdeh, G. (2005) Proteome analysis of sugar beet leaves under drought stress. *Proteomics*, **5**, 950–960.

Heazlewood, J.L., Howell, K.A., Whelan, J. and Millar, A.H. (2003) Towards an analysis of the rice mitochondrial proteome. *Plant Physiol.*, **132**, 230–242.

Hedrich, R. and Schroeder, J.I. (1989) The physiology of ion channels and electrogenic pumps in higher plants. *Annu. Rev. Plant Physiol.*, **40**, 539–569.

Hirano, H., Kawasaki, H. and Sassa, H. (2000) Two-dimensional gel electrophoresis using immobilized pH gradient tube gels. *Electrophoresis*, 21, 440–445.

Hooley, R. (1994) Gibberellins: perception, transduction and responses. *Plant Mol. Biol.*, **26**, 1529–1555.

Imin, N., Kerim, T., Weinman, J.J. and Rolfe, B.G. (2001) Characterization of rice anther proteins expressed at the young microspore stage. *Proteomics*, **1**, 1149–1161.

International Rice Genome Sequencing Project (2005) The map-based sequence of the rice genome. *Nature*, **436**, 793–800.

Islam, N., Lonsdala, M., Upadhyaya, N.M., Higgins, T.J., Hirano, H. and Akhurst, R. (2004) Protein extraction from mature rice leaves for two-dimensional gel electrophoresis and its application in proteome analysis. *Proteomics*, **4**, 1903–1908.

Kerim, T., Imin, N., Weinman, J.J. and Rolfe, B.G. (2003) Proteome analysis of male gametophyte development in rice anthers. *Proteomics*, **3**, 738–751.

Khush, G.S. (2003) Challenges for meeting the global food and nutrient needs in the new millennium. *Proc. Nutr. Soc.*, **60**, 15–26.
Kim, S.T., Cho, K.S., Yu, S., Kim, S.G., Hong, J.C. et al. (2003) Proteomic analysis of differentially expressed proteins induced by rice blast fungus and elicitor in suspension-cultured rice cells. *Proteomics*, **3**, 2368–2378.
Kim, K.M., Kwon, Y.S., Lee, J.J., Eun, M.Y. and Sohn, J.K. (2004) QTL mapping and molecular marker analysis for the resistance of rice to ozone. *Mol. Cells*, **17**, 151–155.
Koller, A., Washburn, M.P., Lange, B.M., Andon, N.L., Deciu, C. et al. (2002) Proteomic survey of metabolic pathways in rice. *Proc. Natl Acad. Sci. USA*, **99**, 11969–11974.
Komatsu, S. (2005) Rice Proteome Database: a step toward functional analysis of the rice genome. *Plant Mol. Biol.*, **59**, 179–190.
Komatsu, S. and Tanaka, N. (2004) Rice proteome analysis: a step toward functional analysis of the rice genome. *Proteomics*, **4**, 938–949.
Komatsu, S., Kajiwara, H. and Hirano, H. (1993) A rice protein library: a data-file of rice proteins separated by two-dimensional electrophoresis. *Theor. Appl. Genet.*, **86**, 935–942.
Komatsu, S., Masuda, T. and Abe, K. (1996) Phosphorylation of a protein (pp56) is related to the regeneration of rice cultured suspension cells. *Plant Cell Physiol.*, **37**, 748–753.
Komatsu, S., Muhammad, A. and Rakwal, R. (1999a) Separation and characterization of proteins from green and etiolated shoots of rice (*Oryza sativa* L.): towards a rice proteome. *Electrophoresis*, **20**, 630–636.
Komatsu, S., Rakwal, R. and Li, Z. (1999b) Separation and characterization of proteins in rice (*Oryza sativa*) suspension cultured cells. *Plant Cell Tiss. Org. Cult.*, **55**, 183–192.
Komatsu, S., Konishi, H., Shen, S. and Yang, G. (2003) Rice proteomics: a step functional analysis of the rice genome. *Mol. Cell. Proteom.*, **2**, 2–10.
Komatsu, S., Kojima, K., Suzuki, K., Ozaki, K. and Higo, K. (2004) Rice Proteome Database based on two-dimensional polyacrylamide gel electrophoresis: its status in 2003. *Nucleic Acid. Res.*, **32**, 388–392.
Konishi, H. and Komatsu, S. (2003) A proteomics approach to investigating promotive effects of brassinolide on lamina inclination and root growth in rice seedlings. *Biol. Pharm. Bull.*, **26**, 401–408.
Konishi, H., Ishiguro, K. and Komatsu, S. (2001) A proteomics approach towards understanding blast fungus infection of rice grown under different levels of nitrogen fertilization. *Proteomics*, **1**, 1162–1171.
Kruft, V., Eubel, H., Jänsch, L., Whehahn, W. and Braun, H.-P. (2001) Proteomic approach to identify novel mitochondrial proteins in *Arabidopsis*. *Plant Physiol.*, **127**, 1694–1710.
Lee, S., Lee, J.E., Yang, E.J., Lee, J.E., Park, A.R., Song, W.H. and Park, O.K. (2004) Proteomic identification of annexins, calcium-dependent membrane binding proteins that mediate osmotic stress and abscisic acid signal transduction in *Arabidopsis*. *Plant Cell*, **16**, 1378–1391.
Lyons, J.M. (1973). Chilling injury in plants. *Annu. Rev. Plant Physiol.*, **24**, 445–446.
Maurel, C. (1997) Aquaporins and water permeability of plant membranes. *Annu. Rev. Plant Physiol. Plant Mol. Biol.*, **48**, 399–429.
Mikami, S., Hori, H. and Mitsui, T. (2001) Separation of distinct components of rice Golgi complex by sucrose density gradient centrifugation. *Plant Sci.*, **161**, 665–675.
Millar, A.H., Sweetlove, L.J., Giege, P. and Leaver, C.J. (2001) Analysis of the *Arabidopsis* mitochondrial proteome. *Plant Physiol.*, **127**, 1711–1727.
O'Farrell, P.H. (1975) High resolution two-dimensional electrophoresis of proteins. *J. Biol. Chem.*, **250**, 4007–4021.
Oguchi, K., Tanaka, N., Komatsu, S. and Akao, S. (2004a) Methylmalonate-semialdehyde dehydrogenase is induced in auxin- and zinc-stimulated root formation in rice. *Plant Cell Rep.*, **22**, 848–858.
Oguchi, K., Tanaka, N., Komatsu, S. and Akao, S. (2004b) Characterization of NADPH-dependent oxidoreductase from rice induced by auxin and zinc. *Plant. Physiol.*, **121**, 124–131.
Okamoto, T., Higuchi, K., Shinkawa, T., Isobe, T., Lorz, H., Koshiba, T. and Kranz, E. (2004) Identification of major proteins in maize egg cells. *Plant Cell Physiol.*, **45**, 1406–1412.
Peltier, J.-B., Friso, G., Kalume, D.E., Roepstorff, P., Nilsson, F., Adamska, I. and van Wijk, K.J. (2000) Proteomics of the chloroplast: systematic identification and targeting analysis of lumenal and peripheral thylakoid proteins. *Plant Cell*, **12**, 319–341.

Peltier, J.-B., Emanuelsson, O., Kalume, D.E., Ytterberg, J., Friso, G. et al. (2002) Central function of the lumenal and peripheral thylakoid proteome of *Arabidopsis* determined by experimentation and genome-wide prediction. *Plant Cell*, **14**, 211–236.

Pinel, A., N'Guessan, P., Bousalem, M. and Fargette, D. (2000) Molecular variability of geographically distinct isolates of rice yellow mottle virus in Africa. *Arch. Virol.*, **145**, 1621–1638.

Rabbani, M.A., Maruyama, K., Abe, H., Khan, M.A., Katsura, K., Ito, Y., Yoshiwara, K., Seki, M., Shinozaki, K. and Yamaguchi-Shinozaki, K. (2003) Monitoring expression profiles of rice genes under cold, drought, and high-salinity stresses and abscisic acid application using cDNA microarray and RNA gel-blot analyses. *Plant Physiol.*, **133**, 1755–1767.

Raison, J.K., Lyons, J.M. and Keith, A.D. (1971) Temperature-induced phase changes in mitochondrial membranes detected by spin labeling. *J. Biol. Chem.*, **246**, 4036–4040.

Rakwal, R. and Komatsu, S. (2000) Role of jasmonate in the rice (*Oryza sativa* L.) self-defense mechanism using proteome analysis. *Electrophoresis*, **21**, 2492–2500.

Salekdeh, G.H., Siopongco, J., Ghareyazie, B. and Bennett, J. (2002) Proteomic analysis of rice leaves during drought stress and recovery. *Proteomics*, **2**, 1131–1145.

Salt, L.J., Robertson, J.A., Jenkins, J.A., Mulholland, F. and Mills, E.N. (2005) The identification of foam-forming soluble proteins from wheat (*Triticum aestivum*) dough. *Proteomics*, **5**, 1612–1623.

Santoni, V., Rouquie, D., Doumas, P., Mansion, M., Boutry, M., Degand, H., Dupree, P., Packman, L., Sherrier, J., Prime, T., Bauw, G., Posada, E., Rouze, P., Dehais, P., Sahnoun, I., Barlier, I. and Rossignol, M. (1998) Use of a proteome strategy for tagging proteins present at the plasma membrane. *Plant J.*, **16**, 633–641.

Sasaki, T., Matsumoto, T., Yamamoto, K., Sakata, K., Baba, T. et al. (2002) The genome sequence and structure of rice chromosome 1. *Nature*, **420**, 312–316.

Sasse, J.M. (1997) Recent progress in brassinosteroids research. *Plant Physiol.*, **100**, 696–701.

Schubert, M., Petersson, U.A., Haas, P.J., Funk, C., Schröder, W.P. and Kiesebach, T. (2002) Proteome map of the chloroplast lumen of *Arabidopsis thaliana*. *J. Biol. Chem.*, **277**, 8354–8365.

Shen, S., Matsubae, M., Takao, T., Tanaka, N. and Komatsu, S. (2002) A proteomic analysis of leaf sheath from rice. *J. Biochem.*, **132**, 613–620.

Shen, S., Sharma, A. and Komatsu, S. (2003) Characterization of proteins responsive to gibberellin in the leaf-sheath of rice (*Oryza sativa* L.) seedling using proteome analysis. *Biol. Pharm. Bull.*, **26**, 129–136.

Staehelin, L.A. and Moore, I. (1995) The plant Golgi apparatus: structure, functional organization and trafficking mechanisms. *Annu. Rev. Plant Physiol. Plant Mol. Biol.*, **46**, 261–288.

Sze, H., Liang, F. and Hwang, I. (2000) Diversity and regulation of plant Ca^{2+} pumps: insights from expression in yeast. *Annu. Rev. Plant Physiol. Plant Mol. Biol.*, **51**, 433–462.

Tanaka, N., Konishi, H., Khan, M. and Komatsu, S. (2004a) Proteome analysis of rice tissues separated and visualized by two-dimensional electrophoresis: approach to investigating the gibberellin regulated proteins. *Mol. Genet. Genome*, **270**, 485–496.

Tanaka, N., Fujita, M., Handa, H., Murayama, S., Uemura, M., Kawamura, Y., Mitsui, T., Mikami, S., Tozawa, Y., Yoshinaga, T. and Komatsu, S. (2004b) Proteomics of the rice cell: systematic identification of the protein population in subcellular compartments. *Mol. Genet. Genome*, **271**, 566–576.

Tsugita, A., Kawakami, T., Uchiyama, Y., Kamo, M., Miyatake, N. and Nozu, Y. (1994) Separation and characterization of rice proteins. *Electrophoresis*, **15**, 708–720.

Valent, B., Farrall, L. and Chumley, F.G. (1991) *Magnaporthe grisea* genes for pathogenicity and virulence identified through a series of backcrosses. *Genetics*, **127**, 87–101.

Vensel, W.H., Tanaka, C.K., Cai, N., Wong, J.H., Buchanan, B.B. and Hurkman, W.J. (2005) Developmental changes in the metabolic protein profiles of wheat endosperm. *Proteomics*, **5**, 1594–1611.

Ventelon-Debout, M., Delalande, F., Brizard, J.-P., Diemer, H., Van Dorsselaer, A. and Brugidou, C. (2004) Proteome analysis of cultivar-specific deregulations of *Oryza sativa indica* and *O. sativa japonica* cellular suspensions undergoing rice yellow mottle virus infection. *Proteomics*, **4**, 216–225.

Vencent, D., Lapierre, C., Pollet, B., Cornic, G., Negroni, L. and Zivy, M. (2005) Water deficits affect caffeate *O*-mrthyltransferase, lignification, and related enzymes in maize leaves. A proteomic investigation. *Plant Physiol.*, **137**, 949–960.

Wong, J.H., Cai, N., Balmer, Y., Tanaka, C.K., Vensel, W.H., Hurkman, W.J. and Buchanan, B.B. (2004) Thioredoxin targets of developing wheat seeds identified by complementary proteomics approaches. *Phytochemistry*, **65**, 1629–1640.

Xu, Y., Buchholz, W.G., DeRose, R.T. and Hall, T.C. (1995) Characterization of a rice gene family encoding root-specific proteins. *Plant Mol. Biol.*, **27**, 237–248.

Yan, S., Tang, Z., Su, W. and Sun, W. (2005) Proteomic analysis of salt stress-responsive proteins in rice root. *Proteomics*, **5**, 235–244.

Yang, G., Inoue, A., Takasaki, H., Kaku, H., Akao, S. and Komatsu, S. (2005) A proteomic approach to analyze auxin and zinc-responsive protein in rice. *J. Proteome Res.*, 4, 456–463.

Yu, J., Hu, S., Wang, J., Wong, G.K.-S., Li, S. et al. (2002) A draft sequence of the rice genome (*Oryza sativa* L. ssp. *indica*). *Science*, **296**, 79–92.

Zhong, B., Karibe, H., Komatsu, S., Ichimura, H., Nagamura, Y., Sasaki, T. and Hirano, H. (1997) Screening of rice genes from a cDNA based on the sequence data-file of proteins separated by two-dimensional electrophoresis. *Breeding Sci.*, **47**, 245–251.

Zhao, C., Wang, J., Cao, M., Zhao, K., Sgao, J., Lei, T., Yin, J., Hill, G.G., Xu, N. and Liu, S. (2005) Proteomic changes in rice leaves during development of field-grown rice plants. *Proteomics*, **5**, 961–972.

7 Proteome analysis for the study of developmental processes in plants

Loïc Rajjou, Karine Gallardo, Claudette Job and Dominique Job

7.1 Introduction

Proteome analysis, which involves the identification and characterization of expressed proteins, is a powerful tool for determining the biological roles and functions of individual proteins. Furthermore, by providing a systematic and with-out any *a priori* mean for large-scale identification of cellular proteins, proteomics is expected to accelerate discoveries in complex processes such as development. Allied to other genome-scale gene expression profiling methods as transcriptomics and metabolomics, proteomics also constitutes a powerful tool toward the development of modern systems biology (Ge et al., 2003; Patterson and Aerbersold, 2003), an emerging approach that seeks to describe (model) biological systems through integration of diverse types of massive data and which will ultimately allow computational simulations of these complex systems, notably in plants (Gutiérrez et al., 2005). With the completion of genome sequencing projects and the constitution of large expressed sequence tags (ESTs) collections for several reference (*Arabidopsis*, rice, *Medicago truncatula*, poplar, *Physcomitrella patens*) and crop (e.g. maize, wheat, barley, soybean) plants and the development of analytical methods for protein characterization (see Hirano et al., 2004; Newton et al., 2004), proteomics has become a major field of functional genomics, allowing large-scale analysis of protein sequences deposited into databases (e.g. *Arabidopsis* genome initiative, 2000; Goff et al., 2002; Yu et al., 2002; Li et al., 2003; Gutiérrez et al., 2004; Sarnighausen et al., 2004; Stauber and Hippler, 2004; Zhang et al., 2004; Giavalisco et al., 2005; http://www.tigr.org/tdb/tgi/plant.shtml; http://mips.gsf.de/proj/sputnik/). For example, the rice proteome database based on two-dimensional gel electrophoresis (2-DE) presently comprises about 12,000 identified proteins from different tissues and organelles, corresponding to more than 4000 separate entries (Komatsu et al., 2003, 2004; http://gene64.dna.affrc.go.jp/RDP/). Also, a 2-DE proteomics reference map comprising 1367 identified proteins has been established from a cell suspension culture of the model legume *M. truncatula* (Lei et al., 2005).

Agrawal et al. (2005b) proposed that developmental proteomics be defined 'as a set of proteins present at particular developmental stage of a plant'. Therefore, two general proteomics approaches have been used to investigate plant development (see Baginsky and Gruissem, 2004). The one, which can be defined as 'Shotgun proteomics' or 'Tissue-specific proteomics', aims at providing an exhaustive overview of the proteins present in various organs and organelles. At the transcript level, there is evidence that each organ or tissue-type representing the life cycle of *Arabidopsis* has a defining genome expression pattern. Therefore, the knowledge of organ-specific expression patterns and their response to the changing environment provides a foundation for dissecting the molecular processes underlying development (Ma et al., 2005). The other approach, which can be referred to as 'Functional proteomics', concentrates on the analysis of specific proteins related to specific biological processes. In particular, this latter approach aims at the identification of proteins that are subject to post-translational modifications or whose accumulation level changes during specific stages of development. To date, this relies heavily on differential 2-DE, a robust technique used since three decades (O'Farrell, 1975). An example is presented in Plate 7.1 in Plate section, showing the tremendous change in the proteome during *Arabidopsis* embryo development. Such gels can typically resolve more than 1500 proteins following staining by silver nitrate, out of which the accumulation level of several hundreds may be observed to vary during seed development (Plate 7.1). Here, it is expected that the identification of all these changes will provide clues toward understanding seed development.

Recent advances in proteomics include the refinement of 2-DE techniques and the development of sensitive methods for protein visualization and mass spectrometry (MS) analysis. Owing to these progresses, proteomic studies in plants have proliferated during the recent years (reviewed by Hirano et al., 2004; Newton et al., 2004; Rose et al., 2004; Agrawal et al., 2005a; Peck, 2005). A number of excellent papers and reviews are available in several areas of plant proteomics, dealing with the establishment of reference protein maps in plant tissues and organelles (Bardel et al., 2002; Koller et al., 2002; Baginsky and Gruissem, 2004; Heazlewood et al., 2004; Sarnighausen et al., 2004; Stauber and Hippler, 2004; Agrawal et al., 2005b), the identification of proteins that would serve as possible markers of different genotypes and phenotypes and their use in determination of phylogenetic relationships (Thiellement et al., 1999, 2002; Chevalier et al., 2004), the importance of protein modifications ('modificomics') in plants (Hirano et al., 2004; Laugesen et al., 2004), the characterization at the protein level of plant–microorganism interactions (Rolfe et al., 2003; Bestel-Corre et al., 2004; Cánovas et al., 2004; Colditz et al., 2004), or the characterization of the influence of the environment on plant proteomes (Agrawal et al., 2005c). Particularly interesting in the context of plant proteomics is the genetic approach based upon comparison of protein profiles from cultivars and recombinant inbreed lines showing diversity for agronomic and developmental traits and enabling the discovery of the genetic factors accounting for the accumulation of the proteins related to these traits (Thiellement et al., 1999, 2002).

The 'Tissue-specific proteomics' approach for the study of plant development has been covered by several reviews (Thiellement et al., 2002; Hirano et al., 2004; Newton et al., 2004; Agrawal et al., 2005b). Therefore, the present analysis will mainly concentrate on the 'Functional proteomics' approach, taking as an example seed development and germination, in which we have been particularly involved.

7.2 Examples of proteome analyses of plant development

7.2.1 Seed development and germination

The new plant formed by sexual reproduction starts as an embryo within the developing seed, which arises from the ovule. The development of the embryo proceeds through histodifferentiation and seed filling and terminates, for most species growing in temperate climates, with a desiccation phase, after which the embryo enters in a quiescent state, thereby permitting its storage and survival for many years in various environmental conditions. The seed, therefore, occupies a central position in the higher plant life cycle. Dry mature seeds are resting organs, having low moisture content (5–15%) with metabolic activity almost at a standstill. In the absence of dormancy (a physiological process conditioning seed germination not considered here), for germination to occur seeds need to be hydrated under conditions that encourage metabolism, for example a suitable temperature and the presence of oxygen. Broad proteomic analyses of seed development and germination have been initiated using reference plants, as well as with seeds of commercial importance as barley, maize, or tomato. The general aim is to identify characteristic proteins involved in seed development and germination, which will help understanding the biochemical and molecular processes underlying seed quality and vigor. Additionally, these specific proteins might help optimizing industrial seed treatments such as germination enhancement (priming) or malting.

7.2.1.1 Proteomics of developing seeds

Seed development in *M. truncatula* (cv. Jemalong J5) was investigated at specific stages of seed filling corresponding to the acquisition of germination capacity and protein deposition (Gallardo et al., 2003). Individual flowers were tagged on the day of flower opening, and the pods were harvested between 8 and 44 days after pollination (DAP). One hundred and twenty proteins differing in kinetics of appearance were subjected to matrix-assisted laser desorption ionization-time-of-flight (MALDI-TOF) MS. These analyses provided peptide mass fingerprint data that identified 84 of them. Some of these proteins had previously been shown to accumulate during seed development in legumes (e.g. legumins, vicilins, convicilins, and lipoxygenases), confirming the validity of *M. truncatula* as a model for analysis of legume seed filling. The study also revealed proteins presumably involved in cell division during embryogenesis (tubulin and annexin). Their abundance decreased before the accumulation of the major storage protein families, which itself occurs in a specific temporal order: vicilins (14 DAP), legumins (16 DAP), and convicilins (18 DAP). Furthermore, the study showed an accumulation of enzymes of carbon metabolism (e.g. sucrose synthase, starch synthase) and of proteins involved in embryonic photosynthesis (e.g. chlorophyll *a*/*b* binding), which may play a role in providing cofactors for protein/lipid synthesis or for CO_2 refixation during seed filling. Correlated with the reserve deposition phase was the accumulation of proteins associated with cell expansion (actin 7 and reversibly glycosylated polypeptide). This study had not only cataloged proteins but has also described their accumulation patterns at specific stages during seed development, before and during protein deposition. These findings, therefore, contribute to our understanding of how metabolic networks are regulated at the protein level during reserve deposition in seeds of a legume species. Finally, this work revealed a differential accumulation of enzymes involved in methionine metabolism (*S*-adenosylmethionine (AdoMet) synthetase and *S*-adenosylhomocysteine (AdoHcy) hydrolase; see Figure 7.2) and proposed a role for these enzymes in the transition from a highly metabolically active to a quiescent state during seed development (see Section 7.2.1.4). This knowledge will support further attempts to engineer legume seed composition for added end user value.

In parallel, Watson et al. (2003) also reported a proteome of seed proteins from a variety of developmental stages (including very young pods to those with maturing seeds) of 3-month-old *M. truncatula* plants (cv. Jemalong A17). Sixty-one proteins were identified from a total of 91 protein spots (67% success rate), and grouped into 10 functional categories: protein destination/storage (60%), energy (15%), disease/defense (7%), signal transduction (5%), unclear (3%), protein synthesis (2%), cell growth/division (2%), transcription (2%), secondary metabolism (2%), and transporters (2%).

A high-throughput proteomic approach was recently employed to determine the expression profile and identity of hundreds of proteins during seed filling in soybean (*Glycine max*) cv. Maverick (Hajduch et al., 2005). Soybean seed proteins were analyzed at 2, 3, 4, 5, and 6 weeks after flowering (WAF) using 2-DE and MS. This led to the establishment of high-resolution proteome reference maps and expression profiles of 679 spots. These corresponded to 422 proteins representing 216 non-redundant proteins, which were classified into 14 major functional categories. Proteins involved in metabolism, protein destination and storage, metabolite transport, and disease/defense were the most abundant. To further detail global expression trends of proteins involved in different processes, the authors established composite expression profiles by summing protein abundance, expressed as relative volume, for each protein in each functional class for the five seed stages (Hajduch et al., 2005). Relative abundances of metabolic proteins decreased during the experimental period, suggesting metabolic activity curtails as seeds approach maturity. Interestingly, proteins involved in lipid and sterol metabolism decreased from 2 to 4 WAF, but after 4 WAF their abundance slightly increased. The protein destination and storage class of proteins increased during late seed filling, and this was due to the preponderance of seed storage proteins. The transporter class of proteins, which includes the ubiquitous Suc-binding proteins, exhibited almost constant but slightly increased expression during the experimental period. Disease- and defense-related proteins were highly abundant at the early stage of seed filling; later their abundance decreased to about 50% and from 4 WAF was stable. Proteins involved in energy production increased in abundance during seed filling, whereas cell growth and division proteins as well as signal transduction proteins each had decreasing expression profiles. Proteins involved in protein synthesis and secondary metabolism decreased in abundance during the experimental period. A user-intuitive database (http://oilseedproteomics.missouri.edu) has been developed to access these data for soybean and other oilseeds currently being investigated. In summary, an overall decrease in metabolism-related proteins versus an increase in proteins associated with destination and storage was characteristic of seed filling in soybean.

Proteomic approaches were also used to investigate embryo development in monocots. Recently, expressions of more than 400 polypeptide spots during rice caryopsis development, in response to temperature treatments or between varieties were monitored (Lin et al., 2005). Among them, more than 70 differentially expressed polypeptides were analyzed, allowing identifying 54 proteins with known functions. Of these, 21 were involved with carbohydrate metabolism, 14 with protein synthesis and sorting, including glutelins and prolamins, the major storage proteins in rice seeds and 9 with stress responses, most of them being heat-shock proteins (HSPs). The data indicate that changes in the expression of these proteins can be used to mark the physiological development stage of rice caryopses. Waxy (Wx) proteins and glutelins were the most significant spots, which increased significantly during development. Allergen-like proteins, pyruvate orthophosphate dikinase (PPDK) and $NADP^+$-dependent sorbitol dehydrogenase (NADH-SDH), also were expressed during development, implying their physiological roles in caryopsis. Concerning PPDK, its prominent and consistent expression suggested that it might be involved in CO_2 fixation or amino acid biosynthesis. The observed accumulation of NADH-SDH was the first report of its expression in rice, although its role in development remains to be clarified. Expression of large isoforms of Wx proteins was correlated with the amylose content of rice caryopses. High temperature (35/30°C) decreased the expression of Wx proteins, allergen-like proteins, and elongation factor 1b, but increased the expression of small heat-shock proteins (sHSP), glyceraldehyde-3-phosphate dehydrogenase (GAPDH), and prolamin. sHSP was positively correlated with the appearance of chalky kernels. During development, four Wx proteins isoforms were phosphorylated, while several glutelins were phosphorylated and glycosylated. Although the reason of this post-translational modification is unknown, this is an interesting observation, since there are only few reports on post-translational regulation of rice seed proteins. It is noted that phosphorylation of rice glutelins, which was not reported before, was directly demonstrated in this careful study using immunological and LC-MS/MS analyses (Lin et al., 2005).

2-DE was used for a time-resolved study of the changes in proteins that occur during seed development in barley (*Hordeum vulgare*) (Finnie et al., 2002). Protein spots were divided into six categories according to the timing of appearance and disappearance during the 5-week period of comparison. Thirty-six selected spots were identified by MALDI-TOF MS or by nano-electrospray tandem MS/MS. Some proteins were present throughout development (e.g. cytosolic malate dehydrogenase), whereas others were associated with the early grain filling (ascorbate peroxidase), desiccation (cold-regulated protein Cor14b) or late (embryo-specific protein) stages. Most noticeably, the development process was characterized by an accumulation of serpin, low-M_r α-amylase inhibitors (presumed to defend the starch reserves of the seed against invading insect pathogens), serine protease inhibitors, and enzymes involved in protection against oxidative stress (glyoxalase I, ascorbate peroxidase,

1cys-peroxyredoxin). The expression of this last class of proteins throughout the development process most presumably reflects the importance of protection against reactive oxygen species (ROS) produced during seed development (see Section 7.2.1.7). The study also presented examples of proteins not previously experimentally observed (e.g. Cor14b), differential extractability of thiol-bound proteins, and possible allele-specific spot variation (e.g. at the level of β-amylase).

A proteomic approach was utilized to identify over 250 proteins of wheat (*Triticum aestivum* L., cv. Butte 86) starchy endosperm that participate in 13 biochemical processes: ATP interconversion reactions, carbohydrate metabolism, cell division, cytoskeleton, lipid metabolism, nitrogen metabolism, protein synthesis/assembly, protein turnover, signal transduction, protein storage, stress/defense, transcription/translation, and transport (Vensel et al., 2005). Endosperm protein populations were compared at early (10 days post-anthesis, dpa) and late (36 dpa) stages of grain development. Analysis of protein number and spot volume revealed that carbohydrate metabolism, transcription/translation, and protein synthesis/assembly were the principal endosperm functions at 10 dpa followed by nitrogen metabolism, protein turnover, cytoskeleton, cell division, signal transduction, and lipid metabolism. Carbohydrate metabolism and protein synthesis/assembly were also major functions at 36 dpa, but stress/defense and storage were predominant.

7.2.1.2 Proteomics of somatic embryogenesis

Somatic embryogenesis is a process analogous to zygotic embryogenesis, in which a single cell or a small group of vegetative (i.e. somatic) cells are the precursors of the embryos. This phenomenon can be divided into four major steps: (i) initiation of proembryogenic masses, (ii) proliferation of embryogenic cultures, (iii) maturation of somatic embryos, and (iv) regeneration of whole plants. Somatic embryogenesis is widely investigated in several plants because, on one hand, it provides useful systems for plant propagation (e.g. conifer biotechnology for reforestation programs) and, on the other hand, it allows fundamental studies on embryo development (Zimmerman, 1993).

A recent proteomic study used leaf explants from the mutant line 2HA of *M. truncatula*, which presents a 500-fold greater capacity to regenerate plants in culture by somatic embryogenesis than the wild-type Jemalong cultivar chosen as reference to produce extensive genomic sequences (Imin et al., 2005). Both 2HA and Jemalong leaf explants were grown on media containing the auxin 1-naphthalene-acetic acid and the cytokinin 6-benzylaminopurine. Proteins were extracted from the cultures at different time points (2, 5, and 8 weeks), separated by 2-DE, and detected by silver staining. More than 2000 proteins could be reproducibly resolved and detected on each gel. Statistical analysis showed that 54 protein spots were significantly changed in expression (accumulation) during the 8 weeks of culture. They were subjected to MALDI-TOF or LC-MS/MS analyses. This allowed identifying 16 differentially expressed proteins. More than 60% of the differentially expressed protein spots had very different patterns of gene expression between 2HA and Jemalong during the 8 weeks of culture.

Among the identified proteins, RuBisCo small-chain proteins were gradually decreased in both Jemalong and 2HA during explant cultures. As such, the decreased trend of RuBisCo small-chain proteins can be used as a marker for the dedifferentiation and proliferation of the mesophyll tissues. Two of the most abundant proteins were an abscisic acid (ABA)-responsive protein with homology to the pathogenesis-related protein PR10-1 and PR10-1 itself in both Jemalong and 2HA. They were not detected in the young leaves from which the explant cultures originated. Interestingly, they changed little throughout the 8 weeks of culture, suggesting a general role for ABA-responsive proteins and PR10 proteins in cell maintenance or cell defense. Chaperone proteins (dnaK-type HSP70 and luminal binding), in general, showed a decrease in the 8-week-old cultures in both Jemalong and 2HA, although their expression levels were different. This may imply that a higher level of expression of the chaperones is required for the maintenance of cells during early culture. It is noteworthy that this study also identified proteins involved in seed formation as being expressed only in the highly embryogenic 2HA of 8-week-old cultures. These included a seed maturation protein and a vicilin. Since the same behavior was observed during seed development in *M. truncatula* (Gallardo et al., 2003), it can be concluded that somatic embryogenesis closely resembles the zygotic embryogenesis, and that the approach of inducing embryogenesis in 2HA was a valid comparison of the two systems. One of the most interesting proteins identified was thioredoxin *h* (Imin et al., 2005). It appeared early in both protoplast and explant cultures and became undetectable at later stages of cell proliferation. Furthermore, its expression was not detected in leaf tissue or isolated protoplasts, suggesting it is induced only after putting the tissues in culture. There was a higher expression in 2HA than in Jemalong in this early induction process. These results suggest that in *M. truncatula* thioredoxin *h* plays an important role during early stages of commitment from the vegetative stage to a pathway of cellular differentiation and proliferation. As discussed below (Section 7.2.1.6), thioredoxin *h* regulates a myriad of post-transcriptional biological functions in plant cells (Buchanan and Balmer, 2005). Another interesting protein found in this study is the 1cys peroxiredoxin, which remained at a minimum level in Jemalong during 8 weeks of culture. By contrast, it showed a slight increase at 5 weeks before reaching a much higher level of expression at 8 weeks in the highly embryogenic line 2HA, clearly indicating a role for this enzyme in embryo development (Imin et al., 2005).

Forests play a vital role in both the stability of global ecosystems and in the economy of countries that retain large stands of timber. Sustainable forest management is therefore of considerable interest worldwide. In North America, white spruce is an important component of the forest landscape, not only for its abundance, but also for its high value for the solid wood industry and for pulp and paper production. Somatic embryogenesis of white spruce has the potential to deliver a stable supply of superior seedlings on a large scale for forest plantations. In a recent study (Lippert et al., 2005), a proteomic approach was employed to quantitatively assess the expression levels of proteins across four stages of somatic embryo maturation in white spruce (0, 7, 21, and 35 days post-ABA treatment). The increasing cellular complexity of the developing embryos was clearly evident at the molecular level. Measurement of protein content in these samples revealed a 5-fold increase in total protein content based on fresh weight between the early and late stages of development. Subsequent analysis of these tissues by 2-DE showed a concomitant increase in the total number of distinct protein species which could be resolved as the embryos developed. This number rose from 696 distinct protein species in the immature stage 0 embryos to 1250 distinct protein species in the mature embryos at day 35, presumably reflecting the added complexity arising as tissues begin to differentiate.

Forty-eight differentially expressed proteins have been identified by LC-MS/MS, which displayed a significant change in abundance as early as day 7 of embryo development. These proteins are involved in a variety of cellular processes, many of which have not previously been associated with embryo development. The identification of these proteins was greatly assisted by the availability of a substantial EST resource developed for white, sitka and interior spruce. It must be stressed that the combined use of these spruce ESTs in conjunction with GenBank accessions for other plants improved the rate of protein identification from 38% to 62% when compared with GenBank alone using automated, high-throughput techniques. This underscored the utility of EST resources in a proteomic study of any species for which a genome sequence is unavailable. Interestingly, as in zygotic embryogenesis (Gallardo et al., 2003), AdoMet synthetase was detected at high level in the immature stage 0 embryos and decreased to background level in mature embryos (Lippert et al., 2005), suggesting common features of metabolic regulation in somatic and zygotic embryogenesis (see Section 7.2.1.4). Furthermore, as in zygotic embryogenesis, developing somatic embryos proved capable of accumulating storage proteins as vicilins (Lippert et al., 2005). One potential application of these proteomic data would be in the assessment of seedling performance during the biotechnological process of somatic seedling production.

7.2.1.3 Proteomes of mature seeds and their evolution during germination

The storage compounds found in most mature seeds accumulate during seed filling. These reserves are of paramount importance for two reasons: (a) they support early seedling growth when mobilized upon germination, and (b) they are widely used for human and animal nutrition. Thus, plants provide over 70% of the protein in the human diet, most of which is constituted by the storage proteins of cereal and legume seeds. Among them, 11–12S globulins are abundant seed storage proteins, being widely distributed in higher plants. They are synthesized during seed maturation on the mother plant in a precursor form consisting of a single protein chain of about 60 kDa. At later stages, the precursor form is cleaved, yielding the mature globulins generally found in dry mature seeds (Figure 7.1). These are composed of six subunit pairs that interact non-covalently, each of which consists of an acidic A-subunit of $M_r \approx 40{,}000$ and a basic B-subunit of $M_r \approx 20{,}000$ covalently joined by a single disulfide group. These storage proteins are subsequently broken down during germination and used by the germinating seedling as an initial food source (Bewley and Black, 1994; Adachi et al., 2001, 2003) (Figure 7.1).

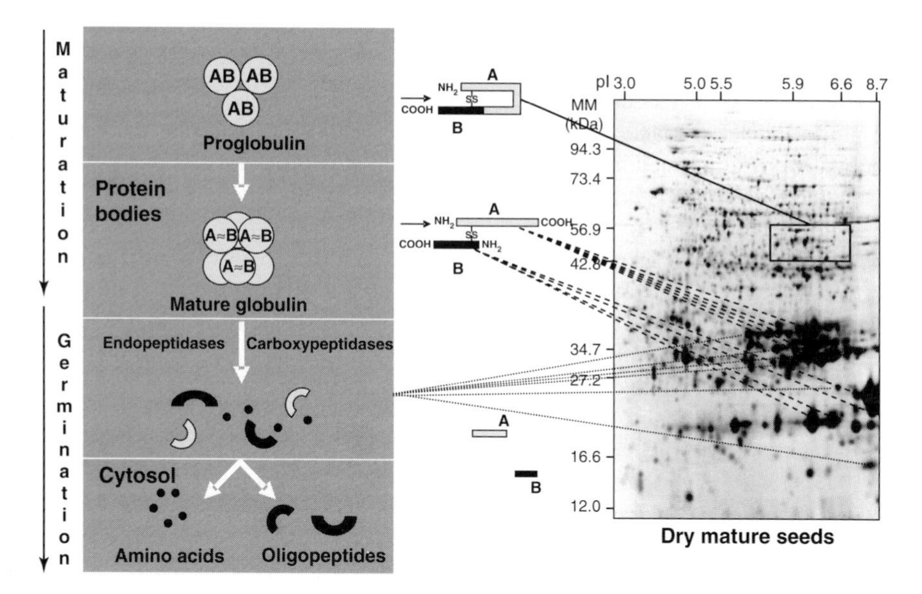

Figure 7.1 Proteomics of the mature dry *Arabidopsis* seed, with emphasis on the identification of 12S cruciferins, the major seed storage proteins in crucifers. Main features of 12S cruciferin synthesis and mobilization during seed maturation and germination are shown on the left. The figure also shows on the right a 2-D gel of total soluble proteins from dry mature seeds. Proteins were first separated by electrophoresis according to charge. Isoelectrofocusing (IEF) was carried out with protein samples with an equivalent to an extract of approximately 100 seeds, corresponding to about 200 μg protein for all samples. Proteins were then separated according to size by SDS-PAGE. Proteins were visualized by silver nitrate staining (for details on protein extractions and proteomic analyses, see Job et al., 2005).

Contrasting with these well accepted findings, a proteomic approach (Gallardo et al., 2001) revealed that total soluble protein extracts from dry mature *Arabidopsis* seeds contain three forms of 12S globulins (cruciferins): (a) residual precursor forms (shown with solid lines in Figure 7.1), (b) A- and B-subunits (shown by dashed lines in Figure 7.1), and (c) proteolysed forms of A- and B-subunits (dotted lines in Figure 7.1). The presence of some residual precursor forms in the dry mature *Arabidopsis* seeds was unexpected (Bewley and Black, 1994). A possible explanation could be that the maturation process giving rise to the formation of the A- and B-chains was not fully completed when developing seeds entered into quiescence. Another unexpected finding was to recover fragments of A- and B-subunits in the soluble protein extracts from the dry mature *Arabidopsis* seeds. This behavior presumably reflects an early mobilization of the cruciferins during the maturation phase. Thus, proteomics enlightened new features of 12S globulin structure in dry seeds, showing that the anabolic processes that occur before germination and the catabolic processes that normally occur during germination are not fully separated developmentally in *Arabidopsis*.

During germination, the residual cruciferin precursor forms disappeared, suggesting that processing of pro-cruciferins during maturation is not required for their mobilization (see also Section 7.2.1.8). Also, a systematic analysis of all cruciferin fragments appearing and disappearing during germination indicated that initial mobilization of the cruciferins preferentially begins with proteolysis of the A-chains (Gallardo et al., 2001). This hypothesis is consistent with the determination of cleavage sites under limited trypsinolysis of 11–12S globulins; all of these sites being located within the A-chains (Adachi et al., 2001, 2003). Likewise, during germination *sensu stricto* (prior to radicle protrusion) and during the radicle emergence step fragments of both A- and B-chains were released. Thus, once initiated at the level of A-subunits, mobilization of cruciferins continues during germination at the level of the B-subunits, indicating that the initial proteolysis of A-chains increased the sensitivity of B-chains toward further proteolytic attacks (Gallardo et al., 2001).

Besides the 12S cruciferins, changes in the abundance (up- and down-regulation) of a number of proteins were observed during germination *sensu stricto* and the radicle protrusion step (Gallardo et al., 2001, 2002b; Rajjou et al., 2004; Job et al., 2005). Based on their accumulation patterns, the seed proteins were classified into 12 types: during germination (types 0–4), during imbibition drying (types 5 and 6), and during priming (types 7–12). Furthermore, these protein types were discussed according to their characteristics, such as those involved in mobilization of stored seed reserves (types 0, 1, 3–5, 7, and 11), germination *sensu stricto* (types 1, 2, 5–7, and 11), radicle emergence (types 3 and 4), imbibition (type 5), desiccation (type 6), and priming (types 7, 9–11). Proteins associated with germination *sensu stricto* correlated with initial events in the mobilization of protein and lipid reserves, and the resumption of cell cycle activity, such as WD-40 repeat protein, tubulin, and cytosolic GAPDH. During radicle emergence, proteins mostly involved in defense mechanisms to protect the future seedlings against herbivores, pathogens, and other stresses were identified, such as myrosinase, jasmonate-induced myrosinase-binding proteins, AdoMet synthetase, late embryogenesis abundant (LEA), and heat shock (HSP70) proteins. In addition, a seed maturation protein, probably involved in sequestering biotin (Alban et al., 2000), and a chloroplast translation elongation factor (EF-Tu) were found. Among a total of 19 imbibition-associated proteins, 7 proteins were identified, including actin 7 (ACT 7) and WD-40 repeat proteins. Accumulation of the WD-40-repeat protein during imbibition provided first evidence for its role in the germination process. Remarkably, all protein changes associated with the desiccated state of seeds corresponded to proteins already present in the dry mature seeds. One of the three detected desiccation-specific proteins was the cytosolic GAPDH, whose induction during desiccation is a conserved feature among different tissues and organs in plants. Two priming treatments, hydro and osmo-priming, were also employed to examine the priming-associated proteins that resulted in the identification of a total of nine proteins, such as tubulin, 12S-cruciferin B-subunits, catalase, and low-molecular-weight HSPs (Gallardo et al., 2001).

In order to further characterize seed vigor, we have submitted *Arabidopsis* seeds to a controlled deterioration treatment (CDT), and analyzed the consequence of this treatment on the seed proteome by differential proteomics based on a comparison of eight seed lots treated for different periods of time, up to 7 days (Rajjou et al., 2006). Germination tests showed a progressive decrease of seed vigor depending on the duration of CDT, as described (Tesnier et al., 2002). Proteomic analyses revealed that CDT-induced qualitative and quantitative protein modifications in dry mature seeds. Moreover, during germination, the proteome evolution with the deteriorated seeds appeared to be very different from that with the non-deteriorated control seeds. Therefore, loss in seed vigor was accounted for by specific protein changes in the dry seeds and by an inability of low-vigor seeds to display a normal proteome during germination. By following [^{35}S]Met incorporation during germination, we also observed that the loss of germination vigor was associated to a strong reduction in the protein neosynthesis capacity. This finding is in agreement with other results showing the importance of protein synthesis in germination (Rajjou et al., 2004; see Section 7.2.1.5). This differential proteome analysis also revealed 183 proteins involved in seed vigor, all of them being identified by MS sequencing, and raised the possibility of using plant seeds as a very general model for the understanding of cellular aging.

A database based on clickable protein reference maps (http://seed.proteome.free.fr) has been developed to access all these *Arabidopsis* data.

Proteomics was also used to identify major proteins in extracts of mature barley seeds and to follow their fate during germination (Østergaard et al., 2004). To overcome the problem of working with total protein extracts which are dominated by storage proteins and to identify new proteins involved in seed germination, a low-salt extraction buffer was used. At this point, it is worth mentioning that the systematic study of plant proteins dates from the nineteenth century and in particular from the work of Osborne (Osborne, 1924). Osborne pioneered the systematic study of plant proteins and introduced the widely used classification into solubility groups based on their sequential extraction in water (albumins), dilute salt solutions (globulins), alcohol–water mixtures (prolamins), and dilute acid or alkali solutions (glutelins). As stressed by Shewry (1995), this classification was, of course, introduced before we had a detailed understanding of protein structure and characterization, and it was therefore inevitable that some modifications have proved necessary (e.g. the use of reducing agents to cleave the interchain disulfide bonds). Nevertheless, it had formed an invaluable framework for the modern study of seed proteins, notably for the establishment of seed proteomes.

The isolation of the albumin fraction of the mature barley seeds resulted in 198 identifications of 103 proteins in 177 spots (Østergaard et al., 2004). These included housekeeping enzymes (glycolysis, starch metabolism, and citric acid cycle), chaperones, defense proteins (including enzyme inhibitors presumably involved in defense of the germinating seed against pathogens), and proteins related to desiccation and oxidative stress. It is noted that a number of the identifications were made using ESTs. Numerous spots in the 2-D gel pattern changed during germination (micromalting) and an intensively stained area which contained large amounts of the serpin (serine protease inhibitor) protein Z appeared centrally on the 2-D gel. Spots containing α-amylase also appeared. Identification of 22 spots after 3 days of germination represented 13 different database entries and 11 functions including hydrolytic enzymes, chaperones, housekeeping enzymes, and inhibitors. Despite the fact that cereal seed proteins have been studied for decades, new proteins have been identified on 2-D gels. Several of the identified proteins have homologs in other organisms but their function in barley seeds is not clear. Yet, their identification will facilitate the analysis of the changes in the proteome that occur during seed development and germination of cereal grains. Furthermore, by combining proteomics with malting quality analysis and genetics, 2-D gel patterns can be related to cultivar characteristics (Finnie et al., 2004a, b).

The above studies have been carried out with whole seeds, which comprise several tissues and organs: the embryo itself containing an embryonic axis (radicle) and one or two cotyledons, a storage tissue (either an endosperm, a perisperm, or the cotyledons), and the seed envelopes (Bewley and Black, 1994). Therefore, it is worth noting that proteome analysis of germinating tomato seeds has only recently been carried out from isolated embryo and endosperm tissues. This revealed several spots with different abundance in these tissues in the mature dry state and during germination. In particular, a number of protein spots present mainly in the endosperm were identified as coat protein and chain-P of tomato mosaic virus, suggesting their possible role in defense against such viruses during germination (Sheoran et al., 2005). It is also noted in this context a detailed proteome analysis of the endoplasmic reticulum (ER) from germinating castor seeds including separation of ER proteins into luminal, peripheral membrane an integral membrane subfractions. Most of the proteins identified are concerned with roles in protein processing and storage, and lipid metabolism which occur in the ER (Maltman et al., 2002). Both proteomic studies on tomato and castor seeds help demonstrating the usefulness of tissue dissection and separation technologies for a better understanding of the germination process.

The defense response of plants against fungal infection has been intensively studied in vegetative organs, but quite surprisingly not in seeds. This question was addressed for the first time in germinating maize embryos by a proteomics approach (Campo et al., 2004). Among the proteins induced in response to infection are proteins involved in protein synthesis, folding and stabilization, proteins involved in oxidative stress tolerance, as well as the specific pathogenesis-related proteins. Altogether, the available proteomic data highlighted the unexpected occurrence of a defense response during germination, which most presumably is required for protection of the germinating seed and the establishment of a vigorous plantlet.

In summary, these facts highlighted the power of proteomics to unravel specific features of complex developmental processes such as germination and to detect protein markers that can be used to characterize seed vigor of commercial seed lots and to develop and monitor priming treatments.

7.2.1.4 Proteomics unravels metabolic control of seed development and germination

Proteomics of *Arabidopsis* and *M. truncatula* seeds revealed the differential accumulation during germination and development of housekeeping enzymes involved in Met metabolism (Gallardo et al., 2001, 2002b, 2003; Rajjou et al., 2004; Job et al.,

2005). Among the essential amino acids synthesized by plants, Met is a fundamental metabolite because it functions not only as a building block for protein but also as the precursor of AdoMet, the primary methyl-group donor and the precursor of polyamines and the plant-ripening hormone, ethylene (Ravanel et al., 1998, 2004). In plants, Met can be synthesized through two pathways (Figure 7.2). In the *de novo* biosynthetic pathway, *O*-phosphohomoSer is transformed to cystathionine in a reaction catalyzed by cystathionine γ-synthase, then to Hcy in a reaction catalyzed by cystathionine β-lyase, and finally to Met in the presence of the cobalamin-independent Met synthase. In the Met recycling pathway, *S*-methylmethionine (SMM), a compound unique to plants, is synthesized by a methyl transfer from AdoMet to Met, in a reaction catalyzed by AdoMet:Met *S*-methyltransferase. SMM can then be reconverted to Met by transferring a methyl group to Hcy in a reaction catalyzed by SMM:Hcy *S*-methyltransferase. These reactions, together with the reactions catalyzed by AdoMet synthetase and AdoHcy hydrolase, constitute the SMM cycle, which may be the main mechanism in plants for short term control of AdoMet level (Ranocha et al., 2001).

During *Arabidopsis* seed germination, two enzymes in this pathway showed differential expression (Gallardo et al., 2002b). The first corresponded to Met synthase. This protein was present at low level in dry mature seeds, and its level was increased

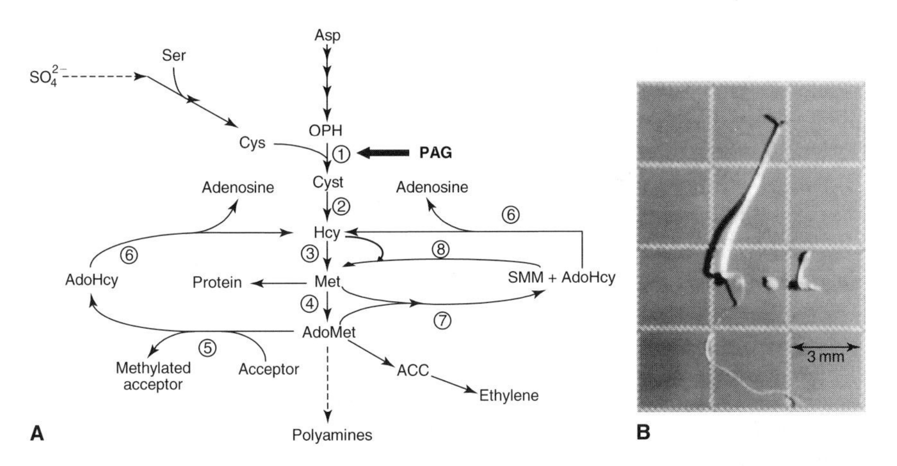

Figure 7.2 Methionine metabolism in seed development and germination (from Gallardo et al., 2002b). **A:** Met metabolism in plants. *Reaction intermediates*: ACC: 1-aminocyclopropane-1-carboxylic acid; AdoHcy: *S*-adenosylhomocysteine; AdoMet: *S*-adenosylmethionine; Cyst: cystathionine; Hcy: homocysteine; OPH: *O*-phosphohomoserine; SMM: *S*-methylmethionine. *Enzymes*: 1: cystathionine γ-synthase; 2: cystathionine β-lyase; 3: cobalamin-independent methionine synthase; 4: AdoMet synthetase; 5: AdoMet-dependent transmethylases; 6: AdoHcy hydrolase; 7: AdoMet: methionine *S*-methyltransferase (MMT); 8: SMM: Hcy *S*-methyltransferase (HMT). DL-propargylglycine (PAG) blocks methionine synthesis by inhibiting cystathionine γ-synthase. **B:** Effect of DL-PAG on seed germination and seedling growth. Morphology of seedlings after 10 days germination on water (Left), 1 mM PAG (Middle) or 1 mM PAG + 10 μM Met (Right). The photograph shows a typical result observed with all seeds in each germination conditions. Square size: 3 mm × 3 mm.

strongly at 1-day imbibition, prior to radicle emergence. Its level was not increased further at 2-day imbibition, coincident with radicle emergence. However, its level in 1-day imbibed seeds strongly decreased upon subsequent drying of the imbibed seeds back to the original water content of the dry mature seeds. The second enzyme corresponded to AdoMet synthetase. In this case, this enzyme was detected in the form of two isozymes with different pI and M_r. Both proteins were absent in dry mature seeds and in 1-day imbibed seeds but specifically accumulated at the moment of radicle protrusion. As transcriptomics, proteomics can only suggest candidate proteins whose function in development remains to be established. To this end, we analyzed the effect of DL-propargylglycine, a specific inhibitor of Met synthesis on *Arabidopsis* seed germination and seedling growth (Gallardo et al., 2002b). Seed germination was strongly delayed in the presence of this compound. Furthermore, this compound totally inhibited seedling growth. These phenotypic effects were largely alleviated upon methionine supplementation in the germination medium (Figure 7.2).

These results therefore validated the proteomics data and established that Met synthase and AdoMet synthetase are fundamental components controlling metabolism in the transition from a quiescent to a highly active state during seed germination. Moreover, the observed temporal patterns of accumulation of these proteins were consistent with an essential role of endogenous ethylene in *Arabidopsis* only after radicle protrusion (Gallardo et al., 2002b).

Strikingly, many proteins whose abundance varied during *M. truncatula* seed filling also corresponded to enzymes involved in Met biosynthesis (Gallardo et al., 2003). Consistent with the high demand for protein synthesis between 12 and 20 DAP, two spots detected in seed extracts throughout this period corresponded to Met synthase. In addition, one spot corresponded to AdoMet synthetase. Interestingly, the level of AdoMet synthetase fell sharply at the 16-DAP stage and remained low up to desiccation. This result was supported by the observation that ESTs corresponding to AdoMet synthetase were only found in cDNA libraries corresponding to early stages of seed development, and is in keeping with the finding that this enzyme is absent from dry mature *Arabidopsis* seeds (Gallardo et al., 2001, 2002a). Recent studies also disclosed a sharp decrease of Met synthase during soybean seed filling (Hajduch et al., 2005) and of both Met synthase and AdoMet synthetase during wheat endosperm development (Vensel et al., 2005). Therefore, proteomics unraveled a characteristic feature of embryo development across plant species.

In summary, these proteomic data revealed a strong correlation between metabolic activity in embryonic cells and accumulation of key enzymes of Met metabolism. Hence, these data provided the first demonstration of a metabolic control of seed development and germination. This type of control might be a more general feature, as inferred from the observation that relative abundances of metabolic proteins steadily decreased during soybean seed filling, this trend being established for 82 identified proteins associated with metabolism (Hajduch et al., 2005). Several studies also documented that Lys metabolism and catabolism are tightly regulated during seed development (Zhu and Galili, 2003).

7.2.1.5 The neosynthesized proteome during germination

Functional proteomics aims at detecting proteins whose expression varies in response to a perturbation. As shown above, this is usually achieved by differential 2-DE. However, the amount of a protein spot in a 2-D gel reflects the accumulation level (turnover) of that protein and not simply its rate of synthesis. The use of labeled precursors of protein synthesis is necessary to clarify this question (e.g. see the paper by Chang et al. (2000) which nicely documents this point). It is noted that combined with 2-DE, this radiolabeling technique has been widely used in the past (Cuming and Lane, 1979; Dure et al., 1981), leading to the foundation of major concepts in seed biology, for example, the role during germination of stored and nascent mRNAs in coding developmentally regulated proteins as Em or germin (reviewed by Lane, 1991) or the mechanisms of cellular desiccation and hydration through the discovery of the LEA proteins (Dure et al., 1989). Since mature dry seeds contain mRNAs stored during maturation there are two mechanisms to account for the synthesis of a protein during germination, which depend on the use of stored or nascent mRNA as template. Earlier studies based on the use of metabolic inhibitors (cycloheximide, actinomycin D) had supported the view that protein synthesis occurs on such stored templates during the early stages of seed germination and that *de novo* transcription is not necessary (Dure and Waters, 1965; Waters and Dure, 1966).

To investigate the role of stored and neosynthesized mRNAs in *Arabidopsis* seed germination, the effect of α-amanitin, a transcriptional inhibitor targeting DNA-dependent RNA polymerase II (de Mercoyrol et al., 1989), was examined on the germination of non-dormant *Arabidopsis* seeds (Rajjou et al., 2004). *Transparent testa* mutants were used of which seed coat is highly permeable (Debeaujon et al., 2000) to better ascertain that the drug can reach the embryo during seed imbibition. Even with the most permeable mutant (*tt2-1*), germination (radicle protrusion) occurred in the absence of transcription, while subsequent seedling growth was blocked. In contrast, germination was abolished in the presence of the translational inhibitor cycloheximide. Taken together, the results highlighted the role of stored proteins and mRNAs for germination in *Arabidopsis* and showed that in this species the potential for germination is largely programmed during the seed maturation process. In the aim of characterizing the *de novo* synthesized proteome during germination, proteomic studies were carried out with this system, in the presence of [^{35}S]Met as a labeled precursor for protein synthesis (Figure 7.3).

The radioactive protein patterns differed remarkably depending on the presence or absence of the transcriptional inhibitor in the germination assay (Figure 7.3). Upon incubation for 24 h on water, seeds synthesized a number of proteins of molecular mass larger than 30 kDa and of pI less than 7.3. Under the same conditions but in the presence of α-amanitin a large number of these protein spots were no more detectable or were only present at a much lower level than in the water control (rectangle in Figure 7.3A). Proteins for which *de novo* synthesis was repressed by α-amanitin are involved in reactivation of metabolic activity during germination; for example, mitochondrial enzymes as the mitochondrial processing peptidase and succinate dehydrogenase, Met biosynthesis, triacylglycerol metabolism, and hexose assimilation, a

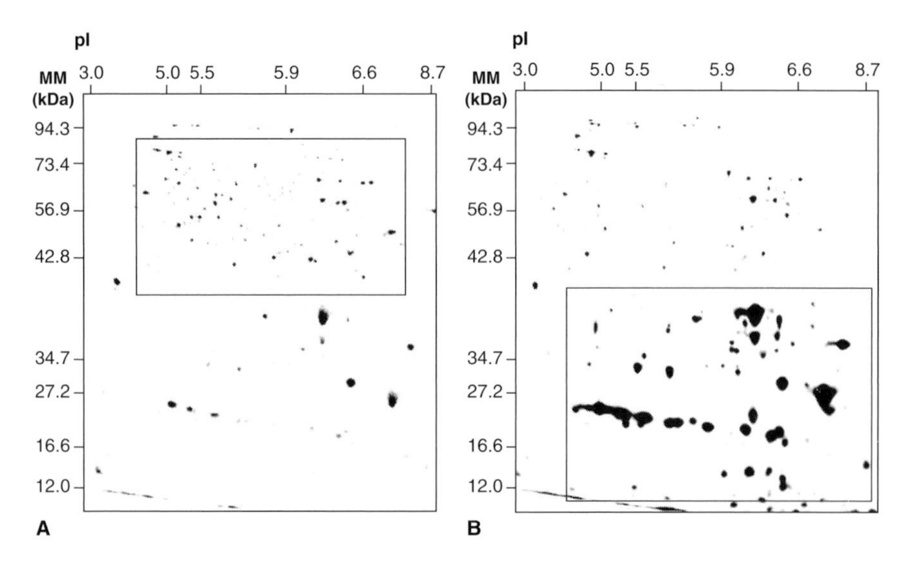

Figure 7.3 Influence of α-amanitin on *de novo* protein synthesis during 24-h germination of the *tt2-1 Arabidopsis* mutant (from Rajjou et al., 2004). Seeds were incubated for 24 h in the presence of [^{35}S]Met, in the absence **(A)** or presence **(B)** of α-amanitin. Proteins were extracted, submitted to 2-DE, and the radiolabeled proteins revealed by Phosphorimager analysis. The labeled protein spots (rectangle in **(A)**, proteins that showed reduced accumulation levels in the α-amanitin-incubated seeds; rectangle in **(B)**, proteins that showed increased accumulation levels in the α-amanitin-incubated seeds) were identified either by MALDI-TOF analysis, by Edman protein sequencing, or by comparison with *Arabidopsis* seed protein reference maps (Gallardo et al., 2001, 2002a,b; Rajjou et al., 2004; Job et al., 2005; http://seed.proteome.free.fr).

finding which provides another indication of the importance of metabolic control of seed germination. Interestingly, a comparison of silver-nitrate-stained gels and radioactive gels revealed a number of spots whose levels apparently remained constant during germination as inferred by quantification from the silver-nitrate-stained gels, although the proteins were labeled with [^{35}S]Met, thereby indicating the occurrence of protein turnover during germination. This finding therefore reveals the existence of regulatory mechanisms to maintain constant the accumulation levels of some enzymes during the germination process, necessary for proper functioning of metabolic pathways. More generally, these data illustrate the power of combining classical proteomics with dynamic proteomics in the interpretation of protein accumulation patterns. Besides these metabolic enzymes, proteins for which *de novo* synthesis was strongly depressed by α-amanitin during germination corresponded to translation elongation and initiation factors a finding which is in agreement with the observation that cycloheximide fully inhibited seed germination and seedling establishment in *Arabidopsis*.

Very surprisingly, in the presence of α-amanitin the seeds massively synthesized proteins of molecular mass smaller than 40 kDa, which were only detected at low level from the radiolabeled proteome obtained in the absence of the fungal toxin (rectangle in Figure 7.3B). The majority of these proteins corresponded to proteins normally synthesized during seed development, such as 12S globulin subunits and members of the dehydrin family. In other words, it seemed that transcriptional inhibition by α-amanitin somehow allowed the germinating seeds to recapitulate part of the maturation program, by favoring the use of stored mRNAs encoding maturation proteins. This finding is in keeping with the work of Lopez-Molina et al. (2002) showing that *Arabidopsis* seed germination comprises a short window of time after the start of imbibition during which the seed can still recruit late maturation programs in order to remain osmo-tolerant. This ABA-mediated developmental checkpoint has physiological significance since control of this developmental transition to auxotrophic growth may enable seeds to monitor environmental water status and to mount appropriate adaptive responses (Lopez-Molina et al., 2002). It must be stressed, however, that our data showing an apparent re-induction of the embryonic maturation program during germination were obtained in the presence of α-amanitin, raising the hypothesis that the fungal toxin repressed the synthesis of factor(s) preventing the use of some of the stored mRNAs during germination (Rajjou et al., 2004). Taken together, these results enlightened the importance of protein synthesis from the pool of stored mRNAs for radicle protrusion. The proteomic data suggested that commitment to seedling growth occurs after radicle protrusion. That stored mRNAs encoding maturation proteins (e.g. cruciferin subunits) can be translated during normal germination is indicated by the data in Figure 7.3 showing that common protein spots are detected in the *novo* protein patterns obtained in the absence or presence of α-amanitin. This last finding raises an old, yet hitherto unanswered, question (Bewley and Black, 1994) that commitment to germination and plant growth requires transcription of genes allowing the imbibed seed to discriminate between mRNAs to be utilized in germination and those to be destroyed.

7.2.1.6 Proteomics of the thioredoxin protein–protein interaction in developing and germinating seeds

The reversible formation of disulfide bonds is important in the regulation of a growing number of enzymes and regulatory proteins such as transcription factors. The thiol-disulfide redox state of these proteins is modulated and maintained by cellular redox agents. Paramount among these is thioredoxin, a 12-kDa protein with a conserved catalytically active disulfide group found in virtually all organisms. The thioredoxin occurring in plant cell cytoplasm (*h*-type) has been found to function in several capacities, including germination and early seedling development (Buchanan and Balmer, 2005).

The role of thioredoxin in wheat starchy endosperm was investigated by proteomics (Wong et al., 2004b). In one approach, thioredoxin targets were isolated from total soluble extracts of endosperm and flour and separated by 2-DE following reduction of the extract by the NADP/thioredoxin system and labeling the newly generated sulfhydryl (SH) groups with monobromobimane (mBBr). In parallel, an elegant method has been developed based upon trapping covalently interacting proteins on an affinity column (Motohashi et al., 2001) prepared with mutant thioredoxin *h* in which one of the active site cysteines was replaced by serine. The two procedures were complementary: of the total targets, one-third was observed with both procedures and one-third was unique to each. Altogether 68 potential targets were identified, for which almost contained conserved cysteines. In addition to confirm known interacting proteins, this study identified 40 potential targets not previously described in seeds. A comparison of the results obtained with young endosperm isolated 10 days after flowering (DAF) to those with mature endosperm, isolated 36 DAF, revealed a unique set of proteins functional in processes characteristics of each developmental stage. Targets of thioredoxin in young endosperm emphasized the importance of biosynthesis (protein synthesis and nitrogen metabolism) and cell division at this stage. Targets of mature endosperm, on the other hand, highlighted functions related to stress, protein storage, and degradation. Indeed, the shift in the processes regulated by thioredoxin reflects the evolution of major biological events characterizing seed development: an emphasis in young endosperm on building the machinery needed for grain filling, growth, cell expansion, and division versus storing and potentially degrading protein in the mature endosperm. The increase in stress- and defense-related proteins in the older tissue could account for the seed's need to survive trash conditions until and when germination commences.

The NADP/thioredoxin system, composed of NADPH, thioredoxin *h* and NADP-thioredoxin reductase (NTR), was also shown to function in the reduction of the major storage proteins of the grain endosperm, gliadins and glutenins in wheat, converting disulfide (S—S) bonds to the reduced (SH) state during the germination process (Kobrehel et al., 1992; Lozano et al., 1996). The soluble proteins (albumins/globulins) and the major insoluble storage proteins (hordeins) of barley endosperm resembled their counterparts in other cereals in undergoing reduction (conversion of S—S to —SH) during germination and early seedling development (Marx and Buchanan, 2003). Recent improvement for identification of thioredoxin targets in mature and germinating barley seeds was achieved by application of a highly sensitive Cy5 maleimide dye and large-format 2-D gels, resulting in a 10-fold increase in the observed number of labeled spots (Maeda et al., 2004). This study confirmed α-amylase/subtilisin inhibitor, several α-amylase/trypsin inhibitors and cyclophilin as thioredoxin targets. Furthermore, lipid transfer protein, embryo-specific protein, three chitinase isoenzymes, a single-domain glyoxalase-like protein and superoxide dismutase were novel identifications of putative target proteins, suggesting new physiological roles of thioredoxins in barley seeds, related to seed defense against pathogens and oxidative stress, or protein folding.

In summary, this work revealed new functions for thioredoxin in seeds, and conclusively documented the general occurrence of a sequence of redox changes taking place in cereal endosperm: proteins are synthesized in the reduced state early in seed development and oxidized during maturation and drying. Upon germination, thioredoxin reduces the oxidized proteins, thereby leading to increased solubility (Wong et al., 2004a), proteolysis and, ultimately, nitrogen and carbon mobilization.

7.2.1.7 The oxidized (carbonylated) proteome in developing and germinating seeds

Increased cellular levels of ROS are known to occur during seed development and germination (Bailly, 2004), but the consequences in terms of protein oxidation and damage are poorly characterized. Protein carbonylation is a widely used marker of protein oxidation and sensitive methods for its detection have been developed (Levine et al., 1994; Nyström, 2005). It occurs by direct oxidative attack on lysine, arginine, proline, or threonine residues of proteins, thus inhibiting or altering their activities and increasing their susceptibility toward proteolytic attack (Levine et al., 1994; Nyström, 2005). A proteomic investigation showed that protein carbonylation can be detected in dry mature *Arabidopsis* seeds and during the first stages of germination (Job et al., 2005) (Figure 7.4), which was a direct demonstration of the accumulation of ROS in seed development and germination.

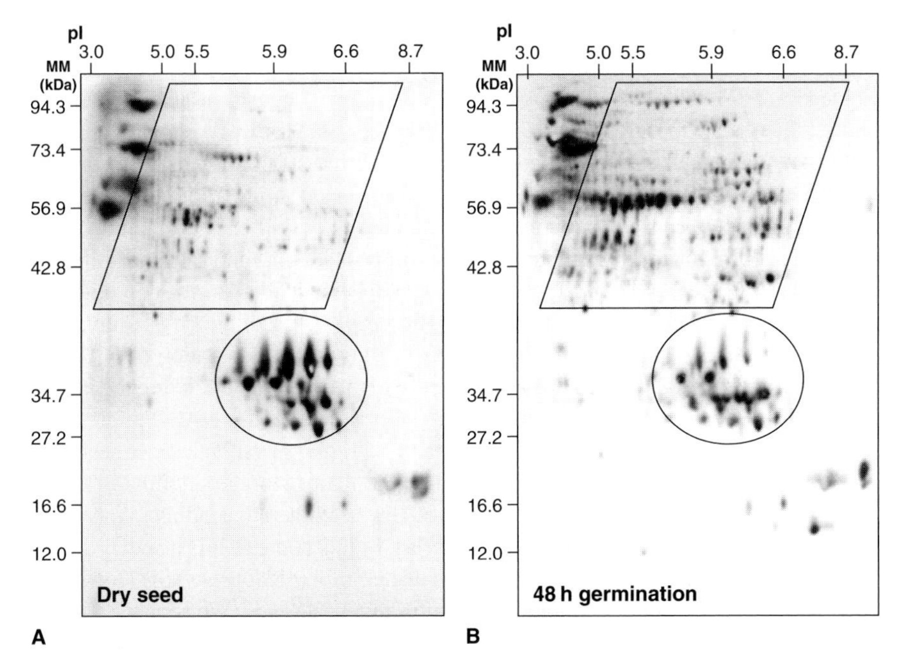

Figure 7.4 The oxidized proteome of *Arabidopsis* seeds (from Job et al., 2005). Characterization of the carbonylated proteins in dry mature seeds **(A)** and in germinating (2-d imbibed) seeds **(B)**. Proteins were separated by 2DE. Following transfer to nitrocellulose, the appearance of carbonyl groups in proteins was analyzed by immunodetection of protein-bound 2,4-dinitrophenylhydrazone (DNP) after derivatization with the corresponding hydrazine, as described (Job et al., 2005). Proteins shown in the oval correspond to 12S globulin (cruciferin) subunits. Those shown in the diamond correspond to various albumins, including a large number of glycolytic enzymes, mitochondrial ATP synthase, chloroplastic ribulose bisphosphate carboxylase large chain, aldose reductase, methionine synthase, translation factors and several molecular chaperones.

In the dry mature seeds, the legumin-type globulins (12S cruciferins) were the major targets (Figure 7.4A). During imbibition various carbonylated proteins accumulated (Figure 7.4B). This oxidation damage was not evenly distributed among seed proteins but targeted specific proteins as glycolytic enzymes, mitochondrial ATP synthase, chloroplastic ribulose bisphosphate carboxylase large chain, aldose reductase, Met synthase, translation factors, and several molecular chaperones. Among the specific targets of protein oxidation detected in *Arabidopsis* seeds (Job et al., 2005), HSP70 chaperones, aconitase, translation initiation and elongation factors, ATP synthase β-subunit, actin, or GAPDHc have also been shown to be highly sensitive to oxidation either in bacterial, yeast, or animal cells, pointing out common features in various living systems. Furthermore, the patterns of protein oxidation observed during *Arabidopsis* seed germination are strikingly reminiscent of those described for *Escherichia coli* or *Saccharomyces cerevisiae* cells exposed to oxidative stress (Tamarit et al., 1998; Cabiscol et al., 2000), lending further support to the contention that oxidative stress normally accompanies seed germination (Bailly, 2004).

Although accumulation of carbonylated proteins is usually considered in the context of aging in a variety of model systems (Berlett and Stadtman, 1997), this was clearly not the case for the *Arabidopsis* seeds since they germinated at a high rate and yielded vigorous plantlets. Thus, the results supported the proposal of Johansson et al. (2004) that the progression of oxidative damage in the life cycle of plants and animals is fundamentally different, and that the observed specific changes in protein carbonylation patterns are probably required for counteracting and/or utilizing the production of ROS caused by recovery of metabolic activity in the germinating seeds (Job et al., 2005). In agreement with a previous proposal (Côme and Corbineau, 1989), the results suggested that blocking glycolysis could be beneficial during conditions of oxidative stress since it would result in an increased flux of glucose equivalents through the pentose phosphate pathway, thus leading to the generation of NADPH. This could provide the reducing power for antioxidant enzymes, including the thioredoxin and GSH/glutaredoxin systems. As mentioned above the NADP/thioredoxin system composed of NADPH, thioredoxin *h* and NTR plays a crucial role in seed germination since it functions in the reduction of the major storage proteins in seeds, converting disulfide S—S bonds to the reduced SH state, thereby leading to increased solubility and mobilization (Wong et al., 2004a). In this context, it is also worth noting that many of the carbonylated proteins detected in *Arabidopsis* seeds (Job et al., 2005) have been identified as thioredoxin targets in wheat seeds (Wong et al., 2004b), lending further support for the existence of a link between ROS and redox regulatory events catalyzed by thioredoxin in seeds (Wong et al., 2004a).

If not correlated to aging, then what might be the role of protein carbonylation in seed physiology? Since carbonylation of proteins increases their susceptibility to proteolytic cleavage, one possibility might be that carbonylation of 12S cruciferin subunits occurring during seed development facilitates their mobilization during germination, which might be advantageous for seedling establishment. Another possibility might be that the 12S cruciferins being the most abundant proteins in the *Arabidopsis* seeds act as scavengers of ROS generated during seed development and germination, thereby counteracting their deleterious effects. A key question is whether the carbonylation events occurring during germination play a role in seedling establishment or if they are merely the consequence of cells undergoing intense metabolism activity in the awakening from quiescence. In view of the specificity of protein oxidation observed and the fact that the patterns of oxidized proteins closely resemble those of bacterial or yeast cells exposed to oxidative stress, it appears more likely that protein carbonylation does not simply reflect secondary epiphenomena but is used as a means to adapt embryo metabolism to the oxidative conditions encountered during germination. The biochemical basis of such specific carbonylation of proteins remains to be elucidated.

7.2.1.8 Proteomics of developmental mutants in seeds

Comparison of the 2-DE pattern of a mutant with that of the wild-type has often been used to evaluate the effects of a mutation, or to characterize the protein(s) encoded or influenced by the mutated gene. Examples of such proteomic analyses can be found in the reviews by Thiellement et al. (1999, 2002) and Cánovas et al. (2004). Here, we discuss some proteomic studies carried out with seed developmental mutants.

The one concerns *Arabidopsis* mutants deficient in (*aba*) or insensitive to (*abi3*) ABA, and the double mutant (*aba,abi3*) (Meurs et al., 1992). 2-D protein patterns demonstrated that in seeds of the double mutant from 14 to 20 DAP, there was only very low amount of various maturation-specific proteins whereas many proteins similar to those occurring during germination were induced, although no germination was apparent. It appears that in the *aba,abi3* double mutant seed development is not completed and the program for seed germination is initiated prematurely. 2-D protein patterns also demonstrated that exogenous ABA can complete the program for seed maturation. Thus, by providing clear and contrasted protein profiling signatures, this study paved the way for further characterization of important mechanisms occurring in seed development as development of desiccation tolerance or the shift from maturation to germination developmental programs, and that are under ABA control. It will be extremely interesting to revisit these mutants with the modern tools of proteomics toward the characterization of the protein targets of ABA during seed development.

In another study, the role of gibberellins (GAs) in germination of *Arabidopsis* seeds using a GA-deficient *ga1* mutant, and wild-type seeds treated with paclobutrazol, a specific GA biosynthesis inhibitor was investigated (Gallardo et al., 2002a). With both systems, radicle protrusion was strictly dependent on exogenous GAs, as previously reported (Koornneef and van der Veen, 1980). The proteome analysis revealed that GAs do not participate in many processes involved in germination *sensu stricto*, that is the initial mobilization of seed protein and lipid reserves. Changes in 46 proteins were detected at this stage in the wild-type and mutant seeds. However, only one protein (α-2,4 tubulin) was suggested to depend on the action of GA, as it was not detected in the *ga1* mutant seeds and accumulated in the mutant seeds incubated in the presence of GAs. In contrast, it was suggested that GAs might be involved, directly or indirectly, in controlling the abundance of several proteins (two isoforms of AdoMet synthetase and β-glucosidase) associated with radicle protrusion and post-germination processes. In conclusion, this proteomic study established for the first time the developmental stage at which GAs exert their action during germination, namely the radicle protrusion step, and unraveled several protein targets that can account for its action at this step.

Recently, the role of specific proteases, the seed-type members of the vacuolar processing enzyme (VPE), in seed protein processing during seed filling and maturation has been investigated by proteomics, using knockout mutant alleles of all four members (α*VPE*, β*VPE*, γ*VPE*, and δ*VPE*) of the VPE gene family in *Arabidopsis* (Gruis et al., 2004). The complete removal of VPE function in the quadruple mutant resulted in a total shift of storage protein accumulation from wild-type processed polypeptides to a finite number of prominent alternatively processed polypeptides cleaved at sites other than the conserved Asn residues targeted by VPE. Although, alternatively proteolyzed legumin-type globulin polypeptides largely accumulated as intrasubunit disulfide-linked polypeptides with apparent molecular masses similar to those of VPE-processed legumin polypeptides, they showed markedly altered solubility and protein assembly characteristics. Instead of forming 11S hexamers (see Figure 7.1), alternatively processed legumin polypeptides were deposited primarily as 9S complexes. However, despite the impact on seed protein processing, plants devoid of all known functional VPE genes appeared unchanged with regard to protein content in mature seeds, relative mobilization rates of protein reserves during germination, and vegetative growth. These findings indicated that VPE-mediated Asn-specific proteolytic processing, and the physiochemical property changes attributed to this specific processing step, are not required for the successful deposition and mobilization of seed storage protein in the protein storage vacuoles of *Arabidopsis* seeds. These results are also in good agreement with the proteomic data of Gallardo et al. (2001) showing that residual precursor forms of 12S cruciferins were mobilized at a high rate during wild-type *Arabidopsis* seed germination (see Section 7.2.1.3).

7.2.1.9 Proteomics of genetic diversity for seed composition

Since the pioneer work of Zivy et al. (1983, 1984), 2-DE and proteomics have been widely used to characterize varieties in several plants (Thiellement et al., 1999, 2002; Chevalier et al., 2004). This is because genetic variations can be evidenced by the position of proteins in 2-D gels, as influenced by sequence polymorphism, and/or by the variation in amount of a given protein in different genotypes (genetically determined quantitative variations). This powerful approach linking plant proteomics and plant genetics can lead to the characterization of PQLs (protein quantity loci) that explain part of the variability in spot intensity (Damerval et al., 1994; de Vienne et al., 1999). Several examples of the use of such an approach can be found in the reviews by Thiellement et al. (1999, 2002). Below we describe a recent proteomic study in this area dealing with genetic diversity for seed protein content and composition in the model legume plant *M. truncatula*. Here, 50 lines of *M. truncatula*, derived from ecotypes or cultivars of diverse geographical origin, were grown under uniform conditions, and variation in seed protein composition and quantity was investigated (Le Signor et al., 2005). Electrophoretic analyses revealed 46 major seed polypeptides, of which 26 were polymorphic within the collection. The polymorphism for the major seed protein classes (the vicilin/convicilin (7S) and the legumin (11S) type proteins) allowed the clustering of the genotypes into four groups (Figure 7.5).

All lines not belonging to either *M. truncatula* ssp. *truncatula* or ssp. *longispina* were clustered in a single group, demonstrating the value of seed protein profiles in delimiting species boundaries. The Jemalong line group was differentiated early in the dendrogram, and thus represents an ancient clade in the seed diversity of *M. truncatula.* Within-accession variation was investigated for 1-D seed profiles, with additional lines obtained from the same ecotypes. As expected for an autogamous species, within-accession variation was low. Seed protein content was highly variable among the 50 lines examined. Lines contrasting for qualitative traits and seed protein content were identified to allow for the genetic determination of these characters, which will be relevant for legume crop improvement (Le Signor et al., 2005).

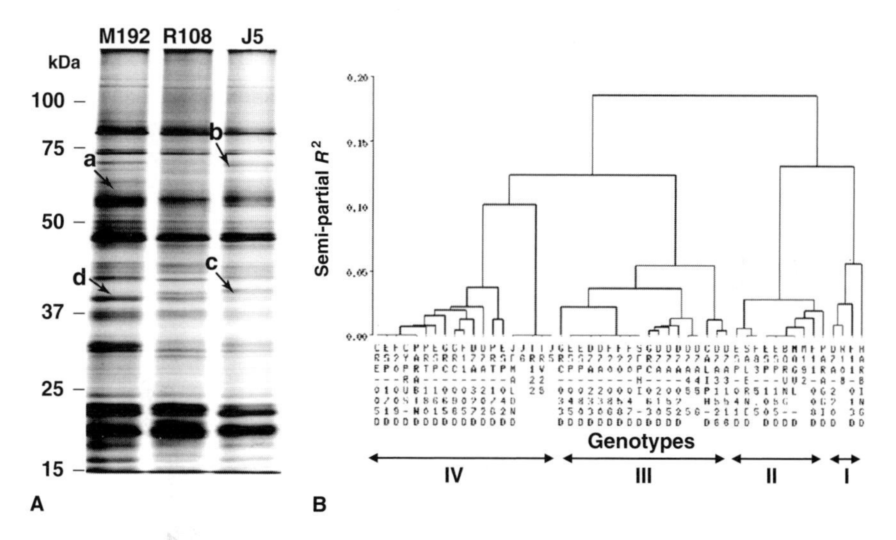

Figure 7.5 Genetic variability for seed protein composition in *M. truncatula* (from Le Signor et al., 2005). **A:** Examples of globulin polymorphism detected in 1-D protein profiles (silver stained) of dry mature seeds from three *M. truncatula* lines (M192, R108, Jemalong J5). Vicilin-like (7S) and legumin-like (11S) storage proteins were detected after Western-blotting using antibodies raised against pea vicilins and pea legumins, respectively. Arrows indicate four polymorphic bands identified by immunoblotting as vicilin-like (a, b) and legumin-like (c, d) proteins. **B:** Dendrogram of the 1-D profiles of 50 *M. truncatula* lines of identical geographical origin (Ward's minimum variance method). Four groups with distinct protein patterns were identified. The lines M192, R108 and J5, shown in **(A)**, fall into the groups II, I, and IV, respectively.

7.2.1.10 A model of seed development and germination based upon proteomics data

The proteomic data described above can be used to elaborate a model of seed development and germination (Figure 7.6). A first salient feature emerging from these studies is the importance of metabolic control to maintain quiescence in the dry state and conversely to allow awakening from that quiescence during germination. As we have discussed here, this finding seems now well established by studies with different plants (*Arabidopsis*, *M. truncatula*, soybean, and wheat) and with different systems (e.g. somatic and zygotic embryos, whole seeds, organelles). In particular, a well-documented metabolic block concerns the absence of important metabolic enzymes in mature quiescent seeds, such as enzymes involved in Met metabolism. Owing to the central role of this sulfur amino acid in metabolism, a control exerted at the level of an enzyme such as AdoMet synthetase will have an impact on a myriad of processes, not only at the level of metabolism (protein synthesis, methylations) but also in regulation of development mediated by ethylene and polyamines. Contrarily to expectation, the mobilization of seed storage protein reserves cannot support germination and *de novo* Met biosynthesis is required for germination to occur. It would be interesting to investigate whether this requirement is specific to Met or holds more generally for other amino acids.

Another central enzyme involved in metabolic control of quiescence appears to be the mitochondrial processing peptidase. This processing peptidase is a fascinating enzyme that catalyzes the specific cleavage of the diverse pre-sequence peptides from hundreds of the nuclear-encoded mitochondrial precursor proteins that are synthesized in the cytosol and imported into the mitochondrion. Knowing the central

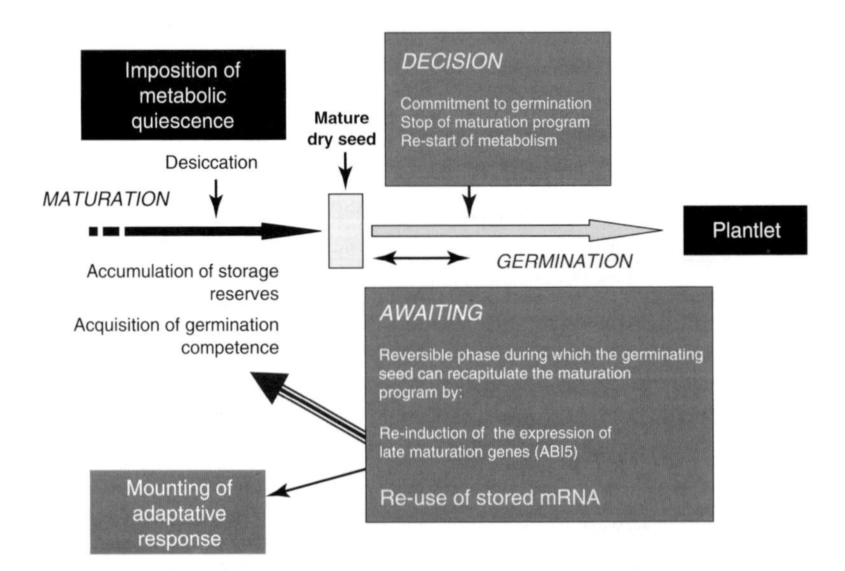

Figure 7.6 A working model of seed development and germination.

role of mitochondria in seed germination and seedling establishment (Bewley and Black, 1994), the possibility of controlling this important enzyme during germination would provide a simple and very efficient means to functionality-control these essential organelles.

A second salient feature put forward by the proteomic data was the general importance of protein modification in seed development and germination. It was somewhat amazing to observe the very large number of proteins that proved to be the specific targets in seeds of thioredoxins and of oxidation through carbonylation. Unexpectedly, the proteomic studies unraveled a new role for protein carbonylation in seed development and germination that has no counterpart in microorganisms and animals. Last, the proteomic data revealed a link between ROS leading to protein carbonylation and redox regulatory events catalyzed by thioredoxin in seeds. The results suggest that a tuning of such protein modifications might exert a dramatic control upon seed development and germination, which will be the object of future research.

Finally, the proteomic data documented the distinct role of stored and nascent mRNA pools in germination and seedling growth. Along with other molecular studies (Lopez-Molina et al., 2002; Li et al., 2005; Nakabayashi et al., 2005), the proteomic studies provided the foundation of a new paradigm in seed germination, in that imbibition of the mature dry seed does not irreversibly commit the seed to the germination program, but rather there exists a possibility to reinduce the maturation program in early phase of seed germination. Presumably, the *PKL* gene, which encodes a CHD3-chromatin-remodeling factor preventing expression of the embryonic development state during *Arabidopsis* germination (Ogas et al., 1999), is involved in such control, suggesting the importance of epigenetic regulation is seed developmental programs. In this context, it is worth noting that a significant portion of the *Arabidopsis* genome shows a co-regulated expression pattern (Ma et al., 2005; Nakabayashi et al., 2005).

As discussed above, and as stressed by Lane (1991), this peculiar 'reversible' behavior observed during early germination may serve as a barometric control that buffers a germinating plant embryo against instant loss of desiccation tolerance. For example, in response to the vagaries of nature's water supply, anhydrobiosis proteins may be reversely lost and restored. Plant cell enlargement through water uptake is of paramount importance in plants because it is an analog of mobility in animals in that it allows higher plants to move through the air or the soil, thereby providing the plant's photosynthetic machinery with access to light (Lane, 1991).

7.2.2 Proteomics of other developmental processes in plants

As mentioned in Section 7.1, most developmental studies in plants based on the proteomic approach concern the establishment of lists of specific proteins in organs and organelles (Thiellement et al., 2002; Hirano et al., 2004; Newton et al., 2004; Agrawal et al., 2005b). By the 'Functional proteomics' approach, seed and embryo are the tissues for which the most extensive proteomic studies have been carried out among all the tissues examined so far in dicots (Agrawal et al., 2005b). There are, however, some studies using this approach in other plant tissues. They are described briefly below.

A flower proteome for all stages of the flower, from buds until petal browning has been established in *M. truncatula* (Watson et al., 2003). Forty-three proteins were identified from a total of 94 protein spots (46% success rate), and grouped into 12 functional categories: energy (38%), protein destination/storage (16%), disease/defense (9%), metabolism (7%), signal transduction (7%), protein synthesis (5%), cell growth/division (5%), transcription (5%), cell structure (2%), secondary metabolism (2%), transporters (2%), and unclear (2%).

A proteomic approach was also used to study petal development in rose (*Rosa hybrida*) (Dafny-Yelin et al., 2005). Using 2-DE, stage-specific (closed bud, mature flower, and flower at anthesis) petal protein maps with ca. 1000 unique protein spots were generated. Expression analyses of all resolved protein spots revealed that almost 30% of them were stage-specific, with ca. 90 protein spots for each stage. Most of the proteins exhibited differential expression during petal development, whereas only ca. 6% were constitutively expressed. Eighty-two of the resolved proteins were identified by MS and annotated. Classification of the annotated proteins into functional groups revealed energy, cell rescue, unknown function (including novel sequences), and metabolism to be the largest classes, together comprising ca. 90% of all identified proteins. Interestingly, a large number of stress-related proteins were identified in developing petals.

A proteomics study of senescence, a final stage of development of leaves, has been carried out in clover (*Trifolium repens*). This study allowed the quantitative analysis of 590 spots separated by 2-DE (Wilson et al., 2002). Of these, about 40% (178 spots) were found to be senescence-related, most of them being localized to chloroplast. More recently, Schiltz et al. (2004) analyzed quantitative variations in leaf proteins during nitrogen remobilization related to seed filling in pea (*Pisum sativum*) and identified 130 spots implicated in a variety of cellular functions. Altogether, the data emphasized the importance of proteolysis and chloroplast degradation, and remobilization of nitrogen reserves in senescing leaves of white clover and pea.

Two recent studies concerned proteomic investigations of leaves under drought stress, a major limitation to agricultural production worldwide. The one concerned sugar beet leaves (Hajheidari et al., 2005). Out of more than 500 protein spots separated by 2-DE, 79 showed significant changes (up- and down-regulation) under drought. Twenty protein spots were analyzed by LC-MS/MS, leading to the identification of RuBisCo and 11 other proteins involved in redox regulation, oxidative stress, signal transduction, and chaperone activities. Interestingly, a protein was identified as putative nascent polypeptide-associated complex α-chain (NAC) whose level showed a significant decrease in response to drought. NAC is associated with ribosomes and binds to emergent polypeptides; it prevents inappropriate interactions of nascent polypeptide chains with other factors (Wiedmann et al., 1994). A decrease in α-NAC levels may reflect a general repression of translation under drought stress and/or result in the mistranslation, mistargeting, and proteolysis of proteins, decreasing overall plant performance. It is worth noting that a recent proteomic study demonstrated that the NAC was down-regulated in rice roots submitted to a salt stress in the presence of 150 mM NaCl (Yan et al., 2005). It is proposed that some of these proteins could constitute a physiological advantage under drought, making them potential targets in marker-assisted selection programs.

The other dealt with a proteome analysis of water deficit in maize leaves (Vincent et al., 2005). Remarkably, this study provided 2-D protein patterns from leaf segments evenly sampled along the maize leaf at different stages of development (2 cm-long fragments of the sixth leaf). This allowed separating direct effects of water deficit from side effects related to the decrease of growth. The proteomic data lend to draw a hypothesis on the localization of lignin synthesis in leaves and on the effect of drought on lignification. This study, by pointing out the question of heterogeneity in tissue samples, should be taken as an example for future proteomic research in plant development and encourage the use of well-defined plant samples.

A proteome analysis of plant programmed cell death (PCD) has been carried out in *Arabidopsis* cell culture (Swidzinski et al., 2004). This study allowed identifying 11 proteins potentially associated with PCD, including antioxidant enzymes (catalase, superoxide dismutase), indicating that oxidative stress is associated with PCD, and several mitochondrial proteins, which confirmed the importance of this organelle during PCD. These mitochondrial proteins may be involved in redox signaling that triggers PCD or in the release of mitochondrial pro-apoptotic proteins into the cytosol.

A proteome analysis was used to investigate the changing patterns of protein synthesis during pollen development in anthers from rice plants grown under strictly controlled growth conditions (Kerim et al., 2003). More than 150 protein spots separated by 2-DE showed changing accumulation levels during development (six discrete microspore developmental stages were studied, ranging from the pollen mother cell to mature pollen), out of which 40 representing 33 unique gene products were identified. These proteins included regulators of light-mediated signal transduction in plants, enzymes of carbohydrate metabolism, cell wall, and cytoskeleton-associated proteins. Furthermore, new Open Reading Frames (ORFs) were defined.

In summary, these studies demonstrated the utility of the proteomic approach in addressing a biological system in which there is little prior knowledge to form the basis of more hypothesis-driven studies.

7.3 Conclusions and perspectives

In conclusion, we have described the usefulness of proteomics to investigate the molecular mechanisms taking place during plant development with an emphasis on two extremely complex processes of the plant life cycle that are seed formation and germination. In this review we have outlined the combination of quantitative proteomics with the comprehensive analysis of the *de novo* synthesized, carbonylated, and reduced (thioredoxin targets) proteins, which leads to additional information about rates of synthesis and degradation, modifications, and protein interactions that might exert a dramatic influence upon seed development and germination. It should be emphasized that such characterization of the functional status of hundreds of proteins at specific stages of seed development and germination, as well as during other aspects of plant growth, represent an important part of attempts to uncover the dynamic complexity of these biological systems and ultimately create working models of entire 'systems biology' networks in plants. It should also be underlined that in this emerging field of 'systems biology', the integration of transcriptomics and metabolomics with proteomics data will lead to substantial information about the regulation of protein accumulation and activity, and contribute to an essential part of understanding biological functions. The comparative analysis of transcript, protein and metabolite profiling during different aspects of plant growth, development and physiological processes, such as responses to biotic or abiotic stresses, seems therefore to be of great importance.

An additional important focus in plant proteomics is directed toward the identification of candidate proteins governing complex agronomic traits, such as yield, hybrid vigor, disease resistance, and stress tolerance. In this context, the combination of quantitative trait loci (QTL) analyses for these traits and the identification of loci affecting the quantity (PQL), the position shift (PSL) or the functional state of the proteins detected in 2-DE, should help elucidating the link between the protein spots and the traits (Thiellement et al., 2002). For example, co-location of a PQL with the structural gene encoding the protein detected in 2-D gels and a QTL controlling an important agronomic trait would point to an association between the protein spot itself and the variation observed for the trait. Furthermore, co-location of PQLs with agronomic trait QTLs but not with the corresponding structural genes would result in identifying those proteins in 2-D gels whose genetic factors control their quantity. The use of plants that have their genome sequenced should help identifying those regulatory genes in the QTL/PQL regions. To further validate the role of candidate proteins, methodologies of reverse genetics are very useful. In particular, a method for Targeting-Induced Local Lesions In Genomes, or TILLING, overcomes the limitations of knocking out an entire essential gene and expands knowledge of active protein domains. This method uses chemical mutagenesis to yield an allelic series of point mutations in the genome. High-throughput TILLING facilities have been established for a number of species, such as *Arabidopsis*, *M. truncatula*, rice, pea, and maize, to provide a screening resource for isolating different mutant alleles corresponding to genes of interest with known sequences (Henikoff et al., 2004). Recently, TILLING has been extended to the improvement of crop plants since favorable alleles detected in mutant collections of crop plants can be used directly in selection (Slade and Knauf, 2005). Therefore, proteomics combined with genetics and TILLING shows great promise as a general method for functional study and validation of key proteins related to diverse aspects of plant growth, development, and physiological processes.

References

Adachi, M., Kanamori, J., Masuda, T., Yagasaki, K., Kitamura, K., Mikami, B. and Utsumi, S. (2003) Crystal structure of soybean 11S globulin: glycinin A3B4 homohexamer. *Proc. Natl Acad. Sci. USA*, **100**, 7395–7400.

Adachi, M., Takenaka, Y., Gidamis, A.B., Mikami, B. and Utsumi, S. (2001) Crystal structure of soybean proglycinin A1aB1b homotrimer. *J. Mol. Biol.*, **305**, 291–305.

Agrawal, G.K., Yonekura, M., Iwahashi, Y., Iwahashi, H. and Rakwal, R. (2005a) Systems, trends and perspectives of proteomics in dicot plants. Part I: technologies in proteome establishment. *J. Chromatogr.* B, **815**, 109–123.

Agrawal, G.K., Yonekura, M., Iwahashi, Y., Iwahashi, H. and Rakwal, R. (2005b) Systems, trends and perspectives of proteomics in dicot plants. Part II: proteomes of the complex developmental stages. *J. Chromatogr. B*, **815**, 125–136.

Agrawal, G.K., Yonekura, M., Iwahashi, Y., Iwahashi, H. and Rakwal, R. (2005c) Systems, trends and perspectives of proteomics in dicot plants. Part III: unraveling the proteomes influenced by the environment, and at the levels of functions and genetic relationships. *J. Chromatogr. B*, **815**, 137–145.

Alban, C., Job, D. and Douce, R. (2000) Biotin metabolism in plants. *Annu. Rev. Plant Physiol. Plant Mol. Biol.*, **51**, 17–47.

Arabidopsis Genome Initiative (AGI) (2000) Analysis of the genome sequence of the flowering plant *Arabidopsis thaliana*. *Nature*, **408**, 796–815.

Baginsky, S. and Gruissem, W. (2004) Chloroplast proteomics: potentials and challenges. *J. Exp. Bot.*, **55**, 1213–1220.

Bailly, C. (2004) Active oxygen species and antioxidants in seed biology. *Seed Sci. Res.*, **14**, 93–107.

Bardel, J., Louwagie, M., Jacquinot, M., Jourdain, A., Luche, S., Rabilloud, T., Macherel, D., Garin, J. and Bourguignon, J. (2002) A survey of the plant mitochondrial proteome in relation to development. *Proteomics*, **2**, 880–898.

Berlett, B.S. and Stadtman, E.R. (1997) Protein oxidation in aging, disease and oxidative stress. *J. Biol. Chem.*, **272**, 20313–20316.

Bestel-Corre, G., Dumas-Gaudot, E. and Gianinazzi, S. (2004) Proteomics as a tool to monitor plant-microbe endosymbioses in the rhizosphere. *Mycorrhiza*, **14**, 1–10.

Bewley, J.D. and Black, M. (1994) *Seeds. Physiology of Development and Germination*. Plenum Press, New York.

Buchanan, B.B. and Balmer, Y. (2005) Redox regulation: a broadening horizon. *Annu. Rev. Plant Biol.*, **56**, 187–220.

Cabiscol, E., Piulats, E., Echave, P., Herrero, E. and Ros, J. (2000) Oxidative stress promotes specific protein damage in *Saccharomyces cerevisiae*. *J. Biol. Chem.*, **275**, 27393–27398.

Campo, S., Carrascal, M., Coca, M., Abián, J. and San Segudo, B. (2004) The defense response of germinating maize embryos against fungal infection: a proteomics approach. *Proteomics*, **4**, 383–396.

Cánovas, F.M., Dumas-Gaudot, E., Recorbet, G., Jorrin, J., Mock, H.-P. and Rossignol, M. (2004) Plant proteome analysis. *Proteomics*, **4**, 285–298.

Chang, W.W.P., Huang, L., Webster, C., Burlingame, A.L. and Roberts, J.M. (2000) Patterns of protein synthesis and tolerance to anoxia in root tips of maize seedling acclimated to a low-oxygen environment, and identification of proteins by mass spectrometry. *Plant Physiol.*, **122**, 295–317.

Chevalier, F., Martin, O., Rofidal, V., Devauchelle, A.-D., Barteau, S., Sommerer, N. and Rossignol, M. (2004) Proteomic investigation of natural variation between *Arabidopsis* ecotypes. *Proteomics*, **4**, 1372–1381.

Colditz, F., Nyamsuren, O., Niehaus, K., Eubel, H., Braun, H.-P. and Krajinski, F. (2004) Proteomic approach: identification of *Medicago truncatula* proteins induced in roots after infection with the pathogenic oomycete *Aphanomyces euteiches*. *Plant Mol. Biol.*, **55**, 109–120.

Côme, D. and Corbineau, F. (1989) Some aspects of metabolic regulation of seed germination and dormancy. In: Taylorson, R.B. (ed) *Recent Advances in the Development and Germination of Seeds*. Plenum Press, New York, pp. 165–179.

Cuming, A.C. and Lane, B.G. (1979) Protein synthesis in imbibing wheat embryos. *Eur. J. Biochem.*, **99**, 217–224.

Dafny-Yelin, M., Guterman, I., Menda, N., Ovadis, M., Shalit, M., Pichersky, E., Zamir, D., Lewinsohn, E., Adam, Z., Weiss, D. and Vainstein, A. (2005) Flower proteome: changes in protein spectrum during the advanced stages of rose petal development. *Planta*, **222**, 37–46.

Damerval, C., Maurice, A., Josse, J.M. and de Vienne, D. (1994) Quantitative trait loci underlying gene product variation-a novel perspective for analyzing regulation of genome expression. *Genetics*, **137**, 289–301.

de Mercoyrol, L., Job, C. and Job, D. (1989) Studies on the inhibition by alpha-amanitin of single-step addition reactions and productive RNA synthesis catalysed by wheat germ RNA polymerase II. *Biochem. J.*, **258**, 165–169.

de Vienne, D., Leonardi, A., Damerval, C. and Zivy, M. (1999) Genetics of proteome variation for QTL characterization: application to drought-stress responses in maize. *J. Exp. Bot.*, **50**, 303–309.

Debeaujon, I., Léon-Kloosterziel, K.M. and Koornneef, M. (2000) Influence of the testa on seed dormancy, germination, and longevity in *Arabidopsis*. *Plant Physiol.*, **122**, 403–413.

Dure III, L., Crouch, M., Harada, J., Ho, T.-H.D., Mundy, J., Quatrano, R., Thomas, T. and Sung, Z.R. (1989) Common amino acid sequence domains among the LEA proteins of higher plants. *Plant Mol. Biol.*, **12**, 475–486.

Dure III, L., Greeway, S.C. and Galau, G.A. (1981) Developmental biochemistry of cottonseed embryogenesis and germination: changing messenger ribonucleic acid populations as shown by *in vitro* and *in vivo* protein synthesis. *Biochemistry*, **20**, 4162–4168.

Dure III, L.S. and Waters, L.C. (1965) Long-lived messenger RNA: evidence from cotton seed germination. *Science*, **147**, 410–412.

Finnie, C., Maeda, K., Østergaard, O., Bak-Jensen, K.S., Larsen, J. and Svensson, B. (2004a) Aspects of the barley seed proteome during development and germination. *Biochem. Soc. T.*, **32**, 517–519.

Finnie, C., Melchior, S., Roepstorff, P. and Svensson, B. (2002) Proteome analysis of grain filling and seed maturation in barley. *Plant Physiol.*, **129**, 1–12.

Finnie, C., Steenholdt, T., Noguera, O.R., Knudsen, S., Larsen, J., Brinch-Pedersen, H., Holm, P.B., Olsen, O. and Svensson, B. (2004b) Environmental and transgene expression effects on the barley seed proteome. *Phytochemistry*, **65**, 1619–1627.

Gallardo, K., Job, C., Groot, S.P.C., Puype, M., Demol, H., Vandekerckhove, J. and Job, D. (2001) Proteomic analysis of *Arabidopsis* seed germination and priming. *Plant Physiol.*, **126**, 835–848.

Gallardo, K., Job, C., Groot, S.P.C., Puype, M., Demol, H., Vandekerckhove, J. and Job, D. (2002a) Proteomics analysis of *Arabidopsis* seed germination. A comparative study of wild-type and GA-deficient seeds. *Plant Physiol.*, **129**, 823–837.

Gallardo, K., Job, C., Groot, S.P.C., Puype, M., Demol, H., Vandekerckhove, J. and Job, D. (2002b) Importance of methionine biosynthesis for *Arabidopsis* seed germination and seedling growth. *Plant Physiol.*, **116**, 238–247.

Gallardo, K., Le Signor, C., Vandekerckhove, J., Thompson, R.D. and Burstin, J. (2003) Proteomics of *Medicago truncatula* seed development establishes the time frame of diverse metabolic processes related to reserve accumulation. *Plant Physiol.*, **133**, 1–19.

Ge, H., Walhout, A.J.M. and Vidal, M. (2003) Integrating 'omic' information: a bridge between genomics and systems biology. *Trend. Genet.*, **19**, 551–560.

Giavalisco, P., Nordhoff, E., Kreitler, T., Klöppel, K-D., Lehrach, H., Klose, J. and Gobom, J. (2005) Proteome analysis of *Arabidopsis thaliana* by two-dimensional gel electrophoresis and matrix-assisted laser desorption/ionization-time of flight mass spectrometry. *Proteomics*, **5**, 1902–1913.

Goff, S.A., Ricke, D., Lan, T.H., Presting, G., Wang, R., Dunn, M., Glazebrook, J., Sessions, A., Oeller, P., Varma, H., Hadley, D., Hutchison, D., Martin, C., Katagiri, F., Lange, B.M., Moughamer, T., Xia, Y., Budworth, P., Zhong, J., Miguel, T., Paszkowski, U., Zhang, S., Colbert, M., Sun, W.L., Chen, L., Cooper, B., Park, S., Wood, T.C., Mao, L., Quail, P., Wing, R., Dean, R., Yu, Y., Zharkikh, A., Shen, R., Sahasrabudhe, S., Thomas, A., Cannings, R., Gutin, A., Pruss, D., Reid, J., Tavtigian, S., Mitchell, J.,

Eldredge, G., Scholl, T., Miller, R.M., Bhatnagar, S., Adey, N., Rubano, T., Tusneem, N., Robinson, R., Feldhaus, J., Macalma, T., Oliphant, A. and Briggs, S. (2002) A draft sequence of the rice genome (*Oryza sativa* L. ssp. *japonica*). *Science*, **296**, 92–100.

Gruis, D., Schulze, J. and Jung, R. (2004) Storage protein accumulation in the absence of the vacuolar processing enzyme family of cysteine proteases. *Plant Cell*, **16**, 270–290.

Gutiérrez, R.A., Green, P.J., Keegstra, K. and Ohlrogge, J.B. (2004) Phylogenetic profiling of the *Arabidopsis thaliana* proteome: what proteins distinguish plants from other organisms? *Genome Biol.*, **5**, R53.

Gutiérrez, R.A., Shasha, D.E. and Coruzzi, G.M. (2005) Systems biology for the virtual plant. *Plant Physiol.*, **138**, 550–554.

Hajduch, M., Ganapathy, A., Stein, J.W. and Thelen, J.J. (2005) A systematic proteomic study of seed filling in soybean. Establishment of high-resolution two-dimensional reference maps, expression profiles, and an interactive proteome database. *Plant Physiol.*, **137**, 1397–1419.

Hajheidari, M., Abdollahian-Noghabi, M., Askari, H., Heidari, M., Sadeghian, S.Y., Ober, E.S. and Salekdeh, G.H. (2005) Proteome analysis of sugar beet leaves under drought stress. *Proteomics*, **5**, 950–960.

Heazlewood, J.L., Tonti-Filippini, J.S., Gout, A.M., Day, D.A., Whelan, J. and Millar, A.H. (2004) Experimental analysis of the *Arabidopsis* mitochondrial proteome highlights signaling and regulatory components, provides assessment of targeting prediction programs, and indicates plant-specific mitochondrial proteins. *Plant Cell*, **16**, 241–256.

Henikoff, S., Till, B.J. and Comai, L. (2004) TILLING. Traditional mutagenesis meets functional genomics. *Plant Physiol.*, **135**, 630–636.

Hirano, H., Islam, N. and Kawasaki, H. (2004) Technical aspects of functional proteomics in plants. *Phytochemistry*, **65**, 1487–1498.

Imin, N., Nizamidin, M., Daniher, D., Nolan, K.E., Rose, R.J. and Rolfe, B.G. (2005) Proteomic analysis of somatic embryogenesis in *Medicago truncatula*. Explant cultures grown under 6-benzylaminopurine and 1-naphtaleneacetic acid treatments. *Plant Physiol.*, **137**, 1250–1260.

Job, C., Rajjou, L., Lovigny, Y., Belghazi, M. and Job, D. (2005) Patterns of protein oxidation in *Arabidopsis* seeds and during germination. *Plant Physiol.*, **138**, 790–802.

Johansson, E., Olsson, O. and Nyström, T. (2004) Progression and specificity of protein oxidation in the life cycle of *Arabidopsis thaliana*. *J. Biol. Chem.*, **279**, 22204–22208.

Kerim, T., Imin, N., Weinmann, J.J. and Rolfe, B.G. (2003) Proteome analysis of male gametophyte development in rice anthers. *Proteomics*, **3**, 738–751.

Kobrehel, K., Wong, J.H., Balogh, A., Kiss, F., Yee, B.C. and Buchanan, B.B. (1992) Specific reduction of wheat storage proteins by thioredoxin. *Plant Physiol.*, **99**, 919–924.

Koller, A., Washburn, M.P., Lange, B.M., Andon, N.L., Decieu, C., Haynes, P.A., Hays, L., Schieltz, A., Ulasek, R., Wei, J., Wolters, D. and Yates, J.R. (2002) Proteomic survey of metabolic pathways in rice. *Proc. Natl Acad. Sci. USA*, **99**, 11969–11974.

Komatsu, S., Kojima, K., Suzuki, K., Ozaki, K. and Higo, K. (2004) Rice proteome database based on two-dimensional polyacrylamide gel electrophoresis: its status in 2003. *Nucleic Acid. Res.*, **32**, D388–D392.

Komatsu, S., Konishi, H., Shen, S. and Yang, G. (2003) Rice proteomics. A step toward functional analysis of the rice genome. *Mol. Cell. Proteom.*, **2**, 2–10.

Koornneef, M. and van der Veen, J.H. (1980) Induction and analysis of gibberellin sensitive mutants in *Arabidopsis thaliana* (L.) Heynh. *Theor. Appl. Genet.*, **58**, 257–263.

Lane, B.G. (1991) Cellular desiccation and hydration: developmentally regulated proteins, and the maturation and germination of seed embryos. *FASEB J.*, **5**, 2893–2901.

Laugesen, S., Bergoin, A. and Rossignol, M. (2004) Deciphering the plant phosphoproteome: tools and strategies for a challenging task. *Plant Physiol. Biochem.*, **42**, 929–936.

Le Signor, C., Gallardo, K., Prosperi, J.M., Salon, C., Quillien, L., Thompson, R. and Duc, G. (2005) Genetic diversity for seed protein composition in *Medicago truncatula*. *Plant Genet. Resour.*, **3**, 59–71.

Lei, Z., Elmer, A.M., Watson, B.S., Dixon, R.A., Mendes, P.J. and Summer, L.W. (2005) A 2-DE proteomics reference map and systematic identification of 1367 proteins from a cell suspension culture of the model legume *Medicago truncatula*. *Mol. Cell. Proteomics*, **4**, 1812–1825.

Levine, R.L., Williams, J.A., Stadtman, E.R. and Shacter, E. (1994) Carbonyl assays for determination of oxidatively modified proteins. *Method. Enzymol.*, **233**, 346–357.

Li, F., Wu, X., Tsang, E. and Cutler, A. (2005) Transcriptional profiling of imbibed *Brassica napus* seed. *Genomics*, **86**, 718–730.

Li, W.W., Quinn, G.B., Alexandrov, N.N., Bourne, P.E. and Shindyalov, I.N. (2003) A comparative proteomics resource: proteins of *Arabidopsis thaliana*. *Genome Biol.*, **4**, R51.

Lin, S.-K., Chang, M.-C., Tsai, Y.-G. and Lur, H.-S. (2005) Proteomic analysis of the expression of proteins related to rice quality during caryopsis development and the effect of temperature on expression. *Proteomics*, **5**, 2140–2156.

Lippert, D., Zhuang, S., Ralph, S., Ellis, D.E., Gilbert, M., Olafson, R., Ritland, K., Ellis, B., Douglas, C.J. and Bohlmann, J. (2005) Proteome analysis of early somatic embryogenesis in *Picea glauca*. *Proteomics*, **5**, 461–473.

Lopez-Molina, L., Mongrand, S., McLachlin, D.T., Chait, B.T. and Chua, N.-H. (2002) ABI5 acts downstream of ABI3 to execute an ABA-dependent growth arrest during germination. *Plant J.*, **32**, 317–328.

Lozano, R.M., Wong, J.H., Yee, B.C., Peters, A., Kobrehel, K. and Buchanan, B.B. (1996) New evidence for a role for thioredoxin h in germination and seedling development. *Planta*, **200**, 100–106.

Ma, L., Sun, N., Liu, X., Jiao, Y., Zhao, H. and Deng, X.W. (2005) Organ-specific expression of *Arabidopsis* genome during development. *Plant Physiol.*, **138**, 80–91.

Maeda, K., Finnie, C. and Svensson, B. (2004) Cy5 maleimide labeling for sensitive detection of free thiols in native protein extracts: identification of seed proteins targeted by barley thioredoxin *h* isoforms. *Biochem. J.*, **378**, 497–507.

Maltman, D.J., Simon, W.J., Wheeler, C.H., Dunn, M.J., Wait, R. and Slabas, A.R. (2002) Proteomic analysis of the endoplasmic reticulum from developing and germinating seed of castor (*Ricinus communis*). *Electrophoresis*, **23**, 626–639.

Marx, J.H. and Buchanan, B.B. (2003) Thioredoxin and germinating barley: target and protein redox changes. *Planta*, **216**, 454–460.

Meurs, C., Basra, A.S., Karssen, C.M. and van Loon, L.C. (1992) Role of abscisic acid in the induction of desiccation tolerance in developing seeds of *Arabidopsis thaliana*. *Plant Physiol.*, **98**, 1484–1493.

Motohashi, K., Kondoh, A., Stumpp, M.T. and Hisabori, T. (2001) Comprehensive survey of protein targeted by chloroplast thioredoxin. *Proc. Natl Acad. Sci. USA*, **98**, 11224–11229.

Nakabayashi, K., Okamoto, M., Koshiba, T., Kamiya, Y. and Nambara, E. (2005) Genome-wide profiling of stored mRNA in *Arabidopsis thaliana* seed germination: epigenetic and genetic regulation of transcription in seed. *Plant J.*, **41**, 697–709.

Newton, R.P., Brenton, A.G., Smith, C.J. and Dudley, E. (2004) Plant proteome analysis by mass spectrometry: principles, problems, pitfalls and recent developments. *Phytochemistry*, **65**, 1449–1485.

Nyström, T. (2005) Role of oxidative carbonylation in protein quality control and senescence. *EMBO J.*, **24**, 1311–1317.

O'Farrell, P.F. (1975) High-resolution two-dimensional electrophoresis of proteins. *J. Biol. Chem.*, **250**, 4007–4021.

Ogas, J., Kaufmann, S., Henderson, J. and Somerville, C. (1999) *PICKLE* is a CHD3 chromatin-remodeling factor that regulates the transition from embryonic to vegetative development in *Arabidopsis*. *Proc. Natl Acad. Sci. USA*, **96**, 13839–13844.

Osborne, T.B. (1924) *The Vegetable Proteins*. Logmans, Green, London.

Østergaard, O., Finnie, C., Laugesen, S., Roepstorff, P. and Svensson, B. (2004) Proteome analysis of barley seeds: identification of major proteins from two-dimensional gels (pI 4–7). *Proteomics*, **4**, 2437–2447.

Patterson, S.D. and Aebersold, R. (2003) Proteomics: the first decade and beyond. *Nat. Genet.*, **33**, 311–323.

Peck, S.C. (2005) Update on proteomics in *Arabidopsis*. Where do we go from here? *Plant Physiol.*, **138**, 591–599.

Ranocha, P., McNeil, S.D., Ziemak, M.J., Li, C., Tarczynski, M.C. and Hanson, A.D. (2001) The *S*-methylmethionine cycle in angiosperms: ubiquity, antiquity and activity. *Plant J.*, **25**, 575–584.

Rajjou, L., Gallardo, K., Debeaujon, I., Vandekerckhove, J., Job, C. and Job, D. (2004) The effect of α-amanitin on the *Arabidopsis* seed proteome highlights the distinct roles of stored and neosynthesized mRNAs during germination. *Plant Physiol.*, **134**, 1598–1613.

Rajjou, L., Lovigny, Y., Job, C., Belghazi, M., Groot, S.P.C. and Job, D. (2006) Seed quality and germination. In: Navie, S. (ed) *Seed Biology. Eighth International Workshop on Seeds*, Brisbane, Australia, 2005. CAB International publishing, Cambridge, USA (in press).

Ravanel, S., Block, M.A., Rippert, P., Jabrin, S., Curien, G., Rébeillé, F. and Douce, R. (2004) Methionine metabolism in plants: chloroplasts are autonomous for *de novo* methionine synthesis and can import *S*-adenosylmethionine from the cytosol. *J. Biol. Chem.*, **279**, 22548–22557.

Ravanel, S., Gakière, B., Job, D. and Douce, R. (1998) The specific features of methionine biosynthesis and metabolism in plants. *Proc. Natl Acad. Sci. USA*, **95**, 7805–7812.

Rolfe, B.G., Mathesius, U., Djordjevic, M., Weinman, J., Hocart, C., Weiller, G. and Bauer, W.D. (2003) Proteomic analysis of legume-microbe interactions. *Comp. Funct. Genom.*, **4**, 225–228.

Rose, J.K.C., Bashir, S., Giovannoni, J.J., Jahn, M.M. and Saravan, R.S. (2004) Tackling the plant proteome: practical approaches, hurdles and experimental tools. *Plant J.*, **39**, 715–733.

Sarnighausen, E., Wurtz, V., Heintz, D., Van Dorsselaer, A. and Reski, R. (2004) Mapping the *Physcomitrella patens* proteome. *Phytochemistry*, **65**, 1589–1607.

Schiltz, S., Gallardo, K., Huart, M., Negroni, L., Sommerer, N. and Burstin, J. (2004) Proteome reference maps of vegetative tissues in pea. An investigation of nitrogen mobilization from leaves during seed filling. *Plant Physiol.*, **135**, 2241–2260.

Sheoran, I.S., Olson, D.J., Ross, A.R. and Sawhney, V.K. (2005) Proteome analysis of embryo and endosperm from germinating tomato seeds. *Proteomics*, **5**, 3752–3764.

Shewry, P.R. (1995) Cereal seed storage proteins. In: Kigel, J. and Gaglili, G. (eds) *Seed development and germination*. Marcel Dekker, Inc., New York, pp. 45–72.

Slade, A.J. and Knauf, V.C. (2005) TILLING moves beyond functional genomics into crop improvement. *Transgenic Res.*, **14**, 109–115.

Stauber, E.J. and Hippler, M. (2004) *Chlamydomonas reinhardtii* proteomics. *Plant Physiol. Biochem.*, **42**, 989–1001.

Swidzinski, J.A., Leaver, C.J. and Sweetlove, L.J. (2004) A proteomic analysis of plant programmed cell death. *Phytochemistry*, **65**, 1829–1838.

Tamarit, J., Cabiscol, E. and Ros, J. (1998) Identification of the major oxidatively damaged proteins in *Escherichia coli* cells exposed to oxidative stress. *J. Biol. Chem.*, **273**, 3027–3032.

Tesnier, K., Strookman-Donkers, H.M., van Pijlen, J.G., van der Geest, A.H.M., Bino, R.J. and Groot, S.P.C. (2002) A controlled deterioration test of *Arabidopsis thaliana* reveals genetic variation in seed quality. *Seed Sci. Technol.*, **30**, 149–165.

Thiellement, H., Bahrman, N., Damerval, C., Plomion, C., Rossignol, M., Santoni, V., de Vienne, D. and Zivy, M. (1999) Proteomics for genetic and physiological studies in plants. *Electrophoresis*, **20**, 2013–2026.

Thiellement, H., Zivy, M. and Plomion, C. (2002) Combining proteomic and genetic studies in plants. *J. Chromatogr. B*, **782**, 137–149.

Vensel, W.H., Tanaka, C.K., Cai, N., Wong, J.H., Buchanan, B.B. and Hurkman, W.J. (2005) Developmental changes in the metabolic protein profiles of wheat endosperm. *Proteomics*, **5**, 1594–1611.

Vincent, D., Lapierre, C., Pollet, B., Cornic, G., Negroni, L. and Zivy, M. (2005) Water deficits affect caffeate *O*-methyltransferase, lignification, and related enzymes in maize leaves. A proteomic investigation. *Plant Physiol.*, **137**, 949–960.

Waters, L.C. and Dure III, L.S. (1966) Ribonucleic acid synthesis in germinating cotton seeds. *J. Mol. Biol.*, **19**, 1–27.

Watson, B.S., Asirvatham, V.S., Wang, L. and Summer, L.W. (2003) Mapping the proteome of barrel medic (*Medicago truncatula*). *Plant Physiol.*, **131**, 1104–1123.

Wiedmann, B., Sakai, H., Davis, T.A. and Wiedmann, M. (1994) A protein complex required for signal-sequence-specific sorting and translocation. *Nature*, **370**, 434–440.

Wilson, K.A., McManus, M.T., Gordon, M.E., Jordan, T.W. (2002) The proteomics of senescence in leaves of white clover, *Trifolium repens* (L.). *Proteomics*, **2**, 1114–1122.

Wong, J.H., Cai, N., Tanaka, C.K., Vensel, W.H., Hurkman, W.J. and Buchanan, B.B. (2004a). Thioredoxin reduction alters the solubility of proteins of wheat starchy endosperm: an early event in cereal germination. *Plant Cell Physiol.*, **45**, 407–415.

Wong, J.H., Cai, N., Balmer, Y., Tanaka, C.K., Vensel. W.H., Hurkman, W.J. and Buchanan, B.B. (2004b) Thioredoxin targets of developing wheat seeds identified by complementary proteomic approaches. *Phytochemistry*, **65**, 1629–1640.

Yan, S., Tang, Z., Su, W. and Sun, W. (2005) Proteomic analysis of salt stress-responsive proteins in rice root. *Proteomics*, **5**, 235–244.

Yu, J., Hu, S., Wang, J., Wong, G.K., Li, S., Liu, B., Deng, Y., Dai, L., Zhou, Y., Zhang, X., Cao, M., Liu, J., Sun, J., Tang, J., Chen, Y., Huang, X., Lin, W., Ye, C., Tong, W., Cong, L., Geng, J., Han, Y., Li, L., Li, W., Hu, G., Huang, X., Li, W., Li, J., Liu, Z., Li, L., Liu, J., Qi, Q., Liu, J., Li, L., Li, T., Wang, X., Lu, H., Wu, T., Zhu, M., Ni, P., Han, H., Dong, W., Ren, X., Feng, X., Cui, P., Li, X., Wang, H., Xu, X., Zhai, W., Xu, Z., Zhang, J., He, S., Zhang, J., Xu, J., Zhang, K., Zheng, X., Dong, J., Zeng, W., Tao, L., Ye, J., Tan, J., Ren, X., Chen, X., He, J., Liu, D., Tian, W., Tian, C., Xia, H., Bao, Q., Li, G., Gao, H., Cao, T., Wang, J., Zhao, W., Li, P., Chen, W., Wang, X., Zhang, Y., Hu, J., Wang, J., Liu, S., Yang, J., Zhang, G., Xiong, Y., Li, Z., Mao, L., Zhou, C., Zhu, Z., Chen, R., Hao, B., Zheng, W., Chen, S., Guo, W., Li, G., Liu, S., Tao, M., Wang, J., Zhu, L., Yuan, L. and Yang, H. (2002) A draft sequence of the rice genome (*Oryza sativa* L. ssp. *indica*). *Science*, **296**, 79–92.

Zhang, H., Sreenivasulu, N., Weschke, W., Stein, N., Rudd, S., Radchuk, V., Potokina, E., Scholz, U., Schweizer, P., Zierold, U., Langridge, P., Varshney, R.K., Wobus, U. and Graner, A. (2004) Large-scale analysis of the barley transcriptome based on expressed sequence tags. *Plant J.*, **40**, 276–290.

Zimmerman, L.J. (1993) Somatic embryogenesis: a model for early development in higher plants. *Plant Cell*, **5**, 1411–1423.

Zivy, M., Thiellement, H., de Vienne, D. and Hofmann, J.P. (1983) Study on nuclear and cytoplasmic genome expression in wheat by two-dimensional electrophoresis I. First results on 18 alloplasmic lines. *Theor. Appl. Genet.*, **66**, 1–7.

Zivy, M., Thiellement, H., de Vienne, D. and Hofmann, J.P. (1984) Study on nuclear and cytoplasmic genome expression in wheat by two-dimensional electrophoresis II. Genetic differences between two lines and two groups of cytoplasms at five developmental stages or organs. *Theor. Appl. Genet.*, **68**, 335–345.

Zu, X. and Galili, G. (2003) Increased lysine synthesis coupled with a knockout of its catabolism synergistically boosts lysine content and also transregulates the metabolism of other amino acids in *Arabidopsis* seeds. *Plant Cell*, **15**, 845–853.

8 Surveying the plant cell wall proteome, or secretome

Tal Isaacson and Jocelyn K.C. Rose

8.1 Introduction

The plant cell wall can be viewed as the physical structure that encases all plant cells and whose architecture determines cellular morphology. Thus, wall synthesis, assembly and remodeling ultimately dictate plant form. The wall matrix also has a well-established defensive function and most microbial pathogens are unable to colonize most plants because of this structural barrier. However, another perspective of the 'cell wall' is as a compartment, comprising an apoplastic continuum that extends throughout the plant. In this sense it is not entirely analogous to a subcellular compartment, as it is not encapsulated by a discrete membrane, or specific to an individual cell. However, it is a multifunctional milieu that represents the site of numerous metabolic pathways and the source and medium for many molecular signals, since the wall is where many biotic and abiotic stimuli are first perceived.

The plant wall is a highly hydrated environment, typically composed of many different polysaccharides (up to 90% of the dry weight), proteins (2–10%) minerals (1–5%) and aromatic substances (less than 2%) (O'Neill and York, 2003). During cell division a new wall is synthesized; a semi-rigid, polysaccharide-rich matrix, referred to as the primary wall, whose composition evolves during development to give rise to the differentiated cells that determine tissue and organ morphology. The thickened walls that are subsequently deposited to confer mechanical strength after cell growth ceases are termed secondary walls and they often have a different composition and structure from primary walls. The molecular composition and organization of the wall differs among species, tissues, cell types and even within microdomains of a wall around a single cell. Accordingly, the repertoires of apoplastic proteins that are associated with different cells and tissues can also be quite distinct, with a high degree of temporal and spatial variation.

This review of the plant cell wall proteome treats the cell wall in a broader 'apoplastic' sense, as the compartment that is found exterior to the plasma membrane, containing the primary or secondary wall and middle lamella. The term 'wall proteome' is thus equivalent to the 'apoplastic' or 'extracellular' proteome. Similarly, the 'secretome' refers to the population of proteins that are secreted to the cell surface and that are either mobile within the apoplast, bound to the extracellular matrix or associated with the plasma membrane, but with apoplastic domains.

8.2 The multifunctional cell wall

The first plant wall proteins were identified only about 40 years ago (Lamport, 1965) and the last few decades have seen the field evolve to the point where recent genome-scale assessments predict the existence of many hundreds of extracellular proteins in the *Arabidopsis* genome. This has occurred in parallel with, and in part due to, a growing appreciation of the dynamic nature of the wall and apoplastic environment and the central role that it plays in most aspects of plant biology.

8.2.1 *Structural proteins*

In an extremely crude sense, plant wall proteins are sometimes divided into two categories: structural proteins and enzymes. Structural proteins are typically immobilized within the wall (Showalter, 1993) and comprise approximately 5–10% of the wall dry weight (Cassab and Varner, 1988). Most are unusually rich in one or two amino acids, contain highly repetitive sequence domains and are glycosylated, to such a degree that some are perhaps more accurately referred to as proteoglycans. They fall into two classes: the hydroxyproline-rich glycoproteins (HRGPs) and the glycine-rich proteins (GRPs). The HRGPs include the extensins, proline-rich proteins (PRPs) and the arabinogalactan proteins (AGPs). Although these protein families have been studied extensively, the functions of most have yet to be determined (Ringli et al., 2001; Showalter, 2001; Wu et al., 2001; Johnson et al., 2003). While many structural proteins are covalently linked to the wall matrix, some are highly soluble and appear to associate only loosely with the wall (Nothnagel, 1997).

8.2.2 *Wall-modifying proteins*

The availability of the first plant genome sequences revealed the existence of hundreds of genes that are predicted to encode wall-localized proteins that function in wall remodeling and disassembly (Henrissat et al., 2001; Davies and Henrissat, 2002; Somerville et al., 2004). Due to the dynamic nature of the wall, these proteins are active not only during cell expansion, but also later when wall restructuring is required in response to external and internal stimuli (Fry, 1995; Rose and Bennett, 1999; Cosgrove, 2003; Rose et al., 2003). Early studies revealed the activities of a numerous endo- and exo-acting polysaccharide hydrolases and lyases, which were relatively easy to detect and which provided a relatively simple mechanistic model to describe wall loosening. More recently, additional classes of wall modifying proteins have been identified, including xyloglucan endotransglucosylase/hydrolases (XTHs) and expansins, whose specific contributions to wall modification or mechanisms of action are less clear, but whose existence emphasizes the complexity of wall restructuring (Rose et al., 2002; reviewed in Cosgrove, 2003; Rose et al., 2004). In almost every instance, these proteins are encoded by large gene families that are collectively represented in all cell and tissue types at every developmental stage, but that individually show a highly specific expression pattern (Imoto et al., 2005; Nishitani, 2005).

8.2.3 *Defense-associated wall proteins*

Plant walls provide an effective barrier to most microbial pathogens and recent reviews of plant–pathogen interactions have described the elaborate arsenal of extracellular defenses and surveillance networks that are localized in the apoplast, or at the cell wall–plasma membrane interface (Gómez-Gómez, 2004; Jones and Takemoto, 2004; Nürnberger et al., 2004; Pignocchi and Foyer, 2003; Schulze-Lefert, 2004; Veronese et al., 2003; Vorwerk et al., 2004). Plant cells secret constitutively expressed or inducible cocktails of defense-related proteins into the apoplast to deter microbial pathogens (Veronese et al., 2003). These include proteins that break down microbe walls, such as chitinases and endo-β-1,3-glucanases, and also inhibitors of proteins that are secreted by the pathogen into the apoplast to aid colonization (De Lorenzo and Ferrari, 2002; York et al., 2004). Many other secreted proteins contribute to defense responses that result in wall reinforcement (Schulze-Lefert, 2004), lipid metabolism (Maldonado et al., 2002; Oh et al., 2005) and an apoplastic oxidative burst (Pignocchi and Foyer, 2003). Other studies have described a broad range of functionally divergent defense-related proteins and peptides (Garcia-Olmedo et al., 1998; Veronese et al., 2003; van der Hoorn and Jones, 2004; Boller, 2005; Narvaez-Vasquez et al., 2005).

8.2.4 *Secreted peptides and developmental regulation*

In addition to their association with defense responses, short peptides are known to function as signaling molecules in the apoplast and to regulate various developmental processes. A well-studied example is the maintenance of meristem identity and size, which is regulated in the cell wall by the *CLAVATA* (CLV) pathway. The CLV3 peptide is a key player in maintaining the apical shoot meristem identity and size (Rojo et al., 2002) and CLE19, a CLV3 homolog, has a similar effect on root meristems (Casamitjana-Martinez et al., 2003). Other examples include the determination of stomatal density and distribution (Von Groll et al., 2002; Nadeau and Sack, 2003; reviewed in Bergmann, 2004), as well as cell growth, which was shown to be influenced by peptides termed RALF (rapid alkalinization factors) (reviewed in Pearce et al., 2001; Ryan et al., 2002) and by a growth factor peptide, phytosulfokine-a (PSK-a) (Matsubayashi et al., 1997, 2002). Extracellular peptides are also associated with pollen–pistil interactions (Kachroo et al., 2001; Franklin-Tong and Franklin, 2003; Tang et al., 2004).

8.2.5 *Extracellular proteins and abiotic stresses*

Cell wall-localized proteins appear to play a role in responses to a broad range of environmental stresses. For example, it has been observed that the abundance and extractability of many apoplastic proteins change in response to heavy metals, such as zinc (Brune et al., 1994), nickel and cadmium (Blinda et al., 1997) or manganese (Fecht-Christoffers et al., 2003), and exposure to aluminum was reported to cause a decrease in extracted apoplast proteins (Kataoka et al., 2003). Apoplastic proteins are also thought to provide a protective role again low-temperature stress and antifreeze proteins have been found in several overwintering plants, where they inhibit the growth and recrystallization of ice that forms in intercellular spaces (Marentes et al., 1993; Stressmann et al., 2004; Griffith et al., 2005). Interestingly, most known secreted antifreeze proteins are homologous to pathogenesis-related (PR) proteins (Griffith and Yaish, 2004), which may indicate a bifunctional role.

8.2.6 Wall proteins and metabolism

Many minerals (Sattelmacher, 2001), assimilates and metabolites are transported through, or generated in, the apoplast and accordingly, apoplastic proteins are involved in their metabolism, trafficking and distribution. For example, wall-localized invertases function in source–sink relations through catabolism of sucrose (Sherson et al., 2003; reviewed in Roitsch and Gonzalez, 2004).

The categories and examples of wall proteins listed above are used to illustrate their functional diversity, rather than to provide exhaustive list. In spite of the critical role the cell wall proteome plays, it remains far less characterized than those of most plant organelles (Lee et al., 2004). This chapter summarizes the unique technical hurdles that are faced in isolating cell wall proteins, describes different approaches for their extraction and identification and the application of functional screens and bioinformatics tools to characterize the cell wall proteome, or secretome.

8.3 Technical and conceptual challenges in defining the cell wall proteome, or secretome

The attractions of examining the subproteome of a particular cellular compartment include the acquisition of vital information regarding the residence of particular proteins or protein complexes, and also the fact that biological fractionation dramatically enhances the detection of lower abundance proteins (Millar, 2004). One of the main challenges of subcellular proteomics is to capture the most comprehensive representation of the protein complement, while minimizing contamination with proteins from other subcellular locations. The cell wall proteome presents some formidable challenges in this regard.

Organelles, such as chloroplasts and mitochondria, are conveniently bounded by membranes, making their isolation relatively simple, usually by controlled tissue disruption and density gradient centrifugation. During such a procedure, some of the organelles may be disrupted or lost, but those that remain intact provide a theoretically complete protein profile. In contrast, the uncompartmentalized nature of the cell wall does not lend itself to easy characterization using similar approaches and 'the apoplast' cannot be isolated as a whole distinct fraction. Any disruption of the plasma membrane, while attempting to extract extracellular proteins will lead to immediate, and often extensive, contamination of the cell wall protein fraction with intracellular proteins. Pellets of cell wall-enriched material can be isolated and washed to remove loosely associated contaminating proteins, but this will also result in the loss of soluble wall proteins and peptides. Moreover, upon tissue disruption, many cytosolic proteins will bind to the wall matrix with extremely high affinity, preventing a simple distinction between an authentic wall protein and a contaminant (see below).

Another technical challenge is the wide spectrum of biochemical characteristics that cell wall proteins exhibit. Some are highly soluble, with no apparent interaction with the polysaccharide matrix, and are therefore easily lost from wall extracts, while others bind to the wall matrix with varying affinities, thus requiring different conditions to detach them. Some proteins, such as extensins, are covalently linked into the wall architecture and are consequently highly resistant to extraction, even with harsh solvents (Johnson et al., 2003). Many wall proteins also undergo extensive post-translational modification (PTM) that can interfere with extraction, isolation and identification. For example, numerous secreted proteins are glycosylated, sometimes heavily: AGPs can comprise more than 90% carbohydrate (Cassab, 1998). Other proteins are known to interact with the plasma membrane via a lipid anchor, such as a GPI (glycosylphosphatidylinositol) anchor (Takos et al., 1997; Borner et al., 2003). In addition, wall proteins may change their solubility as a consequence of environmental changes (Wojtaszek et al., 1995; Otte and Barz, 1996). For instance, it was reported that during osmotic stress, proteins in the apoplast of jack pine (*Pinus banksiana* Lamb.) became more tightly associated with the wall and were harder to extract (Marshall et al., 1999). Therefore, measured differences in protein abundance might reflect changes in solubility and extractability, rather than real quantitative changes *in muro*.

The cell wall matrix also complicates the process of protein extraction since it consists principally of polysaccharides and polyphenolic compounds, both of which are notorious contaminants that interfere with experimental protein separation procedures, such as gel electrophoresis (Saravanan and Rose, 2004).

These problems are such that no perfect technique is available to isolate all, or even most, cell wall proteins, in an uncontaminated form. However, various techniques have been developed to extract different subsets of wall proteins from various tissues and these have provides some insights into the nature of the wall proteome. These techniques can be divided into two broad categories: disruptive and non-disruptive.

8.4 Approaches for plant cell wall protein extraction and identification

8.4.1 Non-disruptive isolation of extracellular proteins

Modest numbers of plant wall proteins have been isolated over the last few decades during efforts to characterize the biochemical activity or properties of specific individual polypeptides. However, the first methodical attempt to examine plant wall protein populations on a larger scale was described by Robertson et al. (1997), in a pioneering comparative study of secreted proteins from plant cell suspension cultures. Suspension cells, unlike cells in a whole plant tissue, are homogeneous individual cells, each surrounded by a separate cell wall. They represent an attractive system for wall proteomics, since aggressive techniques that might cause cell breakage are not needed to disconnect the cells. Many secreted proteins do not bind to the wall matrix and are freely soluble in the culture media while other subsets, that are wall associated to varying degrees, can be isolated by washing the cells with various buffers that may elute the proteins, but not break the plasma membrane, thus avoiding contamination with intracellular proteins. Robertson et al. (1997) looked at freely soluble and wall-bound eluted proteins from cells suspension cultures of five different plant species. Protein populations were isolated from intact cells that were sequentially washed with 200 mM calcium chloride, which has been shown to effectively liberate many wall-bound proteins, followed by 50 mM 1,2-cyclohexanediaminetetraacetic acid (CDTA), a treatment that might be expected to liberate proteins that bind to pectins. A third buffer contained 2 mM dithiothreitol (DTT), to disrupt intermolecular disulfide bonds, followed by 1 M sodium chloride, to extract proteins that were ionically bound to the wall. A final wash was with 200 mM borate buffer, which perturbs interactions between the wall and glycoproteins. The proteins were separated by 1-dimensional gel electrophoresis and more than 200 protein bands subjected to N-terminal amino acid sequencing, which provided sequence information for approximately two-thirds. Many families of known wall proteins were identified, although a large proportion was not assigned a function based on sequence homology. In many cases, this was a consequence of a lack of DNA sequence information for the species that were studied (e.g. carrot). Approximately, one-third of the proteins from *Arabidopsis* were unclassified, since the genome sequence was not available at that time. However, a recent reevaluation of the published sequences now allows a considerably greater proportion, but not all, of the proteins to be assigned a putative biochemical function (S.-J. Lee and J.K.C. Rose, unpublished data). This sequential washing approach was recently reevaluated, using *Arabidopsis* suspension cells, in order to assess the degree of contamination with intracellular proteins and to identify additional loosely bound wall proteins (Borderies et al., 2003). The authors applied two basic protocols: one involving essentially the same solvent series as that described in Robertson et al. (1997) and the other, a protocol that included 2 M lithium chloride but omitted the calcium chloride wash. Microscopic analysis was also used to monitor the integrity of the intracellular compartments. It was shown that the calcium chloride and chelator extractions were particularly likely to cause cell leakage and it was concluded that cellular integrity deceased substantially after more than two sequential extractions. Since the plasma membrane is easily ruptured, great care must be taken, even when using this supposedly non-destructive approach. The authors also noted that individual protein isoforms were not exclusively eluted in one solvent, which further complicates analysis and assessment of protein abundance. Contamination with intracellular proteins was also observed in a similar study of cell wall-associated proteins in the green algae *Haematococcus pluvialis* (Wang et al., 2004), which again used the same reagents as those described in Robertson et al. (1997).

Suspension cells have been used in this way to characterize the secretome associated with both primary and secondary cell wall synthesis and assembly. A recent study used protoplasts derived from *Arabidopsis* suspension cultured cells to look at proteins that are secreted during wall regeneration (Kwon et al., 2005). Potassium chloride was used to extract loosely bound 'wall proteins' from native suspension cultured cells or protoplasts that had been allowed to regenerate their walls for either 1 or 3 h. A comparison of the three protein populations showed that each set was distinct and the authors also suggested that the specific polypeptides showed differing PTMs among the three fractions. The study gave an interesting insight into the dynamic changes of these proteins during cell wall regeneration, and also assessed patterns of phosphorylation and glycosylation. An ingenious approach was used to shed light on proteins associated with secondary cell wall synthesis, using tobacco suspension cells expressing high cytokinin levels (Blee et al., 2001), which consequently had highly thickened walls and exhibited many of the characteristics of cells that are actively synthesizing secondary walls. The suites of proteins that were extracted using either 200 mM calcium chloride or 40 mM CDTA, were substantially different from those in the analogous study of tobacco primary wall proteins reported by Robertson et al. (1997). While many new proteins were identified, other sequences indicated the presence of proteins that are related to secondary wall formation, including peroxidase and polyphenol oxidase/laccase, a lysine-rich protein and extensin. A related analysis of activities in the culture medium and cell wall fractions of Norway spruce suspension cells that were actively synthesizing lignin also inferred the presence of proteins associated with secondary wall synthesis (Karkonen et al., 2002).

Cell cultures have also been used to study wall protein populations in the fungus *Candida albicans* (Pitarch et al., 2002), and cell wall construction and reorganization in yeast by identifying the secreted proteins from *Saccharomyces cerevisiae* protoplasts that were actively regenerating walls (Pardo et al., 2000). The authors reported the identification of several proteins that are known to participate in wall construction.

Suspension cultured cells represent a convenient experimental system to study plant defense by applying an external stimulus, such as plant hormones involved in defense responses, or oligosaccharide elicitors, to the growth medium (Ndimba et al., 2003). In this way, a new secreted pathogen-related protein was identified from tobacco BY2 cell growth medium (Okushima et al., 2000) and a lipase, isolated from *Arabidopsis* cell culture growth medium, was found to be expressed after application of salicylic acid to the cell culture. This lipase was reported to play a role in plant resistance to the fungus *Alternaria brassicicola* (Oh et al., 2005).

Although suspension cells are a useful source of homogenous plant material that can be easily manipulated, rapidly regenerated and from which wall proteins may relatively easily be isolated, they are an artificial biological system. The complement of extracellular proteins in complex plant tissues is likely to be significantly different, given that different tissues and cell types have distinct suites of wall proteins that exhibit substantial spatial and temporal heterogeneity. This was demonstrated by a comparison of the cell wall proteomes of suspension cell cultures and intact rosettes of *Arabidopsis* (Boudart et al., 2005). The authors found that the proteins profiles were remarkably different from each other and of 132 proteins identified from both tissues, only 17 were common to both extracts. While the proteins identified from the rosette tissues had a high proportion of hydrolytic enzymes and only a few oxidases, the reverse was observed in secreted proteins from the suspension cell cultures. In addition, in each condition, different members of same gene families were expressed.

Clearly, the characterization of the wall proteome of intact tissues, rather than cell cultures is likely to provide more biologically relevant information. An experimental approach to extract apoplastic proteins from complex tissues, and one that can be adapted to minimize contamination with cytosolic proteins, is to use pressure rehydration and vacuum infiltration protocols to extract the apoplastic fluid from the target sample. This approach has been adopted successfully to extract extracellular proteins from several tissue or organ types, including roots (Yu et al., 1999; Kataoka et al., 2003), leaves (Blinda et al., 1997; Hiilovaara-Teijo et al., 1999; Zareie et al., 2002; Fecht-Christoffers et al., 2003; Haslam et al., 2003; Gau et al., 2004; Dani et al., 2005), fruit (Ruan et al., 1995, 1996) and tubers (Olivieri et al., 1998). The infiltration of different solutions into the tissues allows the release of different wall protein fractions, such as proteins that are ionically bound to the wall, but that are released by buffers containing relatively high-salt concentrations. Boudart et al. (2005) evaluated the use of different infiltration solutions and found that 0.2 M calcium chloride effectively released large numbers of proteins from intact *Arabidopsis* rosettes without apparently disturbing the plasma membrane. This contrasts with the damage caused by the same salt solution to plasma membranes of suspension cultured cells, which therefore led to high levels of cytoplasmic protein contamination (Borderies et al., 2003). The disadvantages of this technique are that the protein yield is generally low and great care has to be taken to avoid cell lysis. Sample throughput is therefore typically slow and many wall-localized proteins cannot be recovered without rupturing the plasma membrane.

Secreted protein populations have also been identified from both phloem and xylem fluids, which can be isolated by collecting exudates that are forced from plant tissues under positive pressure, or through centrifugation. For example, Kehr et al. (2005) provided a profile of the most abundant proteins present in the xylem sap of *Brassica napus*, including many proteins that had not previously been associated with xylem. A similar study compared proteins that were present in tomato xylem sap during compatible or incompatible interactions with the vascular wilt fungus *Fusarium oxysporum* (Rep et al., 2002). The phloem sap of flowering and non-flowering plants was compared in an effort to find mobile signals that participate in flowering induction and four peptides were found to differ in their abundance (Hoffmann-Benning et al., 2002). Another opportunity to isolate secreted proteins is provided by openings in grass leaves called hydathodes, through which extracellular guttation fluid is secreted. Since hydathodes are constantly open, they represent potential entry points for microbial pathogens and are therefore likely to be associated with an effective set of defense-related secreted proteins. Indeed, most of the proteins identified in a study of guttation fluids from barley were found to belong to PR protein families (Grunwald et al., 2003).

8.4.2 Disruptive isolation of wall-bound proteins

While the non-disruptive techniques described above can effectively release specific sets of soluble proteins, many form tight interactions with the wall matrix and do not dissociate when gentle washes or extraction techniques are applied. An alternative strategy is to homogenize plant tissues and elute protein fractions from cell wall fragments/pellets. The cell wall pellets are typically first washed with a low-salt buffer to remove cytosolic protein contaminants and then with a buffer containing high concentrations of salt (e.g. 1.5 M sodium chloride) to release proteins that are ionically bound to the wall.

This homogenization approach was used in an analysis of wall-associated proteins from *Arabidopsis* suspension cells (Chivasa et al., 2002), where a cell wall fraction from disrupted cells was sequentially extracted with calcium chloride and urea. The authors used 2-dimensional gel electrophoresis followed by mass spectrometry analysis to sequence 69 different proteins. These included numerous known wall proteins with well-established biochemical activities, a number of unclassified proteins and several polypeptides whose location in the wall was unexpected. A similar examination of cell wall-associated proteins from the developing xylem of compression and non-compression wood of Sitka spruce resulted in the identification of several differentially expressed proteins, including oxidases that may contribute to secondary wall formation (McDougall, 2000). Another analysis described wall-associated proteins from alfalfa stems, where about 100 proteins were identified from lithium chloride- and calcium chloride-released protein fractions (Watson et al., 2004). Many corresponded to known extracellular proteins with well-defined functions, such as wall modification or defense, but others were termed non-classical wall proteins, in that they are not typically thought to be secreted. Proteins from cell wall fragments of maize (Chivasa et al., 2005b) and *Arabidopsis* (Ndimba et al., 2003) were analyzed to track changes in the wall proteome in response to elicitors from pathogens. Here too, the authors identified classical cytosolic proteins, such as glyceraldehydes-3-phosphate dehydrogenase (GAPDH), in the cell wall-associated fraction (Chivasa et al., 2005b).

This disruptive 'grind and find' approach is clearly effective for many classes of wall proteins, particularly those that are more tightly associated with the extracellular matrix. The protein classes found by this approach demonstrate its potential value for identifying new cell wall/apoplastic proteins and wall-localized biochemical pathways. Such experiments will provide a platform for subsequent functional studies. However, the identification of supposedly intracellular proteins in wall extracts raises the critical question of how much contamination with intracellular proteins is typically present in these extracts.

A major portion of plant primary walls and the middle lamella is comprised of pectin, a class of polysaccharides that can be highly negatively charged. Pectin polymers can thus act as a polyanionic matrix in cell wall pellets and so positively charged intracellular proteins have the potential to bind to the wall once the plasma membrane has been ruptured, causing extensive contamination of the sample. In some cases a cytosolic protein can associate so strongly with the wall that a high-salt buffer does not disrupt the interaction and a detergent, such as sodium dodecyl sulphate (SDS), is subsequently required to resolubilize the protein from the wall-enriched pellet (R.S. Saravanan and J. Rose, unpublished data). Considerable caution should therefore be exercised when classifying proteins that are isolated using this disruptive technique as cell wall-localized proteins.

8.5 Computational prediction of the plant secretome

The primary route for protein secretion from eukaryotic cells is termed the classical, or endoplasmic reticulum (ER)/Golgi-dependent, secretory pathway (Figure 8.1). Briefly, secreted eukaryotic proteins utilize an N-terminal signal peptide (SP) to direct their co-translation into the ER lumen, after which they progress through the endomembrane system and are ultimately exported to the extracellular environment, or cell surface. It should be noted that these represent a subset of SP-containing proteins, since others are retained in the ER or Golgi or are redirected to other compartments, such as the vacuole (Oufattole et al., 2005). It was recently revealed that some proteins also traffic from the ER to the chloroplast (Villarejo et al., 2005).

A number of computational tools have been developed to predict the presence of SPs (Fariselli et al., 2003; Bendtsen et al., 2004; Hiller et al., 2004; Kall et al., 2004; Small et al., 2004) based on protein composition and structure at the N-terminus, since SPs do not have conserved amino acid sequences. These programs have been greatly refined in recent years and their accuracy has improved with the adoption of machine learning algorithms (Nielsen et al., 1997; Ladunga, 2000; Bendtsen et al., 2004). This bioinformatics approach is a potentially valuable means to predict the secretome and to help validate the localization of proteins whose identification in cell wall protein extracts, as described above, is unexpected. However, as with all prediction-based tools there is a consistent error rate and both false positive and negative predictions should be expected. To date no detailed study of a large plant sequence data set using an SP prediction program has been reported. However, our group screened a recent release of the predicted complete proteome of *Arabidopsis thaliana*, comprising 27,288 open reading frames (ORFs) using the SP prediction program SignalP (www.cbs.dtu.dk/services/SignalP). Almost 10% (2527) of *Arabidopsis* proteins are predicted to have an SP with a 95% probability index and the number was substantially higher using a more liberal probability score (e.g. 2962 at 90% probability and 3739 at 70%). Figure 8.2 shows the probability ranking for all the predicted *Arabidopsis* proteins, based on a Hidden Markov Model

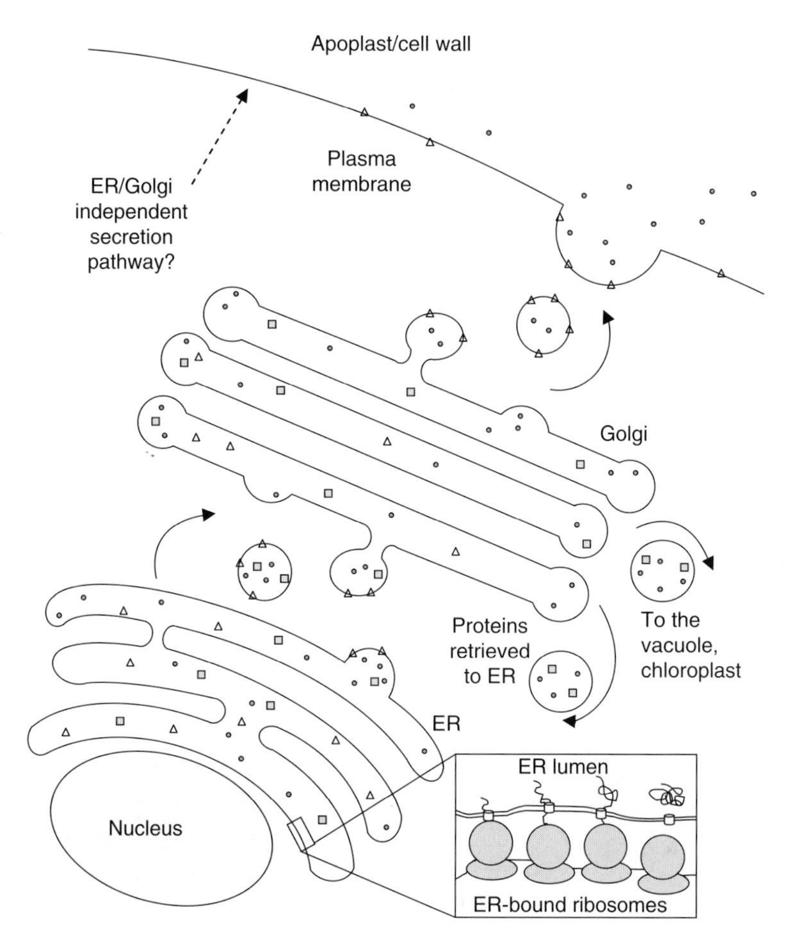

Figure 8.1 The plant secretory pathway. Proteins are synthesized by ribosomes bound to the ER membrane and pass the ER lumen while their SPs cleaved. The proteins traffic through the ER/Golgi and sorting processes occur that direct them to their ultimate destination: retention in the endomembrane system, the plasma membrane, vacuole or chloroplast. Post-translational modifications (PTMs) occur in the ER, Golgi or apoplast.

(HMM) score, with each protein represented as a point on the graph generating the curve. In addition, substantial numbers of proteins were predicted to have signal anchors, or uncleaved SPs, that do not have predicted SP cleavage sites after their hydrophobic transmembrane region and that may be resident in one of the membranes of the endomembrane system. Careful manual inspection of the results revealed, as expected, the presence of both false positives (proteins that are known to be localized in the wall but given a low prediction score) and false negatives (those predicted to be secreted but that have been shown to be located in an intracellular compartment). We estimate that the program has an accuracy of approximately

80–90% with the *Arabidopsis* data set, which is comparable to other assessments. We note that this analysis was performed using SignalP version 2.0 and that version 3.0 has recently been released, which has been reported to be more accurate and one of the most effective predictors of SPs (Bendtsen et al., 2004). Preliminary analyses by our group with version 3.0 generated a similar proportion of *Arabidopsis* proteins with predicted SPs and again, false positives and negatives. In general, we have found that SignalP version 3.0 has a superior performance with plant data compared with other web-based SP prediction programs.

In one sense, 80–90% prediction accuracy is excellent and provides a large data pool for further experimental verification. On the other hand, it could be argued that with 80% accuracy, the prediction is incorrect one time in five and that such results should therefore be interpreted with a great deal of caution. Both perspectives are valid and underline the point that an SP prediction score should not be taken as definitive evidence for or against entry into the secretory pathway, but that it can provide supporting information, or act as a catalyst for further research. In terms of the value of such programs for predicting the wall proteome, it should again be noted that many proteins with functional SPs enter the ER, but are targeted to intracellular compartments and do not ultimately traffic to the cell wall, and also that numerous apoplastic proteins will be overlooked, as a false negatives. However, this bioinformatics approach, when used appropriately, provides a valuable insight into the secretome.

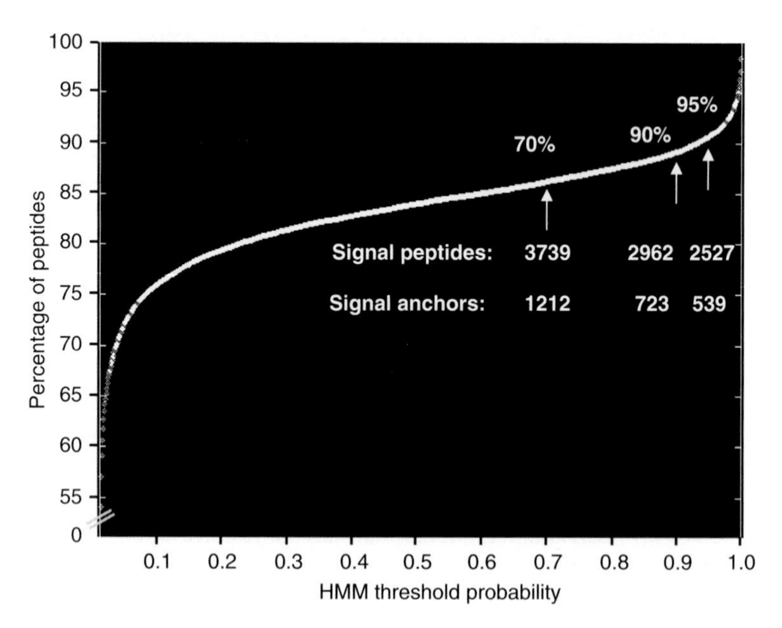

Figure 8.2 Computational prediction of the *Arabidopsis* secretome. The predicted proteome of *Arabidopsis thaliana* (comprising 27,288 ORFs) was screened for the presence of SPs and signal anchors, using the HMM function of SignalP version 2.0.

8.6 Functional screens for extracellular proteins

A number of high throughput screens have been developed to identify populations of secreted plant proteins and their corresponding genes (reviewed in Lee et al., 2004). These include ligating libraries of cDNAs, derived from plant tissues, to DNA sequences encoding reporter proteins, such as green fluorescent protein (GFP), and microscopically determining the subcellular localizations of the resulting chimeric proteins. This has resulted in the identification of new wall-localized proteins that have not been previously identified through protein extraction and has shown that specific proteins can localize to wall microdomains (Escobar et al., 2003). Other screens can be termed 'secretion traps' in that they target proteins that either enter the classical secretory pathway or that are secreted by any route to the cell wall (Lee et al., 2004). An example of the latter is the yeast secretion trap (YST), which is also referred to as the yeast signal sequence trap (YSST) screen (Figure 8.3).

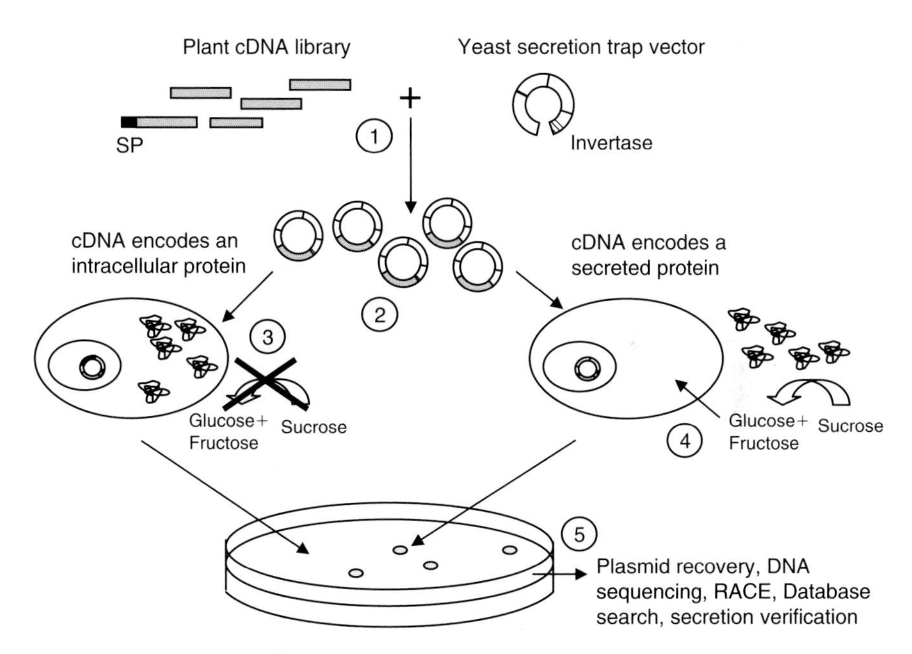

Figure 8.3 The yeast secretion trap screen. (1) Ligation of plant cDNA library into a yeast secretion trap vector containing a truncated yeast invertase gene; (2) Invertase-deficient yeast strain transformation and selection on sucrose medium. Expression of chimeric fusion protein comprising the N-terminus of a plant protein and the C-terminus of invertase; (3) If the plant cDNA encodes an intracellular protein, the translated fusion protein is retained inside the yeast cell and sucrose is not hydrolyzed and so the yeast cells die; (4) If the plant cDNA encodes a secreted protein, the translated fusion protein is secreted to the media, invertase degrades the sucrose and the resulting monosaccharides are taken up and used as an energy source; (5) Plasmid recovery, DNA sequencing, identification of the full length DNA (by database search, cDNA library screening or PCR-based techniques) and verification of secretion (e.g. by expression of the protein fused to GFP in plant cells or immunolocalization).

This high-throughput screen involves ligating cDNAs in-frame to the 5′ end of a yeast invertase gene that has been truncated to remove the native SP sequence. This library is then transformed into an invertase-deficient yeast mutant, which is grown on a medium with sucrose as the sole carbon source. Any yeast transformant containing a plant-derived cDNA with an SP-encoding sequence has the potential to secrete the polypeptide as an invertase fusion protein, resulting in reconstitution of extracellular invertase activity and the rescue of the mutant. The resulting yeast transformants can then be isolated and the gene encoding the secreted protein identified. The YST has now been used to identify secreted proteins from a range of eukaryotes, including plants (Goo et al., 1999; Belanger et al., 2003; Hugot et al., 2004; Yamane et al., 2005). The YST has some limitations, such as redundancy and the possibility of false positives, although in a pilot screen of tomato fruit cDNAs (Lee et al., unpublished), these comprised only about 2% of the clones. Importantly, in studies by our group to date, approximately 40% of the clones/proteins could be classified as 'new' wall proteins, in that they had no functional annotation or had never previously been identified in apoplast protein extracts. The YST screen is an effective means to identify secreted proteins that are not found by traditional protein extraction techniques.

Interestingly, a common feature of these functional screens is the identification of genes encoding secreted proteins that are not predicted to have an SP using SignalP, or computational tools. Additional localization studies by our group have confirmed that some plant proteins that were identified through YST screens clearly have no conventional SP, but are indeed secreted to the wall (Lee and Rose, unpublished data). This underscores a benefit of secretion trap screens of this kind, in that all secreted proteins are potentially identified, rather than specifically those that traffic via the classical secretory pathway. Thus, the YST is referred to here as the yeast secretion trap. It is known that some mammalian and lower eukaryote proteins have non-classical (ER–Golgi independent) secretory pathways (reviewed in Nickel, 2005) but these have yet to be characterized in plants. It remains to be determined whether such pathways are mechanistically similar in all eukaryotes and how many plant proteins are secreted by ER–Golgi independent routes.

8.7 Contamination and validation of the cell wall proteome

The effective isolation of plant cell wall proteins involves a trade-off between extracting the most comprehensive population of wall proteins possible, and minimizing contamination with intracellular proteins. A large portion of the plant wall proteome cannot be extracted without harsh treatments that disrupt the plasma membrane and so a key question is whether disruptive approaches can provide protein fractions that are relatively free from contamination. This is certainly something that has been claimed and several studies have involved careful assessment of apparent contamination levels (Table 8.1). This generally involves determining the presence or absence of a cytosolic 'marker' protein in the extract, using immunological

Table 8.1 Examples of plant cell wall/secretome proteomic projects

Method/tissue	Summary	Contamination assessment	Comments*	Reference
Non-disruptive				
Cell culture	Analysis of wall-bound proteins from cell cultures of *Arabidopsis*, carrot, French bean, tomato and tobacco.	Based on their similarity to known proteins, 2 proteins, out of 125, are estimated to be potentially cytoplasmic.	First systematic attempt to examine the wall proteome. (~125)	Robertson et al. (1997)
	Transformed tobacco cell culture that synthesizes secondary walls.	No comments on contamination.	Transgene leads cytokinin overproduction and to fivefold increase in wall thickness. (21)	Blee et al. (2001)
	Proteomic analysis of loosely bound cell wall proteins of *Arabidopsis* cell-suspension cultures.	Massive contamination claimed, bioinformatic analysis preformed to predict secretion.	Comparison of different protein extraction solutions. (~50)	Borderies et al. (2003)
	Apoplastic proteins involved in cell wall regeneration in *Arabidopsis* protoplasts.	G6PDH assayed and cell integrity assessed microscopically: minor contamination claimed. Bioinformatic analysis predict 56% of identified proteins to contain SPs.	Proteins found with different PTMs, including phosphorylation and glycosylation. (327)	Kwon et al. (2005)
Whole plant/organ	Apoplastic proteins from *Arabidopsis*, rice and wheat leaves.	G6PDH assayed. Estimated 0.3% contamination but dominant presence of RuBisCo in extracts indicates major contamination.	Protein profiles were distinct for each species. The principal proteins identified were defense-related. (23)	Haslam et al. (2003)
	Effect of manganese toxicity on the wall proteome of cowpea leaves.	No comments on contamination.	Concentration of soluble apoplastic proteins increased in response to manganese. Some PR-proteins detected. (31)	Fecht-Christoffers et al. (2003)
	Wall proteins in leaves of susceptible and resistant apple cultivars in response to infection by the fungus *Venturia inaequalis*.	No comments on contamination.	Constitutive presence of PR proteins in the apoplast of the resistant strain. (~20)	Gau et al. (2004)

(*Continued*)

Table 8.1 (*Continued*)

Method/tissue	Summary	Contamination assessment	Comments*	Reference
	Effects of salt stress on the tobacco leaf apoplast proteome.	MDH assayed, major RuBisCo subunit presence assayed by SDS-PAGE analysis: minor contamination claimed.	Salt stress resulted in enhanced accumulation of known defense-associated proteins. (20)	Dani et al. (2005)
	Proteomic analysis of *Arabidopsis* rosettes leaves and comparison to proteome of cells in suspension culture.	MDH assayed and cell integrity assessed microscopically: minor contamination claimed. Bioinformatic analysis predicted 94% of identified proteins to contain SPs.	Of 132 proteins identified from both rosette and cells in culture, 17 were found in both tissues. (93)	Boudart et al. (2005)
Cell culture growth medium	Proteins secreted by tobacco cultured BY2 cells.	No comments on contamination.	Identification of a new member of PR proteins. (6)	Okushima et al. (2000)
	Proteomic analysis of secreted proteins from SA-treated cultured *Arabidopsis* cells.	ADH and G6PDH activity assayed and immunoanalysis of cytosolic proteins. Low contamination claimed. Bioinformatic analysis predicted 54% of identified proteins to contain SPs.	Analysis reveals an *Arabidopsis* lipase involved in defense against *Alternaria brassicicola.* (91)	Oh et al. (2005)
Sap or guttation fluid	Comparison of peptides in the phloem sap of flowering and non-flowering *Perilla* and lupine plants.	No comments on contamination.	Microbore high-performance liquid chromatography used to separate the peptides. (16 peptides)	Hoffmann-Benning et al. (2002)
	Identification of proteins from xylem-sap of fungus-infected tomato.	All four proteins identified are known or predicted to be secreted.	Identification of isoforms of PR proteins. (4)	Rep et al. (2002)
	Identification of guttation fluid proteins in barley plants.	Bioinformatic analysis predicts 6 out of 7 identified proteins to contain SPs.	Identification of PR proteins in non-infected plants. (7)	Grunwald et al. (2003)
	Analysis of xylem sap proteins from *Brassica napus.*	Bioinformatic analysis predicts 97% of identified proteins to contain SPs.	An overview of the most abundant proteins present in xylem sap of *Brassica napus*. (69)	Kehr et al. (2005)

Disruptive				
Cell culture	Cell wall-associated proteins from *Arabidopsis* suspension cells.	Callose synthase assayed and wall fragments examined microscopically. Negligible contamination with plasma membrane claimed. No immunodetection of enoyl-ACP reductase cytosolic protein. Bioinformatic analysis predicts 25% of identified proteins to contain SPs.	Suggests a possible involvement of cell wall kinases in plant responses to pathogen attack. (69)	Chivasa et al. (2002)
	Fungal elicitor-induced changes in extracellular proteins from *Arabidopsis* cell suspension cultures.	No comments on contamination.	Identification of cell wall protein phosphorylation and a putative extracellular receptor-like kinase. (10)	Ndimba et al. (2003)
	Pathogen elicitor-induced changes in the maize secretome.	SDS-PAGE-based comparison of apoplastic protein extract with total cell protein extract.	Detection of changes in phosphorylation status of wall proteins in response to pathogen elicitors. (12)	Chivasa et al. (2005)
	Arabidopsis cell wall proteome defined using MUDPIT.	Bioinformatic analysis predicts 15% of identified proteins to contain SPs.	Sixty-seven percent of reported extracellular proteins found had not previously been reported as wall associated. (89)	Bayer et al. (2005)
	Identification of GPI-anchored proteins (GAPs) in *Arabidopsis*.	No comments on contamination.	Validation of previous bioinformatics predictions of GAPs in *Arabidopsis*. (30)	Broner et al. (2003)
Whole plant/organ	Comparison of wall-associated proteins from developing xylem of compression and non-compression Sitka spruce wood.	No comments on contamination.	Laccase was confirmed as a major oxidase in extracts of lignifying compression xylem. (4)	McDougall (2000)
	Isolation of wall-associated proteins from *Medicago sativa* stems.	Bioinformatic analysis predicts 53% of identified proteins to contain SPs.	Construction of alfalfa cell wall proteome reference map. (97)	Watson et al. (2004)

*Numbers in parentheses correspond to the number of distinct proteins that were identified. SA: salicylic acid; ADH: alcohol dehydrogenase; G6PDH: glucose-6-phosphate dehydrogenase; GPI: glycosylphosphatidylinositol; MDH: malate dehydrogenase; MUDPIT: multidimensional protein identification technology; PR: pathogenesis-related; PTM: post-translational modification; SP: signal peptide.

analysis or measuring enzymatic activity. Examples have included assaying malate dehydrogenase (MDH) (Boudart et al., 2005) or glucose-6-phosphate dehydrogenase (G6PDH) (Haslam et al., 2003; Kwon et al., 2005) activities or immunodetection of the plastid-localized protein ribulose bisphosphate decarboxylase/oxygenase (RuBisCo) or glutathione S-transferase (GST) (Oh et al., 2005). However, it is not clear whether such methods are sufficiently accurate or sensitive to indicate the extent of cytosolic contamination in extracts that are acquired by either disruptive or non-disruptive means. For example, a study of proteins in leaf apoplastic fluids estimated cytoplasmic contamination to be less than 0.3%, based on the enzymatic activity of G6PDH in the extracts, and yet RuBisCo represented a predominant protein in the extracts (Haslam et al., 2003). Another indication that contamination assays of this type may be misleading comes from analyses using disruptive approaches (see Table 8.1). In several cases, great care was taken in sample preparation to ensure optimal tissue/cell homogenization and to evaluate cytosolic contamination, as described above and indeed, the extracts appeared free from intracellular proteins. However, upon sequencing the isolated proteins, an extremely low proportion was reported to have predicted SPs. In one example, despite the absence of G6PDH activity, as many as 44% of the 'extracellular' proteins were not predicted to contain an SP, including proteins, such as 60S ribosomal protein L12 (Kwon et al., 2005). Oh et al. (2005) measured cytoplasmic contamination by both assaying alcohol dehydrogenase and G6PDH activities and by immunodetection of known cytoplasmic proteins, such as GST, the membrane protein annexin1 (AnnAt1), the chloroplast protein oxygen-evolving enhancer protein2 (OEE2) and cyclin D. All assays indicated minimal contamination, but 46% of the proteins were predicted to lack SPs, including some well-characterized cytoplasmic proteins. Similarly, a recent study of *Arabidopsis* wall-associated proteins involved rigorous sample preparation and contamination assessment and yet only 15% of the sequenced proteins had predicted SPs (Bayer et al., 2005). While the bioinformatic tools have a certain error rate, neither this nor the presence of a few 'non-classically secreted proteins' would account for such a dramatic discrepancy.

It seems likely that despite evidence presented to the contrary and every precaution to remove contaminating intracellular proteins, cellular disruption unavoidably results in substantial contamination of the wall protein fraction with intracellular proteins. Data from our lab using antibodies to a range of intracellular proteins support this idea and many such proteins bind with remarkably high affinity to the wall once plant material has been damaged (Saravanan and Rose, unpublished data).

8.8 Conclusions

The field of cell wall proteomics has some unique problems that are not faced by proteomic analysis of plant organelles. This is evident from the relatively small number of wall proteins that have been identified to date (see Table 8.1), compared with the large numbers of known chloroplast and mitochondrial proteins and our

Table 8.2 Advantages and disadvantages of different methodologies used to survey the cell wall proteome or secretome

Method	Advantages	Disadvantages
Protein extraction and identification	Direct access to the cell wall proteins *in muro*. Provide additional information regarding PTMs.	Only a subset of extracellular proteins is extracted and separable with any given protocol. Cell wall proteins samples are invariably contaminated with intracellular proteins. Proteins that are synthesized/secreted under specific condition might not be represented. Laborious and relatively expensive. Lower abundance proteins are more difficult to detect.
Functional screens/secretion traps	High throughput. Does not depend on protein extractability from cell wall.	Identifies genes, rather than proteins directly, so no information obtained about PTMs. Only genes that are transcribed under the conditions of the cDNA library will be represented. Genes expressed at low levels will not be readily selected in the screens.
Bioinformatic prediction	Performed *in silico* so fast and inclusive.	Requires protein sequences and therefore an extensive DNA sequence collection. Prediction relies on existing training data set or known features. Some proteins have signal sequences but are targeted to an internal cellular compartment. Computational prediction inevitably has a finite error rate so experimental confirmation will be required.

knowledge of the identity of the entire wall proteome is still far from approaching saturation. Another obvious challenge is in obtaining highly pure wall protein extracts and thus determining what represents a bona fide extracellular protein. However, on the positive side, great progress has been made over the last few years in developing technologies to identify secreted proteins that complement the standard protein isolation, separation and sequencing by mass spectrometry. Table 8.2 lists some of the advantages and disadvantages of different approaches.

There is no single protocol that can practically be used to identify all cell wall proteins and so techniques must be used that are tailored to a specific subset of proteins of interest, or a combination of strategies employed. Direct protein extraction and isolation captures a relatively small subset and an important consideration is that the subsequent separation technique can also influence the proteins that are identified. Bayer et al. (2005) suggested that traditional 2-dimensional gel electrophoresis biases against proteins with extremes of isoelectric point or molecular weight. They used multidimensional protein identification technology (MUDPIT) to characterize cell wall-bound proteins from *Arabidopsis* and only 33% of the proteins that were identified as secreted proteins had previously been reported in other analyses of the *Arabidopsis* cell wall proteome. The contamination issue, as discussed above, is a major hurdle and it seems that this problem alone will severely limit direct protein isolation as a means to target wall protein populations. On the other hand, bioinformatics and functional screens avoid this limitation, and despite their own particular disadvantages, when used in combination, can certainly provide a large data set of interesting candidates. Whether or not they can give near to full coverage of the wall proteome remains to be seen.

Future important directions in this field will include the characterization of the protein structures themselves, since the function of many wall proteins will be fundamentally dependent on PTMs. An obvious example is glycosylation and indeed a substantial proportion of wall proteins are glycosylated, often heavily. In addition, a few studies of secreted *Arabidopsis* proteins have reported the existence of tyrosine-phosphorylated phosphoproteins (Chivasa et al., 2002; Ndimba et al., 2003; Chivasa et al., 2005b), as well as putative extracellular kinases with no predicted transmembrane domains (i.e. not classical receptor-like kinases). Moreover, the application of fungal elicitors to *Arabidopsis* cells has been shown to induce phosphorylation of a secreted wall modifying protein (Ndimba et al., 2003). These results suggest the existence of an extracellular phosphorylation network and the recent confirmation that adenosine triphosphate (ATP) is present in the plant apoplast (Chivasa et al., 2005a) further supports this idea. Another relatively unexplored area is that of protein–protein interactions and protein complexes in the apoplast.

In conclusion, the plant wall and apoplast still remain relatively uncharted waters in terms of proteomics studies, but tools and technologies have been developed to ask and address many exciting questions and the next few years should see great progress.

References

Bayer, E.M., Bottrill, A.R., Walshaw, J., Vigouroux, M., Naldrett, M.J., Thomas, C.L. and Maule, A.J. (2005) *Arabidopsis* cell wall proteome defined using multidimensional protein identification technology. *Proteomics*, **6**, 301–311.

Belanger, K.D., Wyman, A.J., Sudol, M.N., Singla-Pareek, S.L. and Quatrano, R.S. (2003) A signal peptide secretion screen in *Fucus distichus* embryos reveals expression of glucanase, EGF domain-containing, and LRR receptor kinase-like polypeptides during asymmetric cell growth. *Planta*, **217**, 931–950.

Bendtsen, J.D., Nielsen, H., Von Heijne, G. and Brunak, S. (2004) Improved prediction of signal peptides: SignalP 3.0. *J. Mol. Biol.*, **340**, 783–795.

Bergmann, D.C. (2004) Integrating signals in stomatal development. *Curr. Opin. Plant Biol.*, **7**, 26–32.

Blee, K.A., Wheatley, E.R., Bonham, V.A., Mitchell, G.P., Robertson, D., Slabas, A. R., Burrell, M.M., Wojtaszek, P. and Bolwell, G.P. (2001) Proteomic analysis reveals a novel set of cell wall proteins in a transformed tobacco cell culture that synthesises secondary walls as determined by biochemical and morphological parameters. *Planta*, **212**, 404–415.

Blinda, A., Koch, B., Ramanjulu, S. and Dietz, K.J. (1997) De novo synthesis and accumulation of apoplastic proteins in leaves of heavy metal-exposed barley seedlings. *Plant Cell Environ.*, **20**, 969–981.

Boller, T. (2005) Peptide signalling in plant development and self/non-self perception. *Curr. Opin. Cell Biol.*, **17**, 116–122.

Borderies, G., Jamet, E., Lafitte, C., Rossignol, M., Jauneau, A., Boudart, G., Monsarrat, B., Esquerre-Tugaye, M.T., Boudet, A. and Pont-Lezica, R. (2003) Proteomics of loosely bound cell wall proteins of *Arabidopsis thaliana* cell suspension cultures: a critical analysis. *Electrophoresis*, **24**, 3421–3432.

Borner, G.H., Lilley, K.S., Stevens, T.J. and Dupree, P. (2003) Identification of glycosylphosphatidylinositol-anchored proteins in *Arabidopsis*. A proteomic and genomic analysis. *Plant Physiol.*, **132**, 568–577.

Boudart, G., Jamet, E., Rossignol, M., Lafitte, C., Borderies, G., Jauneau, A., Esquerre-Tugaye, M.T. and Pont-Lezica, R. (2005) Cell wall proteins in apoplastic fluids of *Arabidopsis thaliana* rosettes: identification by mass spectrometry and bioinformatics. *Proteomics*, **5**, 212–221.

Brune, A., Urbach, W. and Dietz, K.J. (1994) Zinc stress induces changes in apoplasmic protein-content and polypeptide composition of barley primary leaves. *J. Exp. Bot.*, **45**, 1189–1196.

Casamitjana-Martinez, E., Hofhuis, H.F., Xu, J., Liu, C.M., Heidstra, R. and Scheres, B. (2003) Root-specific CLE19 overexpression and the sol1/2 suppressors implicate a CLV-like pathway in the control of *Arabidopsis* root meristem maintenance. *Curr. Biol.*, **13**, 1435–1441.

Cassab, G.I. (1998) Plant cell wall proteins. *Annu. Rev. Plant Physiol. Plant Mol. Biol.*, **49**, 281–309.

Cassab, G.I. and Varner, J.E. (1988) Cell-wall proteins. *Annu. Rev. Plant Physiol. Plant Mol. Biol.*, **39**, 321–353.

Chivasa, S., Ndimba, B.K., Simon, W.J., Robertson, D., Yu, X.L., Knox, J.P., Bolwell, P. and Slabas, A.R. (2002) Proteomic analysis of the *Arabidopsis thaliana* cell wall. *Electrophoresis*, **23**, 1754–1765.

Chivasa, S., Ndimba, B.K., Simon, W.J., Lindsey, K. and Slabas, A.R. (2005a) Extracellular ATP functions as an endogenous external metabolite regulating plant cell viability. *Plant Cell*, **17**, 3019–3034.

Chivasa, S., Simon, W.J., Yu, X.L., Yalpani, N. and Slabas, A.R. (2005b) Pathogen elicitor-induced changes in the maize extracellular matrix proteome. *Proteomics*, **5**, 4894–4904.

Cosgrove, D. (2003) Expansion of the plant cell wall. In: Rose, J.K.C. (ed) *The Plant Cell Wall.* Blackwell Publishing Ltd, Oxford, UK.

Dani, V., Simon, W.J., Duranti, M. and Croy, R.R. (2005) Changes in the tobacco leaf apoplast proteome in response to salt stress. *Proteomics*, **5**, 737–745.

Davies, G.J. and Henrissat, B. (2002) Structural enzymology of carbohydrate-active enzymes: implications for the post-genomic era. *Biochem. Soc. Trans.*, **30**, 291–297.

De Lorenzo, G. and Ferrari, S. (2002) Polygalacturonase-inhibiting proteins in defense against phytopathogenic fungi. *Curr. Opin. Plant Biol.*, **5**, 295–299.

Escobar, N.M., Haupt, S., Thow, G., Boevink, P., Chapman, S. and Oparka, K. (2003) High-throughput viral expression of cDNA-green fluorescent protein fusions reveals novel subcellular addresses and identifies unique proteins that interact with plasmodesmata. *Plant Cell*, **15**, 1507–1523.

Fariselli, P., Finocchiaro, G. and Casadio, R. (2003) SPEPlip: the detection of signal peptide and lipoprotein cleavage sites. *Bioinformatics*, **19**, 2498–2499.

Fecht-Christoffers, M.M., Braun, H.P., Lemaitre-Guillier, C., Vandorsselaer, A. and Horst, W.J. (2003) Effect of manganese toxicity on the proteome of the leaf apoplast in cowpea. *Plant Physiol.*, **133**, 1935–1946.

Franklin-Tong, N.V. and Franklin, F.C. (2003) Gametophytic self-incompatibility inhibits pollen tube growth using different mechanisms. *Trends Plant Sci.*, **8**, 598–605.

Fry, S.C. (1995) Polysaccharide-modifying enzymes in the plant cell wall. *Annu. Rev. Plant Physiol. Plant Mol. Biol.*, **46**, 497–520.

Garcia-Olmedo, F., Molina, A., Alamillo, J.M. and Rodriguez-Palenzuela, P. (1998) Plant defense peptides. *Biopolymers*, **47**, 479–491.

Gau, A.E., Koutb, M., Piotrowski, M. and Kloppstech, K. (2004) Accumulation of pathogenesis-related proteins in the apoplast of a susceptible cultivar of apple (*Malus domestica* cv. Elstar) after infection by *Venturia inaequalis* and constitutive expression of PR genes in the resistant cultivar Remo. *Eur. J. Plant Pathol.*, **110**, 703–711.

Goo, J.H., Park, A.R., Park, W.J. and Park, O.K. (1999) Selection of *Arabidopsis* genes encoding secreted and plasma membrane proteins. *Plant Mol. Biol.*, **41**, 415–423.

Gómez–Gómez, L. (2004) Plant perception systems for pathogen recognition and defence. *Mol. Immunol.*, **41**, 1055–1062.

Griffith, M. and Yaish, M.W.F. (2004) Antifreeze proteins in overwintering plants: a tale of two activities. *Trends Plant Sci.*, **9**, 399–405.

Griffith, M., Lumb, C., Wiseman, S.B., Wisniewski, M., Johnson, R.W. and Marangoni, A.G. (2005) Antifreeze proteins modify the freezing process in planta. *Plant Physiol.*, **138**, 330–340.

Grunwald, I., Rupprecht, I., Schuster, G. and Kloppstech, K. (2003) Identification of guttation fluid proteins: the presence of pathogenesis-related proteins in non-infected barley plants. *Physiol. Plant.*, **119**, 192–202.

Haslam, R.P., Downie, A.L., Raveton, M., Gallardo, K., Job, D., Pallett, K.E., John, P., Parry, M.A.J. and Coleman, J.O.D. (2003) The assessment of enriched apoplastic extracts using proteomic approaches. *Ann. Appl. Biol.*, **143**, 81–91.

Henrissat, B., Coutinho, P.M. and Davies, G.J. (2001) A census of carbohydrate-active enzymes in the genome of *Arabidopsis thaliana. Plant Mol. Biol.*, **47**, 55–72.

Hiilovaara-Teijo, M., Hannukkala, A., Griffith, M., Yu, X.M. and Pihakaski-Maunsbach, K. (1999) Snow-mold-induced apoplastic proteins in winter rye leaves lack antifreeze activity. *Plant Physiol.*, **121**, 665–674.

Hiller, K., Grote, A., Scheer, M., Munch, R. and Jahn, D. (2004) PrediSi: prediction of signal peptides and their cleavage positions. *Nucl. Acid. Res.*, **32**, W375–W379.

Hoffmann-Benning, S., Gage, D.A., Mcintosh, L., Kende, H. and Zeevaart, J.A. (2002) Comparison of peptides in the phloem sap of flowering and non-flowering *Perilla* and lupine plants using microbore HPLC followed by matrix-assisted laser desorption/ionization time-of-flight mass spectrometry. *Planta*, **216**, 140–147.

Hugot, K., Riviere, M.P., Moreilhon, C., Dayem, M.A., Cozzitorto, J., Arbiol, G., Barbry, P., Weiss, C. and Galiana, E. (2004) Coordinated regulation of genes for secretion in tobacco at late developmental stages: association with resistance against oomycetes. *Plant Physiol.*, **134**, 858–870.

Imoto, K., Yokoyama, R. and Nishitani, K. (2005) Comprehensive approach to genes involved in cell wall modifications in *Arabidopsis thaliana. Plant Mol. Biol.*, **58**, 177–192.

Johnson, K.L., Jones, B.J., Schultz, C.J. and Bacic, A. (2003) Non-enzymatic cell wall (glycol) proteins. In: Rose, J.K.C. (ed) *The Plant Cell Wall.* Blackwell publishing Ltd, Oxford, UK.

Jones, D. and Takemoto, D. (2004) Plant innate immunity – direct and indirect recognition of general and specific pathogen-associated molecules. *Curr. Opin. Immunol.*, **16**, 48–62.

Kachroo, A., Schopfer, C.R., Nasrallah, M.E. and Nasrallah, J.B. (2001) Allele-specific receptor–ligand interactions in *Brassica* self-incompatibility. *Science*, **293**, 1824–1826.

Kall, L., Krogh, A. and Sonnhammer, E.L.L. (2004) A combined transmembrane topology and signal peptide prediction method. *J. Mol. Biol.*, **338**, 1027–1036.

Karkonen, A., Koutaniemi, S., Mustonen, M., Syrjanen, K., Brunow, G., Kilpelainen, I., Teeri, T.H. and Simola, L.K. (2002) Lignification related enzymes in *Picea abies* suspension cultures. *Plant Physiol.*, **114**, 343–353.

Kataoka, T., Furukawa, J. and Nakanishi, T.M. (2003) The decrease of extracted apoplast protein in soybean root tip by aluminium treatment. *Biol. Plantar.*, **46**, 445–449.

Kehr, J., Buhtz, A. and Giavalisco, P. (2005) Analysis of xylem sap proteins from *Brassica napus. BMC Plant Biol.*, **5**, 11.

Kwon, H.-K., Yokoyama, R. and Nishitani, K. (2005) A proteomic approach to apoplastic proteins involved in cell wall regeneration in protoplasts of *Arabidopsis* suspension-cultured cells. *Plant Cell Physiol.*, **46**, 843–857.

Ladunga, I. (2000) Large-scale predictions of secretory proteins from mammalian genomic and EST sequences. *Curr. Opin. Biotechnol.*, **11**, 13–18.

Lamport, D.T.A. (1965) The protein component of primary cell walls. *Adv. Bot. Res.*, **2**, 151–218.

Lee, S.J., Saravanan, R.S., Damasceno, C.M., Yamane, H., Kim, B.D. and Rose, J.K. (2004) Digging deeper into the plant cell wall proteome. *Plant Physiol. Biochem.*, **42**, 979–988.

Maldonado, A.M., Doerner, P., Dixon, R.A., Lamb, C.J. and Cameron, R.K. (2002) A putative lipid transfer protein involved in systemic resistance signalling in *Arabidopsis. Nature*, **419**, 399–403.

Marentes, E., Griffith, M., Mlynarz, A. and Brush, R.A. (1993) Proteins accumulate in the apoplast of winter rye leaves during cold-acclimation. *Physiol. Plant.*, **87**, 499–507.

Marshall, J.G., Dumbroff, E.B., Thatcher, B.J., Martin, B., Rutledge, R.G. and Blumwald, E. (1999) Synthesis and oxidative insolubilization of cell-wall proteins during osmotic stress. *Planta*, **208**, 401–408.

Matsubayashi, Y., Takagi, L. and Sakagami, Y. (1997) Phytosulfokine-alpha, a sulfated pentapeptide, stimulates the proliferation of rice cells by means of specific high- and low-affinity binding sites. *Proc. Natl Acad. Sci. USA*, **94**, 13357–13362.

Matsubayashi, Y., Ogawa, M., Morita, A. and Sakagami, Y. (2002) An LRR receptor kinase involved in perception of a peptide plant hormone, phytosulfokine. *Science*, **296**, 1470–1472.

Mcdougall, G.J. (2000) A comparison of proteins from the developing xylem of compression and non-compression wood of branches of sitka spruce (*Picea sitchensis*) reveals a differentially expressed laccase. *J. Exp. Bot.*, **51**, 1767.

Millar, A.H. (2004) Location, location, location: surveying the intracellular real estate through proteomics in plants. *Funct. Plant Biol.*, **31**, 563–571.

Nadeau, J.A. and Sack, F.D. (2003) Stomatal development: cross talk puts mouths in place. *Trends Plant Sci.*, **8**, 294–299.

Narvaez-Vasquez, J., Pearce, G. and Ryan, C.A. (2005) The plant cell wall matrix harbors a precursor of defense signaling peptides. *Proc. Natl Acad. Sci. USA*, **102**, 12974–12977.

Ndimba, B.K., Chivasa, S., Hamilton, J.M., Simon, W.J. and Slabas, A.R. (2003) Proteomic analysis of changes in the extracellular matrix of *Arabidopsis* cell suspension cultures induced by fungal elicitors. *Proteomics*, **3**, 1047–1059.

Nickel, W. (2005) Unconventional secretory routes: direct protein export across the plasma membrane of mammalian cells. *Traffic*, **6**, 607–614.

Nielsen, H., Engelbrecht, J., Brunak, S. and Von Heijne, G. (1997) Identification of prokaryotic and eukaryotic signal peptides and prediction of their cleavage sites. *Protein Eng.*, **10**, 1–6.

Nishitani, K. (2005) Division of roles among members of the XTH gene family in plants. *Plant Biosys.*, **139**, 98–101.

Nothnagel, E.A. (1997) Proteoglycans and related components in plant cells. *Int. Rev. Cytol.*, **174**, 195–291.

Nürnberger, T., Brunner, F., Kemmerling, B. and Piater, L. (2004) Innate immunity in plants and animals: striking similarities and obvious differences. *Immunol. Rev.*, **198**, 247–266.

O'Neill, M.A. and York, W.S. (2003) The composition and structure of plant primary cell wall. In: Rose, J.K.C. (ed) *The Plant Cell Wall.* Blackwell publishing Ltd, Oxford, UK.

Oh, I.S., Park, A.R., Bae, M.S., Kwon, S.J., Kim, Y.S., Lee, J.E., Kang, N.Y., Lee, S. M., Cheong, H. and Park, O.K. (2005) Secretome analysis reveals an *Arabidopsis* lipase involved in defense against *Alternaria brassicicola. Plant Cell*, **17**, 2832–2847.

Okushima, Y., Koizumi, N., Kusano, T. and Sano, H. (2000) Secreted proteins of tobacco cultured BY2 cells: identification of a new member of pathogenesis-related proteins. *Plant Mol. Biol.*, **42**, 479–488.

Olivieri, F., Godoy, A.V., Escande, A. and Casalongue, C.A. (1998) Analysis of intercellular washing fluids of potato tubers and detection of increased proteolytic activity upon fungal infection. *Physiol. Plant.*, **104**, 232–238.

Otte, O. and Barz, W. (1996) The elicitor-induced oxidative burst in cultured chickpea cells drives the rapid insolubilization of two cell wall structural proteins. *Planta*, **200**, 238–246.

Oufattole, M., Park, J.H., Poxleitner, M., Jiang, L.W. and Rogers, J.C. (2005) Selective membrane protein internalization accompanies movement from the endoplasmic reticulum to the protein storage vacuole pathway in *Arabidopsis*. *Plant Cell*, **17**, 3066–3080.

Pardo, M., Ward, M., Bains, S., Molina, M., Blackstock, W., Gil, C. and Nombela, C. (2000) A proteomic approach for the study of *Saccharomyces cerevisiae* cell wall biogenesis. *Electrophoresis*, **21**, 3396–3410.

Pearce, G., Moura, D.S., Stratmann, J. and Ryan Jr. C.A., (2001) RALF, a 5-kDa ubiquitous polypeptide in plants, arrests root growth and development. *Proc. Natl Acad. Sci. USA.*, **98**, 12843–12847.

Pignocchi, C. and Foyer, C.H. (2003) Apoplastic ascorbate metabolism and its role in the regulation of cell signalling. *Curr. Opin. Plant Biol.*, **6**, 379–389.

Pitarch, A., Sanchez, M., Nombela, C. and Gil, C. (2002) Sequential fractionation and two-dimensional gel analysis unravels the complexity of the dimorphic fungus *Candida albicans* cell wall proteome. *Mol. Cell. Proteom.*, **1**, 967–982.

Rep, M., Dekker, H.L., Vossen, J.H., De Boer, A.D., Houterman, P.M., Speijer, D., Back, J.W., De Koster, C.G. and Cornelissen, B.J. (2002) Mass spectrometric identification of isoforms of PR proteins in xylem sap of fungus-infected tomato. *Plant Physiol.*, **130**, 904–917.

Ringli, C., Keller, B. and Ryser, U. (2001) Glycine-rich proteins as structural components of plant cell walls. *Cell. Mol. Life Sci.*, **58**, 1430–1441.

Robertson, D., Mitchell, G.P., Gilroy, J.S., Gerrish, C., Bolwell, G.P. and Slabas, A.R. (1997) Differential extraction and protein sequencing reveals major differences in patterns of primary cell wall proteins from plants. *J. Biol. Chem.*, **272**, 15841–15848.

Roitsch, T. and Gonzalez, M.-C. (2004) Function and regulation of plant invertases: sweet sensations. *Trends Plant Sci.*, **9**, 606–613.

Rojo, E., Sharma, V.K., Kovaleva, V., Raikhel, N.V. and Fletcher, J.C. (2002) CLV3 is localized to the extracellular space, where it activates the *Arabidopsis* CLAVATA stem cell signaling pathway. *Plant Cell*, **14**, 969–977.

Rose, J.K. and Bennett, A.B. (1999) Cooperative disassembly of the cellulose-xyloglucan network of plant cell walls: parallels between cell expansion and fruit ripening. *Trends Plant Sci.*, **4**, 176–183.

Rose, J.K., Braam, J., Fry, S.C. and Nishitani, K. (2002) The XTH family of enzymes involved in xyloglucan endotransglucosylation and endohydrolysis: current perspectives and a new unifying nomenclature. *Plant Cell Physiol.*, **43**, 1421–1435.

Rose, J.K.C., Catalá, C., Gonzalez-Carranza, C.Z.H. and Roberts, J.A. (2003) Plant cell wall disassembly. In: Rose, J.K.C. (ed) *The Plant Cell Wall.* Blackwell publishing Ltd, Oxford, UK.

Rose, J.K., Saladie, M. and Catala, C. (2004) The plot thickens: new perspectives of primary cell wall modification. *Curr. Opin. Plant Biol.*, **7**, 296–301.

Ruan, Y.L., Mate, C., Patrick, J.W. and Brady, C.J. (1995) Nondestructive collection of apoplast fluid from developing tomato fruit using a pressure dehydration procedure. *Aust. J. Plant Physiol.*, **22**, 761–769.

Ruan, Y.L., Patrick, J.W. and Brady, C.J. (1996) The composition of apoplast fluid recovered from intact developing tomato fruit. *Aust. J. Plant Physiol.*, **23**, 9–13.

Ryan, C.A., Pearce, G., Scheer, J. and Moura, D.S. (2002) Polypeptide hormones. *Plant Cell*, **14**, S251–S264.

Saravanan, R.S. and Rose, J.K. (2004) A critical evaluation of sample extraction techniques for enhanced proteomic analysis of recalcitrant plant tissues. *Proteomics*, **4**, 2522–2532.

Sattelmacher, B. (2001) The apoplast and its significance for plant mineral nutrition. *New Phytol.*, **149**, 167–192.

Schulze-Lefert, P. (2004) Knocking on heaven's wall: pathogenesis of and resistance to biotrophic fungi at the cell wall. *Curr. Opin. Plant Biol.*, **7**, 377–383.

Sherson, S.M., Alford, H.L., Forbes, S.M., Wallace, G. and Smith, S.M. (2003) Roles of cell-wall invertases and monosaccharide transporters in the growth and development of *Arabidopsis*. *J. Exp. Bot.*, **54**, 525–531.

Showalter, A.M. (1993) Structure and function of plant cell wall proteins. *Plant Cell*, **5**, 9–23.

Showalter, A.M. (2001) Arabinogalactan-proteins: structure, expression and function. *Cell. Mol. Life Sci.*, **58**, 1399–1417.

Small, I., Peeters, N., Legeai, F. and Lurin, C. (2004) Predotar: a tool for rapidly screening proteomes for N-terminal targeting sequences. *Proteomics*, **4**, 1581–1590.

Somerville, C., Bauer, S., Brininstool, G., Facette, M., Hamann, T., Milne, J., Osborne, E., Paredez, A., Persson, S., Raab, T., Vorwerk, S. and Youngs, H. (2004) Toward a systems approach to understanding plant cell walls. *Science*, **306**, 2206–2211.

Stressmann, M., Kitao, S., Griffith, M., Moresoli, C., Bravo, L.A. and Marangoni, A.G. (2004) Calcium interacts with antifreeze proteins and chitinase from cold-acclimated winter rye. *Plant Physiol.*, **135**, 364–376.

Takos, A.M., Dry, I.B. and Soole, K.L. (1997) Detection of glycosyl-phosphatidylinositol-anchored proteins on the surface of *Nicotiana tabacum* protoplasts. *FEBS Lett.*, **405**, 1–4.

Tang, W., Kelley, D., Ezcurra, I., Cotter, R. and Mccormick, S. (2004) LeSTIG1, an extracellular binding partner for the pollen receptor kinases LePRK1 and LePRK2, promotes pollen tube growth *in vitro*. *Plant J.*, **39**, 343–353.

Van Der Hoorn, R.A.L. and Jones, J.D. (2004) The plant proteolytic machinery and its role in defence. *Curr. Opin. Plant Biol.*, **7**, 400–407.

Veronese, P., Ruiz, M.T., Coca, M.A., Hernandez-Lopez, A., Lee, H., Ibeas, J.I., Damsz, B., Pardo, J.M., Hasegawa, P.M., Bressan, R.A. and Narasimhan, M.L. (2003) In defense against pathogens. Both plant sentinels and foot soldiers need to know the enemy. *Plant Physiol.*, **131**, 1580–1590.

Villarejo, A., Buren, S., Larsson, S., Dejardin, A., Monne, M., Rudhe, C., Karlsson, J., Jansson, S., Lerouge, P., Rolland, N., Von Heijne, G., Grebe, M., Bako, L. and Samuelsson, G. (2005) Evidence for a protein transported through the secretory pathway en route to the higher plant chloroplast. *Nat. Cell Biol.*, **7**, 1224–1231.

Von Groll, U., Berger, D. and Altmann, T. (2002) The subtilisin-like serine protease SDD1 mediates cell-to-cell signaling during *Arabidopsis* stomatal development. *Plant Cell*, **14**, 1527–1539.

Vorwerk, S., Somerville, S. and Somerville, C. (2004) The role of plant cell wall polysaccharide composition in disease resistance. *Trend in Plant Sci.*, **9**, 203–209.

Wang, S.B., Hu, Q., Sommerfeld, M. and Chen, F. (2004) Cell wall proteomics of the green alga *Haematococcus pluvialis* (Chlorophyceae). *Proteomics*, **4**, 692–708.

Watson, B.S., Lei, Z., Dixon, R.A. and Sumner, L.W. (2004) Proteomics of *Medicago sativa* cell walls. *Phytochemistry*, **65**, 1709–1720.

Wojtaszek, P., Trethowan, J. and Bolwell, G.P. (1995) Specificity in the immobilisation of cell wall proteins in response to different elicitor molecules in suspension-cultured cells of French bean (*Phaseolus vulgaris* L.). *Plant Mol. Biol.*, **28**, 1075–1087.

Wu, H., De Graaf, B., Mariani, C. and Cheung, A.Y. (2001) Hydroxyproline-rich glycoproteins in plant reproductive tissues: structure, functions and regulation. *Cell. Mol. Life Sci.*, **58**, 1418–1429.

Yamane, H., Lee, S.J., Kim, B.D., Tao, R. and Rose, J.K. (2005) A coupled yeast signal sequence trap and transient plant expression strategy to identify genes encoding secreted proteins from peach pistils. *J. Exp. Bot.*, **56**, 2229–2238.

York, W.S., Qin, Q. and Rose, J.K. (2004) Proteinaceous inhibitors of endo-beta-glucanases. *Biochim. Biophys. Acta*, **1696**, 223–233.

Yu, Q., Tang, C., Chen, Z. and Kuo, J. (1999) Extraction of apoplastic sap from plant roots by centrifugation. *New Phytol.*, **143**, 299–304.

Zareie, R., Melanson, D.L. and Murphy, P.J. (2002) Isolation of fungal cell wall degrading proteins from barley (*Hordeum vulgare* L.) leaves infected with Rhynchosporium secalis. *Mol. Plant Microbe. Interact.*, **15**, 1031–1039.

9 Proteomics of plant mitochondria

Natalia V. Bykova and Ian M. Møller

9.1 Mitochondria as attractive targets for subcellular proteomics

Mitochondria are ubiquitous organelles with a fundamental role in cellular homeostasis. Apart from the primary role in the oxidation of organic acids via the tricarboxylic acid (TCA) cycle and the synthesis of ATP, they also undertake transcription and translation, actively import proteins and metabolites from the cytosol, export organic acid intermediates for cellular biosynthesis, influence programmed cell death (PCD), and respond to cellular signals such as oxidative stress. In plants, mitochondria also perform many important functions such as synthesis of nucleotides, amino acids, lipids and vitamins, and participate in photorespiration.

Mitochondria are thought to have arisen from the endosymbiosis of an α-proteobacterial organism (Andersson et al., 1998; Gray et al., 1999). However, even the most gene-rich mitochondrial genomes encode far fewer genes than the smallest bacterial genome, reflecting the transfer of genes from the presumed mitochondrial ancestor to the nucleus (Burger et al., 2003). Mitochondrial proteome diversity was generated both by gene evolution leading to novel proteins located in the mitochondrion and by relocation of conserved gene products due to altered targeting. The analysis of diversification of mitochondrial proteomes indicates that the number of encoding genes differs among species and the size of the proteome is variable over evolutionary time. Multicellular organisms require tissue-specific isoforms of some mitochondrial proteins, but at the same time a large fraction of species-specific mitochondrial proteins strongly suggests that species evolution has been accompanied by diversification of mitochondrial proteins due to species-related organelle functions (Richly et al., 2003). Prediction of the size and composition of the plant mitochondrial proteome suggests a total of approximately 3000 proteins with 23% of them being homologous with α-proteobacterial organisms, 62% of proteins not shared with the mitochondrial proteome of other species, and one of the largest pool of proteins found to be conserved among all non-parasite mitochondrial proteomes (Richly et al., 2003).

Among the estimated 3000 proteins that accumulate in mitochondria, fewer than 100 are encoded by the mitochondrial genome (Peeters and Small, 2001), the vast majority is encoded by the nuclear genome. These proteins are synthesized by cytosolic ribosomes and are subsequently imported into the organelles via active protein transport systems.

As an aid to better understand the mitochondrial metabolic network, it would be valuable to know the complete set of mitochondrial proteins, the proteome. A number of studies in recent years have addressed this aspect from different angles and to date only about 400 plant mitochondrial proteins have been experimentally identified and the full range of mitochondrial functions in plants is still unknown (Millar et al., 2005). We here give an overview of the present state of knowledge about the plant mitochondrial proteome.

9.2 Methods and approaches briefly for identification of proteins

9.2.1 Gel electrophoresis and immunoblotting

The most direct approach to obtain the complete mitochondrial proteome is to isolate highly purified mitochondria and identify the subset of proteins by mass spectrometry (MS). The use of isoelectric focusing (IEF) and subsequent SDS-PAGE to separate solubilized proteins is the most popular with the major advantage that it permits the visualization of quantitatively comparable sets of proteins while giving information about isoelectric point (pI) and apparent molecular size. Several early reports presented two-dimensional (2D) gel arrays of mitochondrial protein profiles from potato (Colas des Francs-Small et al., 1992, 1993), pea (Humphrey-Smith et al., 1992), maize (Dunbar et al., 1997), and *Arabidopsis* (de Virville et al., 1998). These reports highlighted changes in different tissues from the same plant or within the same tissue during development. The identification of protein spots in these profiles was performed by immunoblotting and N-terminal sequencing. Introduction of MS analysis allowed a much more comprehensive characterization of mitochondrial proteins.

9.2.2 Blue-native PAGE and membrane proteomics

As much as 40% of the proteins in a mitochondrion are found in the outer and inner membrane (Douce, 1985) and it is therefore particularly important to be able to separate and identify membrane proteins when studying the mitochondrial proteome. Many of the membrane proteins are not exposed to the soluble phase but are highly hydrophobic core components of multisubunit complexes. The precipitation of these hydrophobic components is avoided by maintaining the complex structure, for example, by blue-native (BN) PAGE (Schägger et al., 1994). A subsequent denaturing SDS-PAGE dimension separates the complexes into individual subunits ready for identification (Jänsch et al., 1996; Werhahn and Braun, 2002). BN-PAGE was shown to be a suitable procedure for the separation of mitochondrial protein complexes of the inner mitochondrial membrane (Schägger et al., 1994; Jänsch et al., 1996), the outer mitochondrial membrane (Jänsch et al., 1998) as well as the mitochondrial matrix (e.g., Bykova et al., 2003a). This approach has also provided novel insights into the respiratory chain of plant mitochondria, supercomplexes, and protein–protein interactions (Eubel et al., 2003).

9.2.3 One- and two-dimensional LC/MS/MS

There are several major drawbacks and intrinsic limitations of 2D-PAGE-based method. These include the difficulty of identifying low-solubility hydrophobic membrane proteins, highly basic and acidic proteins outside the pI ranges used as well as outside the mass range of 10–100 kDa. In addition, low-abundance proteins are significantly underrepresented, particularly when a relatively small number of very abundant proteins predominate, because of the lack of resolution for proteins of similar mass and pI. Gel arraying also prevents analysis of peptides that are not easily eluted from polyacrylamide gel spots.

One possible way to overcome these problems is based on gel-free separation of peptides from complex lysates using nano- or micro-scale one-dimensional (1D) or 2D liquid chromatography (LC) tandem MS (nano-LC-MS/MS) (Heazlewood et al., 2003a, 2004; Kristensen et al., 2004).

A comparison of different proteomic approaches confirmed that more hydrophobic, large, and basic proteins were identified by BN/SDS-PAGE and the non-gel LC-MS/MS analysis than by IEF/SDS-PAGE separations (Heazlewood et al., 2003a). Another way to solubilize and separate hydrophobic proteins is 1D SDS-PAGE (Millar and Heazlewood, 2003).

These analyses clearly demonstrate that to fully characterize the mitochondrial proteome a series of experimental approaches are required to maximize the representation of all proteins without bias based on physical characteristics of size, charge, and hydrophobicity. Although LC-MS/MS eliminates many problems, it does not appear to be an ideal and complete solution. A great deal of information about the parent protein from which peptides are derived, about possible posttranslational modifications (PTM) (such as multiple spots on IEF-PAGE) and protein–protein interactions (such as ETC complexes on BN-PAGE) is not provided.

9.2.4 Prediction of mitochondrially imported proteins from genomic information

In silico predictions is another way of identifying mitochondrially located proteins. Several commonly used algorithms are based on the fact that a large number of proteins targeted to the matrix or the inner membrane contain a positively charged, amphipathic N-terminal signal sequence (Schneider and Fechner, 2004). However, proteins targeted to the endoplasmic reticulum or highly positively charged proteins such as ribosomal subunits often also give high scores, indicating the insufficient specificity of these approaches. Mitochondrial proteins containing targeting signals with other characteristics, in particular internal signals, are missed (Reichert and Neupert, 2004). Overall, these prediction programs (TargetP, Predotar, MitoProt II, SubLoc, and iPSORT) identified about 40–70% of the experimental mitochondrial proteome from *Arabidopsis* (Heazlewood et al., 2004).

A comparison of six different mitochondrial prediction programs showed that the programs predicted 3000–4500 proteins to be mitochondrial while the best could only predict 47% of the proteins actually found (Heazlewood et al., 2004).

There are several possible explanations for this poor performance: the absence of targeting presequences, cryptic presequences not recognized by these programs, lack of presequences due to incorrect annotation (Heazlewood et al., 2003a; Millar and Heazlewood, 2003), hitherto unrecognized targeting methods or simply that contaminating proteins are incorrectly annotated as mitochondrial in origin. The development of new improved prediction algorithms will be facilitated by access to an increasingly complete catalogues of organellar proteins. Alternative prediction approaches have recently been suggested based either on combination of ortholog cross-species matches of experimentally proven mitochondrially located proteins and N-terminal targeting predictors (Richly et al., 2003), or on more basal complementary information from the primary sequence such as functional domain and the pseudo-amino acid composition, and protein family domain composition (Millar et al., 2005, and references therein).

The *Arabidopsis* Mitochondrial Protein Database (http://www.ampdb.bcs.uwa.edu.au/) was created based on the predicted and experimentally confirmed protein complement of mitochondria from the model plant (Heazlewood and Millar, 2005). The database was formed using the total non-redundant nuclear and organelle encoded sets of protein sequences and allows relational searching of published mitochondrial proteomic analyses, a set of predictions from six independent subcellular-targeting prediction programs, and orthology predictions based on pairwise comparison of the *Arabidopsis* protein set with known yeast and human mitochondrial proteins and with the proteome of *Rickettsia*.

9.3 Mitochondrial proteins identified

9.3.1 Total proteins

The first studies of the plant mitochondrial proteome were concentrated on *Arabidopsis* as the first plant with a fully sequenced genome. Mitochondria were isolated from *Arabidopsis* stems and leaves or from *Arabidopsis* suspension cell cultures (Kruft et al., 2001; Millar et al., 2001). Cell cultures grown in the dark were used to obtain highly purified and functional mitochondria with good yield. After separation of total protein extracts on 2D IEF/SDS-PAGE electrophoresis, 52 protein spots were identified by immunoblotting, Edman degradation and two types of MS analysis, matrix-assisted laser desorption/ionization-time of flight (MALDI-TOF) peptide mass fingerprinting (PMF) and electrospray ionization-quadropole time-of-flight (ESI-QTQF) (Kruft et al., 2001). Thirty percent of the proteins have a role in respiration and 25% in primary metabolism such as pyruvate decarboxylation, TCA cycle, and amino acid and nucleotide metabolism. Other proteins are chaperones or involved in molecular transport, mitochondrial protein biosynthesis and protection against oxidative damage. However, more than 20% of the proteins identified were not previously described for plant mitochondria, indicating novel mitochondrial functions. A set of proteins analysed could not be unambiguously identified on the basis of homology sequence comparisons but their mitochondrial localization was supported by computer prediction programs. Although about 50% of the identified *Arabidopsis* mitochondrial proteins were moderately hydrophobic and represented proteins localized in the outer or inner mitochondrial membrane, a class of highly hydrophobic membrane proteins with more than one membrane-spanning helix could not be detected under the conditions used for solubilization and IEF resolution (Kruft et al., 2001).

Combination of subcellular fractionation into soluble, peripheral and integral membrane proteins with comparison of subfractions on 2D gels and MALDI-TOF PMF identification allowed experimental characterization of a total of 91 protein spots, 81 of which had defined functions based on sequence comparisons (Millar et al., 2001). A set of 43 protein spots were identified as soluble proteins, 21 identified as peripheral membrane proteins, and 18 spots were deemed integral membrane. The characterized proteins represented components of TCA cycle, ETC, membrane carriers, RNA metabolism, translational and protein processing apparatus, chaperonins and HSPs, and enzymes of carbon metabolism. An evaluation of the degree of membrane association demonstrated that many of the major proteins shown to be soluble were subunits of the TCA cycle, HSP60 and HSP70, protein synthesis, and antioxidative stress. The subunits of the F_0F_1-ATP synthase, some subunits of ETC complexes as well as some enzymes of carbon metabolism were in the peripheral membrane class. As expected, membrane carriers and protein processing apparatus had a high degree of association with the integral protein fraction. However, a variety of mitochondrial proteins that are known to be present in significant abundance were not observed in these studies due to the hydrophobicity and basic nature of the protein sequences which was outside the resolving ability of current 2D electrophoresis. These included nearly all of the proteins encoded in the mitochondrial genome and a range of inner membrane carriers. In addition, nearly 40% of the mass spectra acquired could not be matched to predicted open reading frames in the *Arabidopsis* genome. In large part this could be due to protein PTMs which could not be characterized (predicted at best) by PMF and requires tandem MS analysis. The presence of such modifications was evident from the differences in apparent molecular masses and pIs of experimental protein spots and the matched gene products. In addition, the low overall hit rate could be explained by inaccuracies of intron–exon boundaries in translated open reading frames of *Arabidopsis* sequences.

The identified set of *Arabidopsis* mitochondrial proteins was expanded to a total of 416 proteins using a gel-free approach based on LC-MS/MS (Heazlewood et al., 2004). This approach can potentially alleviate the problems of highly hydrophobic, basic, small, or very large molecular mass proteins (more than 80 kDa) as well as low-abundance proteins that are often excluded from IEF 2D gels. An overall comparison of gel-based and LC-MS/MS methods indicated that more than 95% of the proteins identified previously by 2D gel analysis of whole mitochondrial samples (Kruft et al., 2001; Millar et al., 2001; Werhahn and Braun, 2002) were confirmed by LC-MS/MS. Of the total set of 416 non-redundant proteins 236 were identified by LC-MS/MS only. When the proteins were grouped by function the most highly represented were the ETC complexes and TCA cycle components and those involved in

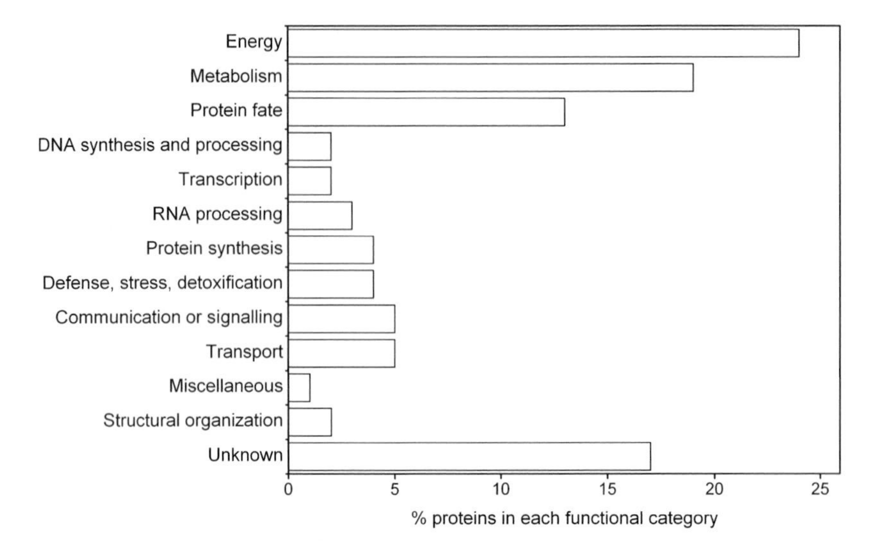

Figure 9.1 Functional distribution of the *Arabidopsis* mitochondrial proteome. The proportion of proteins in 13 functional groups representing a total set of 416 mitochondrial proteins identified in Heazlewood et al. (2004).

metabolism and protein fate. A large set of proteins (17%) could not be confidently assigned a function and were grouped as unclassified proteins (see Figure 9.1).

The analysis of protein representations in EST databases allowed an assessment of transcript abundance, and thus possible protein abundance, for different gene products (Heazlewood et al., 2004). Nearly 45% of the novel LC-MS/MS-based identifications were represented by four or fewer ESTs. The authors suggest that the large number of low-EST-number members could be viewed as evidence of low-abundance paralogs, or even pseudogenes with high levels of identity to major protein isoforms. On the other hand, the low number of ESTs among the novel LC-MS/MS data broadly reflected low-abundance proteins within the mitochondrial proteome. Many of the identified low-abundance components were involved in important areas of mitochondrial information transfer, regulation, and signalling.

9.3.2 Membrane proteins and supercomplexes

BN 2D PAGE has turned out to be a very powerful technique for membrane proteomics studies (Jänsch et al., 1996). The combination of three different electrophoresis procedures (BN-PAGE, IEF and SDS-PAGE) resulted in high-resolution capacity for protein complexes and facilitated detection of extremely hydrophobic proteins (Werhahn and Braun, 2002). The subunit composition of protein complexes of the oxidative phosphorylation (OXPHOS) system has been extensively studied.

Complex I was chromatographically purified from different plants and resolved into about 30 different subunits (Leterme and Boutry, 1993; Herz et al., 1994;

Rasmusson et al., 1994; Jänsch et al., 1996; Combettes and Grienenberger, 1999). Recently, Heazlewood et al. (2003b) identified 30 subunits of *Arabidopsis* and 24 different proteins of rice complex I after separation on BN 2D PAGE and MS analysis. This revealed that plants contain the 14 highly conserved complex I subunits found in other eukaryotic and related prokaryotic enzymes and a set of 9 proteins widely found in eukaryotic complexes. A significant number of proteins present in bovine complex I, but absent from fungal complex I, are also absent from plant complex I and are not encoded in plant genomes. A series of plant-specific nuclear-encoded complex I associated subunits were identified, including ferripyochelin-binding protein-like subunits and a range of small proteins of unknown function. Interestingly, L-galactono-1,4-lactone dehydrogenase, which catalyses the terminal step of ascorbic acid biosynthesis in plant mitochondria (Millar et al., 2003), was found to be associated with complex I. A putative gamma carbonic anhydrase also co-purifies with complex I (Parisi et al., 2004).

Complex II (SDH) was isolated from different sources (Burke et al., 1982; Hattori and Asahi, 1982; Igamberdiev and Falaleeva, 1994) with only two soluble subunits visible after gel electrophoresis. Further clarification of the protein compositions of the complex was done on *Arabidopsis* mitochondria by BN 2D PAGE separation and systematic identification of individual subunits by MS/MS analysis and Edman sequencing (Millar et al., 2004a). It was found to consist of the four classical SDH subunits as well as four subunits unknown in mitochondria from other eukaryotes. Similar to complex II from other heterotrophic eukaryotes, all eight subunits of plant complex II are nuclear-encoded, which makes it an exception because all other OXPHOS complexes include at least one mitochondrially-encoded subunit.

Complex III from potato was purified by cytochrome *c* affinity chromatography and shown to comprise 10 different subunits (Braun and Schmitz, 1992). All potato proteins were partially sequenced by Edman degradation and genes encoding these proteins were characterized. Plant mitochondrial complex III uniquely includes the mitochondrial processing peptidase (MPP) (Braun and Schmitz, 1995).

In *Arabidopsis*, complex IV comprised 9–10 protein bands and co-migrated in part with the translocase of the outer membrane (TOM) complex. Differential analysis of TOM and complex IV revealed that the latter could probably be resolved into eight subunits with similarity to known complex IV subunits from other eukaryotes and a further six putative subunits of unknown function (Millar et al., 2004a).

Purified complex V of plant mitochondria could be resolved into about 10–15 different subunits by electrophoresis (Hamasur and Glaser, 1992; Jänsch et al., 1996). Heazlewood et al. (2003c) characterized the protein components of *Arabidopsis* complex V and linked these to specific gene products. BN-PAGE separation revealed intact F_1F_0, and separated F_1 and F_0 components. The F_1 complex contained five well-characterized subunits, while the four subunits in the F_0 complex were subunit 9, d subunit, and *orfB* that encodes the plant ATP8 and *orf25* that encodes the plant ATP4. One complex V component, the nuclear-encoded F_Ad subunit, appears to be plant-specific and does not have a clear mammalian or yeast counterpart. Chimerics of *orf25*, *orfB*, subunit 9 and subunit 6 have been associated with cytoplasmic male sterility (CMS) in a variety of plant species (see Heazlewood et al., 2003c, for references). Their common physical localization as subunits of plant F_0 may be important in unraveling a common mechanism for CMS.

A systematic investigation of supramolecular structure of the plant mitochondrial respiratory chain revealed an ordered association of protein complexes forming larger structures (Eubel et al., 2003). The existence of respiratory supercomplexes was also reported for yeast and mammalian mitochondria and the term 'respirasome' was suggested for supercomplexes containing the complexes I, III_2 (dimeric), and IV, which autonomously can carry out respiration in the presence of cytochrome *c* and ubiquinone (Schägger and Pfeiffer, 2000). Using gentle protein solubilizations with non-ionic detergents, BN 1D and 2D PAGE, and MS identification, three supercomplexes could be found in mitochondrial fractions from *Arabidopsis*, potato, bean, and barley. These included dimeric ATP synthase, a supercomplex formed by dimeric complex III and complex I, and a supercomplex containing two copies of dimeric complex III and two copies of complex I. Complex II, the alternative oxidase (AOX) and the rotenone-insensitive NAD(P)H dehydrogenases were not part of supercomplexes. Additional supercomplexes of lower abundance and including complex IV, were recently discovered in mitochondria isolated from potato tubers and stems (Eubel et al., 2004). Visualization by in-gel activity staining for COX revealed novel supercomplexes that had III_2IV_{1-2}, and $I_1III_2IV_{1-4}$ compositions. The functional significance of the supercomplexes remains to be elucidated. In yeast, supercomplexes were reported to enhance activity rates of respiratory electron transport (Schägger and Pfeiffer, 2000). It was also suggested that supercomplex formation increases the capacity of the inner mitochondrial membrane for protein insertion (Arnold et al., 1998).

Besides the supercomplexes and the respiratory complexes I–V, several additional protein complexes were identified on BN gels: the prohibitin, HSP60, TOM, and formate dehydrogenase (FDH) complexes (Eubel et al., 2003). In contrast, several protein complexes known to be present in the soluble fraction of plant mitochondria (e.g. PDH and the glycine dehydrogenase complex), were never detected on BN gels possibly due to their instability under conditions used for solubilization and separation. The presence of protein complexes comprising mitochondrial dehydrogenases of the TCA cycle was reported on the basis of diffusion rate measurements of individual enzymes of this metabolic pathway in mammalian mitochondria (Haggie and Verkman, 2002). Therefore, a significant part of all mitochondrial proteins, membrane-associated as well as soluble, may be part of protein complexes, and these protein complexes may be involved in the formation of even larger supermolecular structures. This offers several physiological advantages including substrate channelling, metabolic pathway regulation, and the realization of complicated biochemical reactions with reactive intermediates (Ovadi and Srere, 2000).

Subfractionation of the mitochondrial proteome into a soluble fraction and fractions containing peripheral and integral membrane proteins has been used by several groups to enhance the resolution and thereby improve the basis for our understanding of the biochemical machinery (Millar et al., 2001; Bardel et al.,

2002; Brugière et al., 2004). Bardel et al. (2002) focused their attention on the soluble complement of plant mitochondria and reported 2D-maps of soluble proteins of mitochondria isolated from different tissues and organisms thus revealing the impact of tissue differentiation at the mitochondrial level (see below for detailed review). To get a more complete array of membrane proteins from *Arabidopsis* mitochondria, LC-MS/MS analysis of the hydrophobic proteome was combined with various extraction procedures, such as chloroform/methanol extraction, alkaline, and saline treatments (Brugière et al., 2004). These fractionations allowed the differential extraction of proteins based on their physico-chemical properties and increased the representation of low-abundance proteins. A set of 114 proteins was identified with about 40% identifications not listed in other proteomic studies. The identified proteins were classified according to their known (inner membrane, outer membrane, and matrix) or putative localization within the mitochondria. It was concluded that about two-thirds (75/114) of the identified proteins were genuine mitochondrial membrane proteins.

Components of all the respiratory complexes were identified as well as several members of a large mitochondrial carrier family (MCF) including two ABC transporters and several porins (Brugière et al., 2004). Members of the porin family included a series of voltage-dependent anion channels (VDACs). Several subunits of TIM and TOM complexes and MPP subunits were identified as members of the protein import machinery. Interestingly, several proteins were suggested as candidates for the protein export machinery that is required for the insertion of some integral membrane proteins into the mitochondrial inner membrane.

Four chaperones and a calnexin homologue were found in mitochondrial membrane fractions. Five prohibitins were identified that form another mitochondrial inner membrane complex. Three cytochrome P450 proteins contained at least one predicted transmembrane domain. Oxidative stress-responsive proteins also associated with mitochondrial membrane were phospholipid hydroperoxide glutathione peroxidase, an ascorbate peroxidase, a superoxide dismutase (SOD), thioredoxin *h3*, and a reticuline oxidase-like protein. The high proportion (almost one-third) of putative proteins that were identified in this study requires further validation by relevant functional and localizational studies (Brugière et al., 2004).

9.3.3 Proteins with counterparts in other organisms and plant-specific proteins

Biochemical investigations revealed that isolated plant mitochondria share many similarities with those from animals and fungi. However, plant mitochondria also contain additional features such as non-phosphorylating bypasses in the ETC (Møller, 2001a; Rasmusson et al., 2004), specialized metabolite carriers (Picault et al., 2004), and enzymes involved in the synthesis of folate, lipoic acid, and vitamin C (Rebeillé et al., 1997; Bartoli et al., 2000; Gueguen et al., 2000). Only a few of the nuclear genes encoding proteins that maintain these unique functions have been identified (Heazlewood et al., 2004).

Putative orthology networks were mapped between yeast, human, and *Arabidopsis* mitochondrial proteomics and the *Rickettsia prowazekii* proteome to provide detailed insights into the divergence of the plant mitochondrial proteome from those of other eukaryotes (Heazlewood et al., 2004). Cross-kingdom similarity comparisons were used to identify putative orthologous proteins with conserved function based on the fact that the endosymbiosis of the mitochondrial progenitor by the ancestral eukaryotic cell occurred before the divergence of plants and animals (Gray et al., 1999). A total of 117 yeast mitochondrial proteins were identified as putative orthologs of the *Arabidopsis* mitochondrial protein set. More than 75% of the putative orthologs fall into the major functional groups of energy, metabolism, and protein fate. Similar comparisons with the human mitochondrial proteome and the *Rickettsia* protein set identified within the *Arabidopsis* mitochondrial population 148 and 85 putative orthologs, respectively. A central cluster was found with strict common overlap of 45 putative orthologs between the *Rickettsia* set and the current mitochondrial proteomes of these three divergent organisms. This common set represented proteins of energy (70%), metabolism (13%), and protein fate (15%) (Figure 9.2A).

A very different putative orthology pattern was found for low-abundance proteins involved in mitochondria–cellular interaction (signalling, transport, and structure) and information-transfer processes (DNA synthesis and processing, transcription, and RNA binding/processing) (Heazlewood et al., 2004). Although putative orthologous proteins were present in most of these functional classes, the overlap between different organisms was very low (Figure 9.2B). The kinases observed in human and plant mitochondria were absent from yeast mitochondrial sets of proteins, whereas plant transporters and transcriptional regulators present in the yeast set were not found in the human set. The large set of *Arabidopsis* proteins of unknown function does not have many putative orthologs in other species, suggesting that these proteins are plant-specific.

Thus, although the central bioenergetic and metabolic function of mitochondria appears to be highly conserved during evolution, the regulation of the mitochondrial genome, cellular signalling, and transport between the mitochondrion and the cell have diverged more widely in sequence and/or may have been more readily exchanged with proteins encoded in host (nucleus) and co-endosymbiont genomes, recruiting novel proteins to play roles that are unique to plant mitochondria. One particularly large family of such proteins is the pentatricopeptide repeat proteins. In *Arabidopsis* this gene family contains >400 genes most of them encoding mitochondrial proteins whereas in humans it contains only six genes (Lurin et al., 2004). Ten of these proteins have been identified in *Arabidopsis* mitochondria (Heazlewood et al., 2004), but we can clearly expect to find several hundred more. It has been proposed that they are involved in organellar gene expression by playing the role of sequence-specific adaptors for a variety of other RNA-associating proteins (Lurin et al., 2004).

9.3.4 *Intracellular communication*

Plant mitochondria maintain intracellular communication with the cytosol. Biochemical characterization of plant mitochondrial carrier function over the last

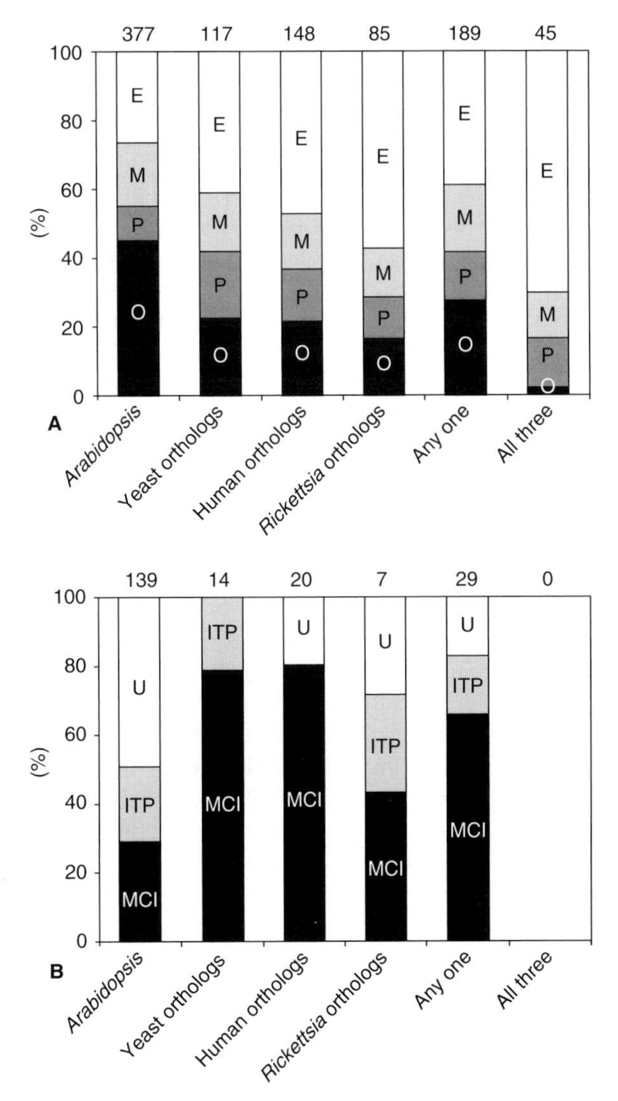

Figure 9.2 Orthologous comparison of mitochondrial proteomes from *Arabidopsis*, yeast, human, and *Rickettsia*. A: Distributions of high-abundance functional categories of matching putative orthologs. Orthologs were determined through high-stringency sequence homology matching as described in Heazlewood et al. (2004). The distribution of putative orthologs to the *Arabidopsis* mitochondrial set of 377 proteins (excluding 26 possible paralogs) is presented for the functional classes of energy (E, white); metabolism (M, light grey); protein fate (P, dark grey) and for all other identifications (O, black). (Reprinted with permission from Heazlewood et al. (2004). Copyright 2003 American Society of Plant Biologists.) **B:** Low-abundance functional group distributions of the matching sequence orthologs. The 139 proteins in the stringent mitochondrial set of 377 proteins in the functional classes involved in mitochondria–cellular interaction (MCI, black) including communication and signalling, transport, and structure; involved in information-transfer processes (ITP, light grey) including DNA synthesis and processing, transcription, and RNA binding/processing; and unknown function (U, white). The total number of putative orthologs in each species, in at least one set, and in all three sets is shown above each bar.

30 years has revealed the operation of carriers for phosphate, adenine nucleotides, mono-, di-, and tri-carboxylates, amino acids, and cofactors such as NAD^+ and coenzyme A (CoA) (Day and Wiskich, 1984; Douce et al., 1997). Recently, up to 58 *Arabidopsis* genes encoding putative members of the mitochondrial carrier protein family have been suggested (Picault et al., 2004). Most of the putative plant carriers appear to be expressed at relatively low levels, as estimated by the number of expressed sequence tags (ESTs) in The Institute for Genomic Research (TIGR) database. However, the EST number is only an estimate of expression level and largely depends on the state of the tissues from which RNAs were isolated. Therefore, there is still the major challenge of identifying the function of all MCF members and determining their substrate specificity, expression pattern, and role in plant metabolism. Recombinant protein technology and functional reconstitution techniques developed for animal and yeast carrier proteins are being combined with a wide range of functional-genomics tools to identify and characterize mitochondrial carriers in plants (for a review, see Picault et al., 2004).

The MCF proteins in *Arabidopsis* were also analysed by a combination of genomic and proteomic approaches (Millar and Heazlewood, 2003). Analyses of microarray experiments reveal differential expression profiles of the more highly expressed members of this gene family in different plant organs and in response to plant hormone application and environmental stresses. The actual presence of carrier proteins in *Arabidopsis* mitochondria was analysed using a proteomic approach based on isolation of integral membrane proteins, 1D SDS-PAGE, and ESI-Q-TOF tandem MS analysis. Six of the nine highly expressed carriers were identified by this approach: adenine nucleotide translocator, dicarboxylate/tricarboxylate carrier, phosphate carrier, uncoupling protein, and a carrier gene of unknown function (Millar and Heazlewood, 2003). More members of MCF were detected in further gel-free LC-MS/MS analysis of the *Arabidopsis* mitochondrial proteome (Heazlewood et al., 2004) and hydrophobic proteome of mitochondrial membranes (Brugière et al., 2004). One of these new identifications has strong similarity to carnitine-acylcarnitine carriers, and the other is a mitochondrial carrier of unknown function. Several other important carrier proteins were identified in the same studies, including two putative ABC transporters, the mitochondrial Fe–S transporter, and a Cys desulphurase which in yeast was shown to act in concert for the synthesis and transport of Fe/S clusters essential for both intramitochondrial and extramitochondrial Fe/S proteins (Kispal et al., 1999). Note that both Heazlewood et al. (2004) and Brugière et al. (2004) worked with cell cultures and that further MCF members may be present in whole plants.

The presence of seven of the ten enzymes of glycolysis was recently discovered in a mitochondrial fraction of *Arabidopsis* cells (Giegé et al., 2003). Four of these enzymes were also identified in an inter-membrane space/outer mitochondrial membrane fraction. The association was confirmed *in vivo* by the expression of enolase- and aldolase-yellow fluorescent protein fusions in *Arabidopsis* protoplasts (Giegé et al., 2003). Enzyme activity assays and experiments on the metabolic conversion of ^{13}C-glucose into ^{13}C-labelled intermediates of the TCA cycle confirmed that the entire glycolytic pathway was present in isolated *Arabidopsis* mitochondria. Sensitivity of enzyme activities to protease treatments indicated that the glycolytic enzymes were present on the outside of the mitochondria. It was proposed that the entire glycolytic pathway is associated with plant mitochondria by attachment to the cytosolic face of the outer mitochondrial membrane which allows pyruvate to be provided directly to the mitochondrion and used as a respiratory substrate. This is in agreement with the recent identification of glycolytic enzymes in a proteomics survey of human mitochondria (Taylor et al., 2003a), although the functional significance of this observation was not investigated further.

Mitochondrial proteomic analysis was also used to study CMS in *Brassica* (Mihr et al., 2001). CMS is a widespread phenomenon in the plant kingdom, caused by an incompatibility between the nucleus and the cytoplasm. Alterations of mitochondrial genome organization are associated with nearly every type of CMS examined so far (Conley and Hanson, 1995). They arise from recombination events through direct or inverted repeats, and cause the expression of chimeric mitochondrial genes which have deleterious effects on respiratory function. Comparison of the protein composition of mitochondria from etiolated CMS seedlings and a near isogenic fertile line revealed distinct differences on 2D gels (Mihr et al., 2001).

Cytoplasmic regulation of the accumulation of nuclear-encoded proteins was recently demonstrated in differential display analysis of the mitochondrial proteomes of normal and T-type CMS maize (Hochholdinger et al., 2004). This study also identified 27 nuclear-encoded proteins that were not previously known to be imported into the mitochondria, which could be a reflection of species- and/or tissue-specific differences. Whereas some of the novel proteins are involved in nucleotide or carbon metabolism or other processes, many of the proteins have only predicted functions or completely lack functional assignments. Thirteen percent of the protein spots analysed by 2D IEF/SDS-PAGE (25 out of 197) exhibited at least a 3-fold difference in accumulation between the mitochondrial proteomes of normal and CMS-T plants that had essentially identical nuclear genomes. As most of these proteins were nuclear-encoded, these findings demonstrate that the genotype of a mitochondrion can regulate the accumulation of the nuclear-encoded fraction of its proteome. Thus, mitochondrial dysfunction signals the expression of particular nuclear genes, thereby altering metabolism to compensate for the mitochondrial dysfunction. This process, retrograde regulation of nuclear genes, occurs by yet undefined mitochondrial signals. The nuclear-encoded AOX is up-regulated by mitochondrial signals when electron transfer through the cytochrome pathway is blocked (McIntosh et al., 1998). In another study, electron transfer through the cytochrome pathway was blocked by antimycin A, which allowed for the identification of seven additional nuclear genes induced by blocking electron flow through complexes III and IV (Maxwell et al., 2002).

The mitochondrial protein import apparatus is one of the major components of intracellular communication system that specifically recognizes and imports hundreds, and possibly up to 3000, nuclear-encoded cytosolically synthesized protein precursors (Sjöling and Glaser, 1998; Zhang et al., 2001). Biochemical studies have characterized the TOM and MPP components of the plant import system (Braun and Schmitz, 1995; Glaser and Dessi, 1999; Werhahn et al., 2001). MS analysis of purified *Arabidopsis* mitochondria and subfractions thereof identified 17 components of the import apparatus and revealed their localization within the mitochondria. Using a combination of transcriptomic and proteomic characterization it was demonstrated that the *Arabidopsis* mitochondrial protein import apparatus responds to mitochondrial dysfunction (Lister et al., 2004). Treatment of *Arabidopsis* cell culture with mitochondrial ECT inhibitors rotenone and antimycin A resulted in a significant increase in transcript level of import components, especially for the minor isoforms. Rotenone-treated cells showed 50% reduced protein import rate and an up-regulation of gene sets involved in mitochondrial chaperone activity, protein degradation, respiratory chain assembly, and division. This indicates that perturbation of complex I activity may have widespread effects on gene expression and, thus, could be an important point for retrograde signalling between the mitochondrion and the nucleus.

9.3.5 Mitochondria from different species, tissues, and organs

The differences in polypeptide composition between mitochondria from different plant organs were highlighted in the work by Colas des Francs-Small et al. (1992, 1993) using potato mitochondria. The impact of tissue differentiation at the mitochondrial level was further investigated by comparative analysis of soluble mitochondrial proteomes of pea mitochondrial purified from green and etiolated leaves, roots, and seeds (Bardel et al., 2002). The 2D map of soluble proteins of green leaf mitochondria revealed 433 Coomassie-stained spots, 78 spots were analysed and 37 proteins identified by Edman degradation, MALDI-MS PMF and nano-ESI-Q-TOF MS/MS. Further fractionation of soluble proteins by gel filtration chromatography prior to 2D IEF/SDS-PAGE analysis was introduced to increase resolution for low-abundance proteins and provide additional information concerning quaternary structure of proteins. This study also revealed intrinsic limitations of the subcellular proteome analysis approach such as unavoidable contamination by other fractions and difficulties in identification of low-abundance proteins. It was concluded that aldehyde dehydrogenase (ALDH) is one of the major proteins expressed in plant mitochondria with the highest level in leaves and roots where at least nine isoforms and/or posttranslationally modified proteins represent a large portion of the soluble proteome ($\sim$7.5%). The possible role of ALDH might be the involvement in the detoxification of reactive aldehydes formed by peroxidation of polyunsaturated fatty acids under stress conditions and/or mitochondrial dysfunction (Møller, 2001b). The most striking difference between green leaf and etiolated leaf mitochondrial proteome was the very low abundance of proteins of the glycine cleavage system (GDC P, T, H, and L subunits) and serine hydroxymethyltransferase (SHMT) when plants were grown in darkness. This dramatic accumulation of the photorespiratory pathway enzymes in mitochondria from mesophyll cells of C_3 plant leaves is attributed to a light-dependent transcriptional control of the genes encoding these proteins (Oliver, 1994). Another protein that seems to be induced by light was a short-chain alcohol dehydrogenase-like protein. On the other hand, some identified spots such as subunits of ATP synthase, E1β subunit of PDH, and chaperonin 60 were more abundant in mitochondria from etiolated leaves. Proteins specifically induced or overexpressed in root mitochondria were FDH, the E1α subunit of PDH and the enzymes of amino acid metabolism cysteine synthase, arginase, and putative glutamate dehydrogenase. The proteins whose abundance suggests an important physiological function in seed mitochondria were thiosulphate sulphur transferase (rhodanese), FDH, and HSP22. Although FDH is known to be highly expressed in non-green tissues and induced by environmental stress in leaves (Colas des Francs-Small et al., 1993; Hourton-Cabassa et al., 1998), its presence as a major protein in seed mitochondria was demonstrated for the first time. Thiosulphate sulphur transferase was recently characterized in plants and has been postulated to have several potential functions such as the formation of iron-sulphur centres, cyanide detoxification, and assimilation of sulphur (Papenbrock and Schmidt, 2000a, b).

9.4 Stress-responsive proteins and redox regulation

9.4.1 Impact of environmental stress

Environmental stresses such as drought, high salinity, extremes of temperature, heavy metal toxicity, ultraviolet radiation, nutrient deprivation, and hypoxia can significantly alter plant metabolism, growth, and development. Perturbations of metabolism by a variety of abiotic stresses trigger increased production of reactive oxygen species (ROS) which can act as important signal molecules and interact with other signalling molecules to achieve integrated responses. Plant mitochondria are not only potential targets for damage during stress but they are a significant source of cellular ROS (Møller, 2001a). Concomitant with imposed oxidative damage, specific proteins are either synthesized or lost from mitochondria. Since the coding capacity of mitochondria is limited, many changes in their protein complement in response to environmental stress require that the proteins be imported from the cytosol where they are synthesized.

Macronutrient (P or N) limitation of wild-type (wt) suspension-cultured tobacco cells resulted in a large induction of AOX, which constitutes a non-energy-conserving branch of the respiratory electron transport chain (Sieger et al., 2005). At the same time, growth of wt cell cultures was dramatically reduced and carbon use efficiency decreased by 42–63%. However, when transgenic (AS8) cells lacking AOX were grown under the same nutrient-deficient conditions, their growth was reduced moderately and carbon use efficiency values remained the same as under nutrient-sufficient conditions. A comparison of the mitochondrial protein profiles of wt and AS8 cells indicated that the lack of AOX in AS8 under P limitation was associated with increased levels of proteins commonly associated with oxidative stress and/or stress injury. These included catalase, ascorbate peroxidase, glutathione *S*-transferase, glyceraldehyde-3-phosphate dehydrogenase, several molecular chaperones, and aldehyde dehydrogenase (ADH). In addition, northern analyses showed increased transcript levels of catalase and glutathione peroxidase, two ROS-detoxifying enzymes (Mittler, 2002), and PR-1a protein whose expression is sensitive to ROS production (Green and Fluhr, 1995). The level of nine ETC components, including subunits of complex I, II, and IV, was consistently reduced in AS8 while TCA cycle enzymes did not show a universal trend in abundance in comparison to the wt. A total set of 72 proteins changed in abundance between low P-grown wt and AS8 cells. Alternatively, the lack of AOX in AS8 cells under N limitation resulted in enhanced carbohydrate accumulation (Sieger et al., 2005). These results support the idea that AOX can essentially act as an 'energyoverflow' during nutrient limitation (Lambers, 1982). In the absence of this 'energy overflow' mechanism during nutrient limitation, a strong imbalance between growth and nutrient availability develops. With the inability to correct the imbalance between carbohydrate supply and demand (by utilizing AOX), AS8 cells maintain anabolism and growth despite the mineral deficiency. This study revealed, therefore, that at least in suspension-cultured cells AOX had nutrient-specific roles, maintaining redox balance during P limitation and carbon balance during N limitation.

The impact of drought, chilling, or herbicide treatment on the capacity of pea leaf mitochondria to import precursor proteins was investigated using *in vitro* protein import and processing assay, and immunoblotting analysis (Taylor et al., 2003b). Drought treatment stimulated import and processing of various precursor proteins via the general import pathway. In contrast, both chilling and herbicide treatment of plants caused inhibition of import with all precursors tested without an observed decrease in processing of imported proteins. Western blot analysis indicated that the steady-state level of several mitochondrial components, including TOM20 receptor and the core subunit of complex III responsible for processing, remained largely unchanged. These stresses also damage several mitochondrial proteins by modification of lipoic acid moieties resulting in loss of enzyme activity (Taylor et al., 2002, 2004a). All three stresses induced the synthesis of AOX. Thus, while different environmental stresses may lead to general oxidative damage in the cell, not all mitochondrial processes are equally affected.

Branched-chain amino acid catabolism has been suggested to be a possible cellular detoxification mechanism (Taylor et al., 2004b). Under conditions of rapid protein turnover, it could be important to keep a balance between maintenance of a branched-chain amino acids pool for protein synthesis and removal of cytotoxic branched-chain amino acids and 2-oxo acids via respiratory oxidation to prevent their accumulation. In the branched-chain 2-oxo acid catabolism pathway, the three branched-chain amino acids, Val, Leu, and Ile, are initially transaminated to their respective branched-chain 2-oxo acid by the branched-chain amino acid transaminase (BCAT). This reversible reaction is also the final step of the biosynthesis of these amino acids. After the transamination step, the 2-oxo acids are decarboxylated and esterified to CoA by the branched-chain oxo-acid dehydrogenase complex (BCKDC). This complex is very similar to the widely studied pyruvate and 2-oxoglutarate dehydrogenase complexes (Lutziger and Oliver, 2001). The CoA esters generated by BCKDC are then oxidized by an acyl-CoA dehydrogenase delivering electrons to the electron transfer flavoprotein (ETF) that directly donates electrons into the respiratory chain at ubiquinone. After this step, the three pathways diverge to a series of separate reaction leading to propionyl-CoA in the case of Val metabolism, propionyl-CoA and acetyl-CoA in the case of Ile metabolism and to acetyl-CoA and acetoacetate in the case of Leu metabolism (Graham and Eastmond, 2002). In total, 13 different enzymes or enzyme complexes are involved in the pathways for Val, Leu, and Ile metabolism, and sequence similarity searches of the *Arabidopsis* genome revealed that multigene families exist for most of these enzymes. Recently, identification of lipoic acid-dependent oxidative catabolism of 2-oxo acids in mitochondria provided evidence for branched-chain amino acid catabolism in *Arabidopsis* (Taylor et al., 2004b). Proteins containing covalently bound lipoic acid were immunodetected on 2D IEF/SDS-PAGE gels with mitochondrial proteins from pea, rice, and *Arabidopsis* using antibodies raised against this cofactor and identified by MS. Lipoic acid-containing acyltransferases from PDH complex and 2-oxoglutarate dehydrogenase complex were identified in all three species. In addition, acyltransferases from BCKDC were identified in both *Arabidopsis* and rice mitochondria. Activity of the BCKDC was only measurable in *Arabidopsis* mitochondria using substrates that represented 2-oxo acids derived by deamination of branched-chain amino acids Val, Leu, and Ile.

Analysis of the mitochondrial proteome by LC-MS/MS provided direct evidence for enzyme components of the Val, Leu, and Ile pathways and identified 11 members of gene families that putatively encode mitochondrial enzymes in *Arabidopsis*. Five identified enzymes are common to the initial stages of catabolism of all three branched-chain amino acids: BCAT, BCKDC, isovaleryl-CoA dehydrogenase (IVD), enoyl-CoA hydratase, and the ETF accepting electrons from IVD. Some of these proteins were reported previously for *Arabidopsis* mitochondria (Daschner et al., 2001; Millar et al., 2001), and some have been found in mitochondria from other plant species (Miernyk et al., 1991; Daschner et al., 2001), while six represented novel identifications. Two enzymes of further Leu metabolism were identified by LC-MS/MS analysis, β subunit of methylcrotonyl-CoA carboxylase and hydroxymethylglutaryl-CoA lyase in agreement with Che et al. (2002). Although there was no convincing MS evidence for the presence of the next step in Ile metabolism or the next two steps in Val metabolism in mitochondria, MS matches were obtained corresponding to the enzymes catalysing the final step in each pathway, methylmalonate-semialdehyde dehydrogenase for Val and acetyl-CoA C-acetyltransferase for Ile. The latter enzyme was also identified as a very significant protein in *Arabidopsis* mitochondrial preparations by Kruft et al. (2001). The absence of key enzymes of Ile and Val metabolism in *Arabidopsis* mitochondria, together with their apparent presence in peroxisomes, suggests a further possible complexity in these catabolic pathways involving trafficking of CoA esters between organelles and the possible presence of carnitine shuttles (Masterson and Wood, 2001).

Differential impact of environmental stresses on the plant mitochondrial proteome was analysed using pea plants subjected to drought, cold, and herbicide treatments (Taylor et al., 2005). The results suggested that herbicide treatment placed a severe oxidative stress on mitochondria, whereas chilling and drought were milder stresses. Mitochondria isolated from the stressed pea plants maintained their ETC activity, but changes were evident in the abundance of uncoupling proteins, non-phosphorylating respiratory pathways, and oxidative modification of lipoic acid moieties on mitochondrial proteins. The H protein of GDC showed dramatically decreased lipoic acid content after all treatments in agreement with previous observations that environmental stress leads to inhibition of the photorespiratory pathway (Taylor et al., 2002). Detailed analysis of the soluble proteome of mitochondria by 2D IEF/SDS-PAGE and MS revealed differential degradation of key matrix enzymes during treatments with chilling being significantly more damaging than drought. Differential induction of HSPs and specific losses of other proteins illustrated the diversity of responses to these stresses at the protein level. BN/SDS-PAGE separation of intact ETC complexes revealed little if any change in response to environmental stresses.

9.4.2 Oxidative stress and anoxia

Most or all stress phenomena – biotic and abiotic – are accompanied by an increased production of ROS. The interaction between ROS and proteins is complex and can result in a large variety of modifications from oxidation of single side groups to chain breakage (Berlett and Stadtman, 1997; Ghezzi and Bonetto, 2003; Møller and Kristensen, 2004; Taylor et al., 2004a).

An analysis of the effects of oxidative stress and respiratory inhibitors on the *Arabidopsis* mitochondrial proteome found changes in protein abundance and also documented losses of mitochondrial function (Sweetlove et al., 2002). Using 2D IEF/SDS gel electrophoresis in combination with tandem MS analysis inducible components of the mitochondrial antioxidant system were identified. A set of 25 protein spots increased more then 3-fold in mitochondria isolated from *Arabidopsis* cell culture after treatment with H_2O_2 or menadione (a redox-active quinone that generates intracellular superoxide), a subset of these increased in antimycin A-treated samples. Nine increased proteins were not found in control mitochondria. One is directly involved in antioxidant defense, a mitochondrial thioredoxin-dependent peroxidase, while another, a thioredoxin reductase-dependent protein disulphide isomerase, is required for protein disulphide redox homeostasis. Other proteins induced in mitochondria after oxidative stress included calreticulin, glyceraldehyde-3-phosphate dehydrogenase and a glutathione *S*-transferase. Interestingly, these proteins are not generally considered to be mitochondrial based on targeting prediction analysis and the available literature, but all three are known to be induced by stress conditions in other organisms (Sweetlove et al., 2002).

Ten protein spots decreased significantly in abundance during oxidative stress. A specific set of mitochondrial proteins were damaged and degraded by stress treatment. The results suggest that sensitive components of the TCA cycle, together with ATP synthesis subunits and the Fe–S centres of complex I are likely to be the most susceptible to oxidative stress. Furthermore, 16 proteins that increased in abundance during all treatments were identified as breakdown products of larger proteins. Using

H_2O_2 as a model stress, further work revealed induction of a specific ATP-dependent protease activity in isolated mitochondria, putatively responsible for the degradation of oxidatively damaged mitochondrial proteins (Sweetlove et al., 2002).

Irreversible modification of proteins by formation of carbonyl groups is also a valuable marker for oxidative stress (Halliwell and Gutteridge, 1999; Møller and Kristensen, 2004). Protein oxidation and its consequences were investigated by a mild oxidative treatment of the soluble matrix fraction of mitochondria from green rice leaves followed by immunoprecipitation and 2D-LC-MS/MS analysis (Kristensen et al., 2004). Carbonyl groups formed were tagged by dinitrophenylhydrazine (DNP), immunoprecipitated using anti-DNP antibodies and digested with trypsin prior to gel-free MS analysis. Using this approach, 20 oxidized proteins were identified in the untreated control sample and these represent proteins oxidized under *in vivo* conditions. A further 32 carbonylated proteins were identified in the *in vitro* oxidized matrix fraction. These susceptible proteins include all of the soluble Krebs cycle enzymes as well as a number of enzymes, like ALDH, that could be termed stress proteins.

Tryptophan oxidation to *N*-formylkynurenine can readily be identified by tandem MS as the addition of 32 mass units/charge to the peptide mass. Almost 200 protein spots from 2D gels of potato tuber and rice leaf mitochondria were analysed and *N*-formylkynurenine was detected in 17 different proteins (Møller and Kristensen, 2005). With one exception, the oxidation-sensitive aconitase, all these proteins were either redox active themselves or subunits in redox-active enzyme complexes. The same site was modified in (i) several adjacent spots containing the P-protein of GDC, (ii) two different isoforms of the MPP in complex III, and (iii) the same tryptophan residues in Mn-SOD in both rice and potato mitochondria. This indicates that Trp oxidation is a selective process.

Significant similarities were found between the mitochondrial proteins observed to decrease in abundance during environmental stresses and those observed to change during chemical oxidative stress of mitochondria from *Arabidopsis* suspension cells (Sweetlove et al., 2002), proteins containing oxidized tryptophan in rice and potato mitochondria, and carbonylated proteins in rice mitochondria (Kristensen et al., 2004) (Figure 9.3). Notably matrix carbon metabolism enzymes appear to be major targets for oxidative modification and breakdown *in vitro* and *in vivo*. GDC subunits and SHMT are highly susceptible to damage and breakdown during environmental stresses in plants which is in agreement with the identification of these proteins as the sites of carbonyl formation following an *in vitro* oxidation in rice (Kristensen et al., 2004), as proteins containing oxidized tryptophan (Møller and Kristensen, 2005) and as the major oxidized proteins in *Arabidopsis* mitochondria (Johansson et al., 2004).

In animals anoxia/hypoxia leads to damage typically during the reoxygenation phase where the reintroduced oxygen comes into contact with the overreduced cellular components (Halliwell and Gutteridge, 1999). Similar responses are seen in plant cells (Blokhina et al., 2003). Rice is one of the most anoxia-tolerant plant crop species, is able to germinate and sustain early seedling growth under strictly anoxic conditions, and can readily return to atmospheric or aerobic conditions

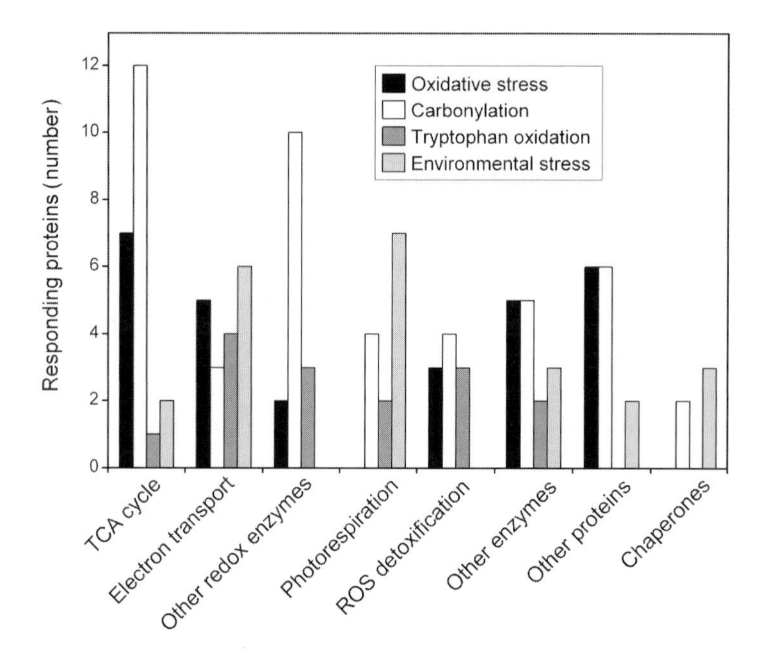

Figure 9.3 Stress-responsive proteins identified in various studies. Identifications include proteins changed in abundance due to the impact of oxidative stress on mitochondria from *Arabidopsis* cell cultures (Sweetlove et al., 2002); carbonylated proteins in the oxidized matrix of rice leaf mitochondria (Kristensen et al., 2004); proteins with oxidized tryptophans detected in rice leaf and potato tuber mitochondria (Møller and Kristensen, 2005); and pea mitochondrial proteins changed in abundance in response to environmental stresses (Taylor et al., 2005).

without substantial damage. Alterations in the mitochondrial proteome during anoxia to air transition in rice were followed to study the stages of mitochondrial biogenesis triggered by oxygen availability (Millar et al., 2004b). Rice coleoptiles grown under anoxia, and the mitochondria isolated from them, respired very slowly compared to air-adapted and air-grown seedlings. Immunodetection of key mitochondrial protein markers, 2D IEF/SDS-PAGE proteome mapping of the soluble mitochondrial fraction, and a shotgun LC-MS/MS analysis revealed similar patterns of the major functional categories of mitochondrial proteins from both anoxic and air-adapted samples. 2D BN/SDS-PAGE, in contrast, showed the very low abundance of assembled complex III and IV in the mitochondrial membrane in anoxic samples and the dramatic increase in the abundance of these complexes on air adaptation. Total heme content, cytochrome absorbance, and cytochrome *c* content also increased markedly on air adaptation. These results likely reflect limited heme synthesis for cytochrome assembly in the absence of oxygen and represent a discrete and reversible blockage of full mitochondrial biogenesis in this anoxia-tolerant species. Interestingly, the adaptation of anoxically grown rice plants to oxygen involved the synthesis and/or assembly of both the cytochrome and alternative respiratory pathways, whereas germination and growth in air only involve the cytochrome pathway. Thus, the heme-independent AOX could be induced following anoxia to provide a rapid response to meet the requirements for aerobic respiration and prevent overreduction of the respiratory chain.

9.4.3 Metal-binding proteins

Like all eukaryotic mitochondria, plant mitochondria contain specific transporters for divalent metal ions that are required for a large number of metal ion-dependent reactions involved in primary respiratory metabolism and signalling (Lill and Kispal, 2001). The structural rearrangements induced by metal ion binding confer stability to proteins by restricting the mobility of domains via non-covalent cross-linking of charged amino acid side-groups. Such interactions stabilize the active conformation of proteins, facilitate enzymatic reactions by active site binding of substrates or products, or participate in the reaction by electron acceptance and donation (Holm et al., 1996). The exact protein subunits that bind metal ions in these reactions, many involving complex multisubunit enzymes, are often unknown. A metal affinity shift assay developed by Kameshita and Fujisawa (1997) relies on the observation that conformational or charge differences induced by metal ion binding to SDS-denatured polypeptides can shift the apparent molecular mass significantly during 2D diagonal SDS-PAGE. Proteomic identification of divalent metal cation-binding proteins in plant mitochondria from *Arabidopsis* cell culture was performed using a combination of the divalent cation-shift electrophoresis approach and tandem MS analysis (Herald et al., 2003). This analysis identified a total of 23 distinct protein spots as the products of at least 11 different genes. A subset of proteins known to be divalent cation-binding, or to catalyse divalent cation-dependent reactions included the β subunit of succinyl-CoA ligase, Mn-SOD, the Fe–S centre-binding 23 kDa subunit of complex I and the Rieske iron-sulphur protein of complex III. A further seven mitochondrial proteins were identified as putatively metal binding. A range of known metal-binding proteins in the mitochondrial ETC, metabolism and proteolysis machinery were not identified in this study possibly due to the absence of metal-binding sites after detergent denaturation or to the lack of mobility shift during electrophoresis. However, it still remains to be established just how the metal binding affects the properties of the target proteins.

9.4.4 Proteins associated with PCD

Oxidative stress can lead to cell death in both animals and plants either by a necrotic route after substantial cell damage or via an ordered and controlled pathway termed PCD. Similar to the key role of animal mitochondria in PCD, this process in plants can also occur via mitochondria-dependent pathways. Cytochrome *c* release is known in plant PCD (Balk and Leaver, 2001), but the mechanism of PCD in plants and its regulation are still unresolved (Yao et al., 2004). The study of transcriptomic changes associated with heat- and senescence-induced PCD in an *Arabidopsis* cell suspension culture revealed several commonly up- and down-regulated candidate genes (Swidzinski et al., 2002). As a further step towards identifying posttranscriptional mechanisms that function during plant PCD, a proteomic analysis of total cellular protein content during both heat- and senescence-induced PCD in an *Arabidopsis* cell culture was undertaken and a range of mitochondrial protein changes were shown to be associated with PCD in plants. Four mitochondrial proteins increased in relative abundance following the induction of PCD (Swidzinski et al., 2004).

The increased relative abundance of two forms of mitochondrial Mn-SOD in both heat- and senescence-induced PCD is consistent with the observation that oxidative stress is implicated in the induction/execution of PCD (Swidzinski et al., 2002). Two isoforms could be a result of PTM of Mn-SOD. The increased lipoamide dehydrogenase content was not accompanied by an increase in abundance of other subunits of PDH and 2-oxoglutarate dehydrogenase complexes indicating an alternative function for lipoamide dehydrogenase during PCD, possibly as a part of redox signalling mechanism. Mitochondrial aconitase may also be involved in redox signalling that triggers PCD. In addition to its role in the TCA cycle, mammalian aconitase functions as an iron regulatory protein (Cairo and Pietrangelo, 2000). Plant aconitases are known to be inactivated in the presence of ROS (Navarre et al., 2000) and may be involved in the regulation of free iron content, or in regulating the expression of other genes. Excess free iron has been implicated in oxidative damage and was suggested to be required for DNA damage, resulting in the production of DNA 'ladders' which are diagnostic signatures of many PCD (Eaton and Qian, 2002).

The notion that VDAC Hsr2 may be involved in plant PCD is intriguing since this protein, along with inner membrane AAC, ADP/ATP carrier is a central component of the mitochondrial permeability transition pore that functions during animal PCD (Martinou and Green, 2001). Interestingly, overexpression of rice VDAC in mammalian cells is sufficient to cause cell death, and inhibition of VDAC function protects against heat-induced cell death in cucumber (Godbole et al., 2003). The presence of a mitochondrial permeability transition has been reported in plant cells in response to stress leading to cell death (Arpagaus et al., 2002; Tiwari et al., 2002). Therefore, the increased relative abundance of VDAC Hsr2 during plant PCD may be a key feature of this permeability transition and potentially involved in the release of pro-apoptotic proteins from plant mitochondria.

9.4.5 Thioredoxins and their target proteins

The regulation of protein function in plant mitochondria by oxidation and/or reduction via thioredoxin (Trx) system might also be part of an oxygen-sensing system to enable mitochondria to adjust their function in accordance with the prevailing cellular redox state and metabolic requirements (Millar et al., 2005). The AOX dimer was the first plant mitochondrial enzyme shown to be regulated by the redox state of inter-molecular disulphide bond (Vanlerberghe and McIntosh, 1997). The mitochondrial thioredoxin system can be distinguished by the electron donor NADPH, an NADPH-Trx reductase, and Trx (Laloi et al., 2001). Two distinct isoforms of

Trxs were found in plant mitochondria Trxs *o* and Trxs *h* (Schürmann and Jacquot, 2000; Gelhaye et al., 2004). The interaction of Trx *h* with two potential targets, AOX and glutathione peroxidase, was recently demonstrated (Gelhaye et al., 2004). In addition, Trx *h* displayed a potential site for glutathionylation and the glutathione adduct had an altered redox potential suggesting that its function might also be affected. However, the specificity and the precise functions of these two isoforms of Trxs remain largely unknown.

Using proteomic approaches in conjunction with specific enrichment by mutant Trx affinity chromatography and direct in-gel fluorescent detection, 50 potential Trx-linked proteins were recently identified in the soluble fraction of mitochondria from both photosynthetic (spinach and pea leaves) and heterotrophic (potato tubers) sources (Balmer et al., 2004). All fundamental processes of plant mitochondria were found to be Trx targeted, including photorespiration, TCA cycle and associated reactions, lipid metabolism, ETC, ATP synthesis/transformation, membrane transport, translation, protein assembly/folding, nitrogen and sulphur metabolism, hormone synthesis, and stress-related reactions. Interestingly, VDAC protein possibly involved in formation of the permeability transition pore complex (Godbole et al., 2003) and PCD trigger was also identified in agreement with previous findings that in animal mitochondria the permeability transition appears to be under redox control (Petronilli et al., 1994). Stress-related proteins were among the major targets found to be linked to Trx in plant mitochondria. The eight identified candidates ranged from alcohol and aldehyde dehydrogenases, enzymes believed to function under anaerobic stress, to proteins such as catalase, SOD, and peroxiredoxin, which are active under oxidative conditions. The other stress-related proteins were FDH and a glutaredoxin-like protein.

Considering the multiplicity of redox-regulated processes in plant mitochondria, the potential extent of the regulatory redox network is not entirely unexpected.

9.5 Protein phosphorylation and signal transduction

Protein phosphorylation is one of the most common PTMs and it can affect virtually any property of a protein. It is often involved in signal transduction cascades and the identification of a new phosphoprotein and/or a protein kinase or phosphatase is therefore a first indication of a new signal transduction pathway.

9.5.1 Phosphoproteins

About 30 labelled polypeptides were detected by autoradiography in mitochondria purified from potato tubers after incubation with [γ-^{32}P]-ATP (Sommarin et al., 1990). Most of the polypeptides including a Ca^{2+}-dependent kinase were located outside the inner membrane. The degree of protein phosphorylation was strongly reduced by the presence of respiratory substrates (succinate, pyruvate, and NADH), mainly the labelling of bands of 40 and 42 kDa (Petit et al., 1990) later shown to be predominantly FDH and PDH (Bykova et al., 2003a, b). In mitochondria from etiolated oat shoots the phosphorylation of two endogenous polypeptides of the molecular masses of 67 and 43 kDa were observed (Pike et al., 1991).

Fractionation experiments of the cells of *Vicia faba* leaves revealed that the Ca^{2+}-dependent phosphorylated proteins with molecular masses of 41 and 25 kDa were present in mitochondria (Kinoshita et al., 1993). Subfractionation of highly purified potato tuber mitochondria into the matrix fraction, inner and outer membrane fractions allowed the detection of more than 20 phosphoproteins and at least one Ca^{2+}-dependent protein kinase located in the outer membrane fraction (Pical et al., 1993). Håkansson and Allen (1995) identified three Tyr-phosphoproteins (28, 27, and 12 kDa) whose phosphorylation was influenced by redox conditions in pea leaf mitochondria. A 37 kDa phosphoprotein with phosphorylation on a histidine residue was proposed to be the α-subunit of succinyl-CoA synthase (Håkansson and Allen, 1995).

More than 20 phosphorylated polypeptides were detected in inside-out submitochondrial particles from potato tuber mitochondria. Two autophosphorylated putative protein kinases were identified, one at 16.5 kDa required divalent cations for autophosphorylation, while the other at 30 kDa did not (Struglics et al., 1999). Two phosphoproteins of 22 and 28 kDa were N-terminally sequenced and identified as the δ'-subunit of the F_1-ATPase and the δ'-subunit of the F_0-ATPase, respectively (Struglics et al., 1998).

Using [γ-^{32}P]-ATP labelling, submitochondrial fractionation and 2D gel electrophoresis followed by MS analysis, 14 new phosphoproteins were identified in potato tuber mitochondria (Bykova et al., 2003a) (Figure 9.4). Seven of them are involved in the TCA cycle or associated reactions. Four identified enzymes are subunits of respiratory complexes and involved in electron transport, ATP synthesis and protein processing. Two are HSPs and one is an antioxidative enzyme Mn-SOD. The characterization of the phosphorylation sites of the E1-α subunit of PDH and FDH in potato tuber mitochondria advances our understanding of the mechanism of the reversible phosphorylation in plant mitochondria (Bykova et al., 2003a, b). PDH is phosphorylated on Ser294 while FDH is phosphorylated on Thr76 and Thr333 (Figure 9.5). Both of the latter sites are on the outer accessible surface of the protein according to structural homology modelling (Bykova et al., 2003b).

PDH and FDH phosphorylation appear to be regulated in the same way by NAD^+, formate and pyruvate. FDH activity is greatly enhanced at low oxygen concentrations where – together with a postulated pyruvate formate-lyase – it might perform the same function as PDH (Bykova et al., 2003b).

9.5.2 *Kinases and phosphatases*

The detailed biochemical studies of the phosphorylation/dephosphorylation mechanism in inside-out inner mitochondrial membranes from potato tuber mitochondria showed that two distinct phosphoprotein phosphatases were involved in the protein dephosphorylation. It was proposed that an inner membrane protein phosphatase was required for activation of the inner membrane protein kinase(s) and that the mitochondrial matrix contained a phosphatase responsible for dephosphorylation of inner membrane phosphoproteins (Struglics et al., 2000).

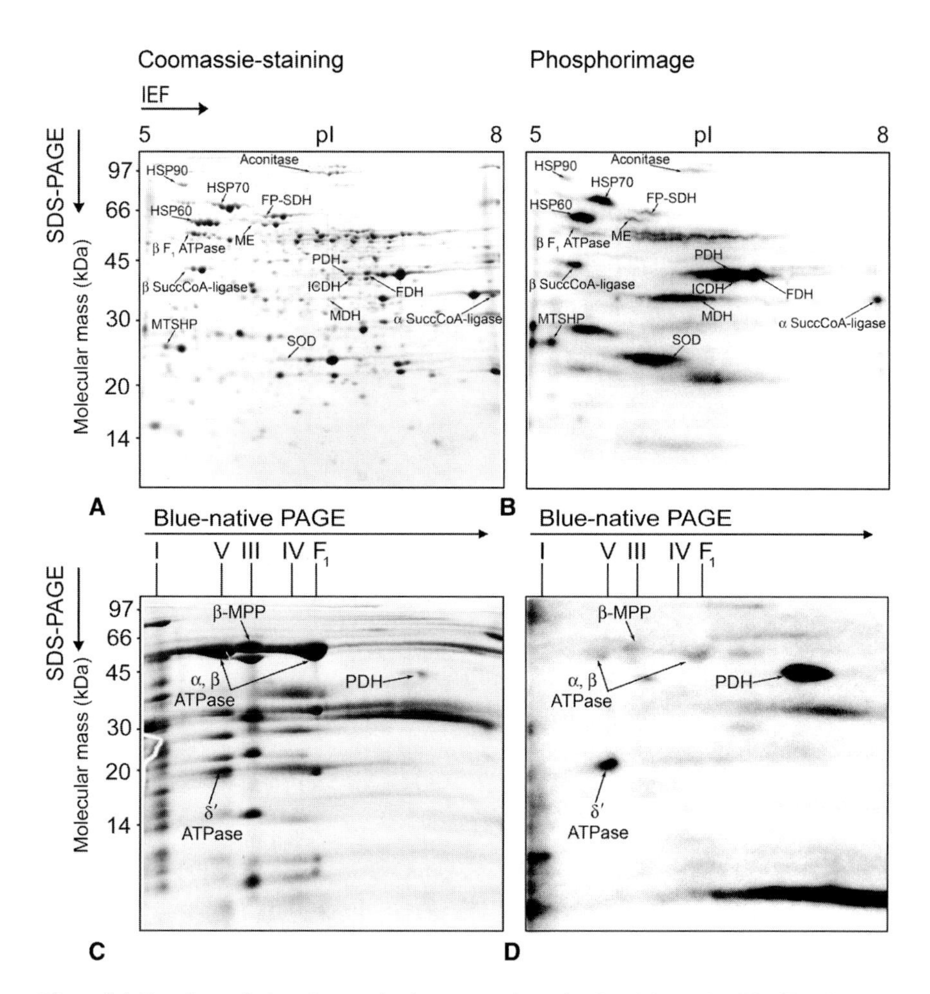

Figure 9.4 Proteins and phosphoproteins in potato tuber mitochondria resolved by 2D gel electrophoresis and visualized by phosphorimaging. **A** and **C**, Coomassie-stained matrix proteins and inner membrane protein complexes, respectively; **B** and **D**, corresponding phosphorimages. The arrows indicate protein spots identified by tandem MS analysis. The designations on top of the BN 2D gels indicate the identity of the membrane protein complexes: I, NADH dehydrogenase; V, F_0F_1–ATP synthase complex; III, bc_1 complex; F_1, F_1 part of the ATP synthase complex. The positions and sizes of standard proteins are given on the left. (Reprinted by permission of Federation of the European Biochemical Societies from Bykova et al. (2003a). Copyright 2003.)

Using [γ-^{32}P]-ATP as phosphate donor, Struglics and Håkansson (1999) found that purified pea mitochondrial NDPK was more heavily labelled on a serine than on a histidine residue, which were both shown to be conserved and phosphorylated in other species. The enhanced labelling of the NDPK in the presence of EDTA in intact mitochondria suggested that the protein is located in the inter-membrane space.

Kinases and phosphatases represent mostly low-abundance proteins with potentially key roles in mitochondrial function. Ten protein kinases, including Leu-rich

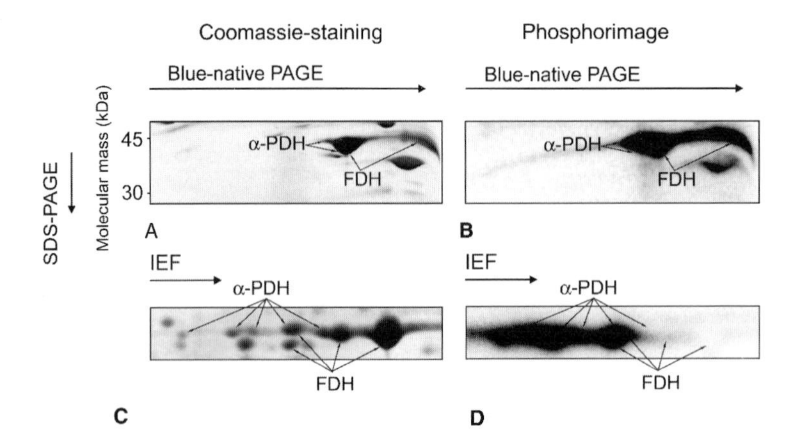

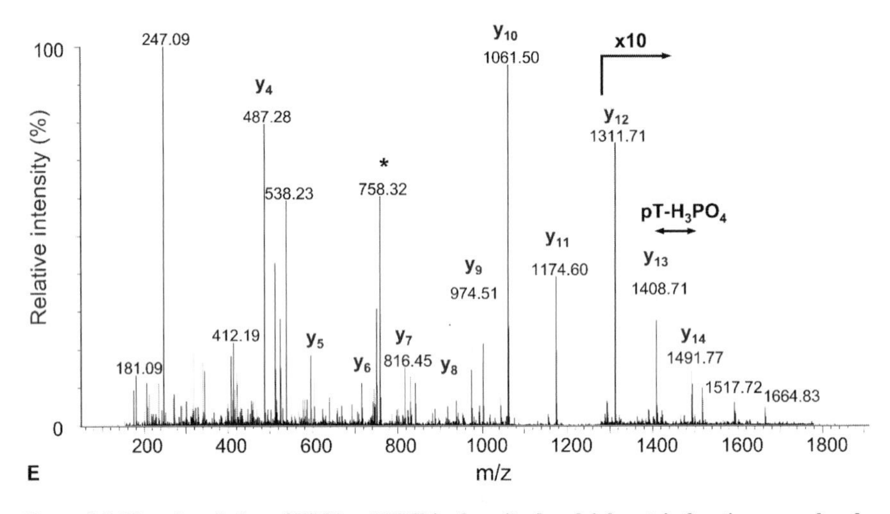

Figure 9.5 Phosphorylation of FDH and PDH in the mitochondrial matrix fraction as analysed by two types of 2D PAGE, phosphorimaging and MS/MS identification of one of the phosphorylation sites. **A** and **C**, Coomassie staining of BN/Tricine SDS-PAGE and IEF/Tricine SDS-PAGE, respectively; **B** and **D**, corresponding phosphorimages. Portions of images containing PDH and FDH are shown. The numbers on the left refer to the molecular masses of standard proteins. The arrows indicate protein spots identified by MS. **E**, Identification of a site of phosphorylation in mitochondrial FDH by automated nano-flow LC-MS/MS analysis. The graph represents the data collected during the MS/MS analysis of the triply charged precursor ion (*m/z* value 820.3), which generated a y-ion series (y4–y14) and confirmed the identity of the peptide as 326-YMPNQAMTPHISGTTIDAQLR-346. The phosphorylation site Thr-333 was assigned by the 83 Da mass difference between the y13 and y14 ions of C-terminal peptide fragments, corresponding to dehydro-2-aminobutyryl residue, which was formed by β-elimination of phosphoric acid from phosphothreonine-333. The peak at *m/z* value 758.32 (indicated with an asterisk) represents the internal fragment ion PNQAMTP that has lost its H_3PO_4 group (98.0 Da). (From Bykova et al. (2003b)).

repeat transmembrane protein kinases, receptor-like protein kinases, Ser–Thr kinases, and a mitogen-activated protein-like kinase, were identified by Heazlewood et al. (2004). Two of the Leu-rich repeat transmembrane receptor protein kinases showed high sequence similarity to a protein kinase of unknown function found recently in human mitochondria (Taylor et al., 2003a). These kinases belong to large subclasses of protein kinases that are poorly represented in EST databases. None of them has a clear mitochondrial targeting sequence.

9.6 Concluding remarks

While the majority of mitochondrial housekeeping enzymes were logically found in all studies, each study brought to light new mitochondrial proteins, thus demonstrating the potential of the proteomics approach to decipher mitochondrial functions. However, most of the low-abundance proteins are still unknown and various types of systematic subfractionation of the mitochondria will be necessary to make their identification possible. Work with different tissues and organs, such as flowers, should be done to identify tissue-specific proteins and their functions. Even when all mitochondrial proteins have been identified and their sub-mitochondrial location established, the largest challenge of them all is still left – the posttranslational modifications! More than 300 different PTMs are known and they can affect a given protein alone or in combination. Thus, we can expect dramatic developments in functional proteomics of plant mitochondria in the coming decade.

References

Andersson, S.G., Zomorodipour, A., Andersson, J.O., Sicheritz-Ponten, T., Alsmark, U.C., Podowski, R.M., Näslund, A.K., Eriksson, A.S., Winkler, H.H. and Kurland, C.G. (1998) The genome sequences of *Rickettsia prowazekii* and the origin of mitochondria. *Nature*, **396**(6707), 133–140.

Arnold, I., Pfeiffer, K., Neupert, W., Stuart, R.A. and Schägger, H. (1998) Yeast mitochondrial F_1F_0-ATP synthase exists as a dimer: identification of three dimer-specific subunits. *EMBO J.*, **17**(24), 7170–7178.

Arpagaus, S., Rawyler, A. and Braendle, R. (2002) Occurrence and characteristics of the mitochondrial permeability transition in plants. *J. Biol. Chem.*, **277**(3), 1780–1787.

Balk, J. and Leaver, C.J. (2001) The PET1-CMS mitochondrial mutation in sunflower is associated with premature programmed cell death and cytochrome *c* release. *Plant Cell*, **13**(8), 1803–1818.

Balmer, Y., Vensel, W.H., Tanaka, C.K., Hurkman, W.J., Gelhaye, E., Rouhier, N., Jacquot, J.P., Manieri, W., Schurmann, P., Droux, M. and Buchanan, B.B. (2004) Thioredoxin links redox to the regulation of fundamental processes of plant mitochondria. *Proc. Natl Acad. Sci. USA*, **101**(8), 2642–2647.

Bardel, J., Louwagie, M., Jaquinod, M., Jourdain, A., Luche, S., Rabilloud, T., Macherel, D., Garin, J. and Bourguignon, J. (2002) A survey of the plant mitochondrial proteome in relation to development. *Proteomics*, **2**(7), 880–898.

Bartoli, C.G., Pastori, G.M. and Foyer, C.H. (2000) Ascorbate biosynthesis in mitochondria is linked to the electron transport chain between complexes III and IV. *Plant Physiol.*, **123**(1), 335–344.

Berlett, B.S. and Stadtman, E.R. (1997) Protein oxidation in aging, disease, and oxidative stress. *J. Biol. Chem.*, **272**(33), 20313–20316.

Blokhina, O., Virolainen, E. and Fagerstedt, K.V. (2003) Antioxidants, oxidative damage and oxygen deprivation stress: a review. *Ann. Bot.*, **91**(2), 179–194.

Braun, H.P. and Schmitz, U.K. (1992) Affinity purification of cytochrome *c* reductase from potato mitochondria. *Eur. J. Biochem.*, **208**(3), 761–767.

Braun, H.P. and Schmitz, U.K. (1995) The bifunctional cytochrome *c* reductase/processing peptidase complex from plant mitochondria. *J. Bioenerg. Biomembr.*, **27**(4), 423–436.

Brugière, S., Kowalski, S., Ferro, M., Seigneurin-Berny, D., Miras, S., Salvi, D., Ravanel, S., d'Herin, P., Garin, J., Bourguignon, J., Joyard, J. and Rolland, N. (2004) The hydrophobic proteome of mitochondrial membranes from *Arabidopsis* cell suspensions. *Phytochemistry*, **65**(12), 1693–1707.

Burger, G., Gray, M.W. and Lang, B.F. (2003) Mitochondrial genomes: anything goes. *Trends Genet.*, **19**(12), 709–716.

Burke, J.J., Siedow, J.N. and Moreland, D.E. (1982) Succinate dehydrogenase. A partial purification from mung bean hypocotyls and soybean cotyledons. *Plant Physiol.*, **70**(6), 1577–1581.

Bykova, N.V., Egsgaard, H. and Møller, I.M. (2003a) Identification of 14 new phosphoproteins involved in important plant mitochondrial processes. *FEBS Lett.*, **540**(1–3), 141–146.

Bykova, N.V., Stensballe, A., Egsgaard, H., Jensen, O.N. and Møller, I.M. (2003b) Phosphorylation of formate dehydrogenase in potato tuber mitochondria. *J. Biol. Chem.*, **278**(28), 26021–26030.

Cairo, G. and Pietrangelo, A. (2000) Iron regulatory proteins in pathobiology. *Biochem. J.*, **352**(2), 241–250.

Che, P., Wurtele, E.S. and Nikolau, B.J. (2002) Metabolic and environmental regulation of 3-methylcrotonyl-coenzyme A carboxylase expression in *Arabidopsis*. *Plant Physiol.*, **129**(2), 625–637.

Colas des Francs-Small, C., Ambard-Bretteville, F., Darpas, A., Sallantin, M., Huet, J.C., Pernollet, J.C. and Rémy, R. (1992) Variation of the polypeptide composition of mitochondria isolated from different potato tissues. *Plant Physiol.*, **98**(1), 273–278.

Colas des Francs-Small, C., Ambard-Bretteville, F., Small, I.D. and Remy, R. (1993) Identification of a major soluble protein in mitochondria from nonphotosynthetic tissues as NAD-dependent formate dehydrogenase. *Plant Physiol.*, **102**(4), 1171–1177.

Combettes, B. and Grienenberger, J.M. (1999) Analysis of wheat mitochondrial complex I purified by a one-step immunoaffinity chromatography. *Biochimie*, **81**(6), 645–653.

Conley, C.A. and Hanson, M.R. (1995) How do alterations in plant mitochondrial genomes disrupt pollen development? *J. Bioenerg. Biomembr.*, **27**(4), 447–457.

Daschner, K., Couee, I. and Binder, S. (2001) The mitochondrial isovaleryl-coenzyme A dehydrogenase of *Arabidopsis* oxidizes intermediates of leucine and valine catabolism. *Plant Physiol.*, **126**(2), 601–612.

Day, D.A. and Wiskich, J. (1984) Transport processes of isolated plant mitochondria. *Physiol. Veg.*, **22**(2), 241–261.

de Virville, J.D., Alin, M.F., Aaron, Y., Remy, R., Guillot-Salomon, T. and Cantrel, C. (1998) Changes in functional properties of mitochondria during growth cycle of *Arabidopsis thaliana* cell suspension cultures. *Plant Physiol. Biochem.*, **36**(5), 347–356.

Douce, R. (1985) Functions of plant mitochondrial membranes. In: *Mitochondria in Higher Plants.* American Society of Plant Physiologists Monograph Series, Academic Press, London, pp. 77–153.

Douce, R., Aubert, S. and Neuburger, M. (1997) Metabolite exchange between the mitochondrion and the cytosol. In: Dennis, D.T., Turpin, D.H., Lefebvre, D.D. and Layzell, D.B. (eds) *Plant Metabolism.* Addison Wesley Longman, Essex, UK, pp. 234–251.

Dunbar, B., Elthon, T., Osterman, J., Whitaker, B. and Wilson, S. (1997) Identification of plant mitochondrial proteins: a procedure linking two-dimensional gel electrophoresis to protein sequencing from PVDF membranes using a fastblot cycle. *Plant Mol. Biol. Report.*, **15**(1), 46–61.

Eaton, J.W. and Qian, M. (2002) Molecular bases of cellular iron toxicity. *Free Radical Biol. Med.*, **32**(9), 833–840.

Eubel, H., Jänsch, L. and Braun, H.P. (2003) New insights into the respiratory chain of plant mitochondria. Supercomplexes and a unique composition of complex II. *Plant Physiol.*, **133**(1), 274–286.

Eubel, H., Heinemeyer, J. and Braun, H.P. (2004) Identification and characterization of respirasomes in potato mitochondria. *Plant Physiol.*, **134**(4), 1450–1459.

Gelhaye, E., Rouhier, N., Gerard, J., Jolivet, Y., Gualberto, J., Navrot, N., Ohlsson, P.I., Wingsle, G., Hirasawa, M., Knaff, D.B., Wang, H., Dizengremel, P., Meyer, Y. and Jacquot, J.P. (2004) A specific form of thioredoxin *h* occurs in plant mitochondria and regulates the alternative oxidase. *Proc. Natl Acad. Sci. USA*, **101**(40), 14545–14550.

Ghezzi, P. and Bonetto, V. (2003) Redox proteomics: identification of oxidatively modified proteins. *Proteomics*, **3**(7), 1145–1153.

Giegé, P., Heazlewood, J.L., Roessner-Tunali, U., Millar, A.H., Fernie, A.R., Leaver, C.J. and Sweetlove, L.J. (2003) Enzymes of glycolysis are functionally associated with the mitochondrion in *Arabidopsis* cells. *Plant Cell*, **15**(9), 2140–2151.

Glaser, E. and Dessi, P. (1999) Integration of the mitochondrial-processing peptidase into the cytochrome bc1 complex in plants. *J. Bioenerg. Biomembr.*, **31**(3), 259–274.

Godbole, A., Varghese, J., Sarin, A. and Mathew, M.K. (2003) VDAC is a conserved element of death pathways in plant and animal systems. *Biochim. Biophys. Acta*, **1642**(1–2), 87–96.

Graham, I.A. and Eastmond, P.J. (2002) Pathways of straight and branched chain fatty acid catabolism in higher plants. *Prog. Lipid Res.*, **41**(2), 156–181.

Gray, M.W., Burger, G. and Lang, B.F. (1999) Mitochondrial evolution. *Science*, **283**(5407), 1476–1481.

Green, R. and Fluhr, R. (1995) UV-B-Induced PR-1 accumulation is mediated by active oxygen species. *Plant Cell*, **7**(2), 203–212.

Gueguen, V., Macherel, D., Jaquinod, M., Douce, R. and Bourguignon, J. (2000) Fatty acid and lipoic acid biosynthesis in higher plant mitochondria. *J. Biol. Chem.*, **275**(7), 5016–5025.

Haggie, P.M. and Verkman, A.S. (2002) Diffusion of tricarboxylic acid cycle enzymes in the mitochondrial matrix *in vivo*. Evidence for restricted mobility of a multienzyme complex. *J. Biol. Chem.*, **277**(43), 40782–40788.

Håkansson, G. and Allen, J.F. (1995) Histidine and tyrosine phosphorylation in pea mitochondria: evidence for protein phosphorylation in respiratory redox signalling. *FEBS Lett.*, **372**(2–3), 238–242.

Halliwell, B. and Gutteridge, J.M.C. (1999) *Free Radicals in Biology and Medicine*, 3rd edition. Oxford University Press, Oxford.

Hamasur, B. and Glaser, E. (1992) Plant mitochondrial F_0F_1 ATP synthase. Identification of the individual subunits and properties of the purified spinach leaf mitochondrial ATP synthase. *Eur. J. Biochem.*, **205**(1), 409–416.

Hattori, T. and Asahi, T. (1982) The presence of two forms of succinate dehydrogenase in sweet potato root mitochondria. *Plant Cell Physiol.*, **23**(2), 515–523.

Heazlewood, J.L. and Millar, A.H. (2005) AMPDB: the *Arabidopsis* mitochondrial protein database. *Nucleic Acid. Res.*, **33**, D605–D610.

Heazlewood, J.L., Howell, K.A., Whelan, J. and Millar, A.H. (2003a) Towards an analysis of the rice mitochondrial proteome. *Plant Physiol.*, **132**(1), 230–242.

Heazlewood, J.L., Howell, K.A. and Millar, A.H. (2003b) Mitochondrial complex I from *Arabidopsis* and rice: orthologs of mammalian and fungal components coupled with plant-specific subunits. *Biochim. Biophys. Acta*, **1604**(3), 159–169.

Heazlewood, J.L., Whelan, J. and Millar, A.H. (2003c) The products of the mitochondrial *orf25* and *orfB* genes are F_0 components in the plant F_1F_0 ATP synthase. *FEBS Lett.*, **540**(1–3), 201–205.

Heazlewood, J.L., Tonti-Filippini, J.S., Gout, A.M., Day, D.A., Whelan, J. and Millar, A.H. (2004) Experimental analysis of the *Arabidopsis* mitochondrial proteome highlights signaling and regulatory components, provides assessment of targeting prediction programs, and indicates plant-specific mitochondrial proteins. *Plant Cell*, **16**(1), 241–256.

Herald, V.L., Heazlewood, J.L., Day, D.A. and Millar, A.H. (2003) Proteomic identification of divalent metal cation binding proteins in plant mitochondria. *FEBS Lett.*, **537**(1–3), 96–100.

Herz, U., Schroder, W., Liddell, A., Leaver, C.J., Brennicke, A. and Grohmann, L. (1994) Purification of the NADH:ubiquinone oxidoreductase (complex I) of the respiratory chain from the inner mitochondrial membrane of *Solanum tuberosum*. *J. Biol. Chem.*, **269**(3), 2263–2269.

Hochholdinger, F., Guo, L. and Schnable, P.S. (2004) Cytoplasmic regulation of the accumulation of nuclear-encoded proteins in the mitochondrial proteome of maize. *Plant J.*, **37**(2), 199–208.

Holm, R.H., Kennepohl, P. and Solomon, E.I. (1996) Structural and functional aspects of metal sites in biology. *Chem. Rev.*, **96**(7), 2239–2314.

Hourton-Cabassa, C., Ambard-Bretteville, F., Moreau, F., de Virville, J.D., Remy, R. and Colas des Francs-Small, C. (1998) Stress induction of mitochondrial formate dehydrogenase in potato leaves. *Plant Physiol.*, **116**(2), 627–635.

Humphrey-Smith, I., Colas des Francs-Small, C., Ambart-Bretteville, F. and Remy, R. (1992) Tissue-specific variation of pea mitochondrial polypeptides detected by computerized image analysis of two-dimensional electrophoresis gels. *Electrophoresis*, **13**(3), 168–172.

Igamberdiev, A.U. and Falaleeva, M.I. (1994) Isolation and characterization of the succinate-dehydrogenase complex from plant mitochondria. *Biochemistry (Moscow)*, **59**(8), 895–900.

Jänsch, L., Kruft, V., Schmitz, U.K. and Braun, H.P. (1996) New insights into the composition, molecular mass and stoichiometry of the protein complexes of plant mitochondria. *Plant J.*, **9**(3), 357–368.

Jänsch, L., Kruft, V., Schmitz, U.K. and Braun, H.P. (1998) Unique composition of the preprotein translocase of the outer mitochondrial membrane from plants. *J. Biol. Chem.*, **273**(27), 17251–17257.

Johansson, E., Olsson, O. and Nyström, T. (2004) Progression and specificity of protein oxidation in the life cycle of *Arabidopsis thaliana*. *J. Biol. Chem.*, **279**(21), 22204–22208.

Kameshita, I. and Fujisawa, H. (1997) Detection of calcium binding proteins by two-dimensional sodium dodecyl sulfate-polyacrylamide gel electrophoresis. *Anal. Biochem.*, **249**(2), 252–255.

Kinoshita, T., Shimazaki, K. and Nishimura, M. (1993) Phosphorylation and dephosphorylation of guard-cell proteins from *Vicia faba* L. in response to light and dark. *Plant Physiol.*, **102**(3), 917–923.

Kispal, G., Csere, P., Prohl, C. and Lill, R. (1999) The mitochondrial proteins Atm1p and Nfs1p are essential for biogenesis of cytosolic Fe/S proteins. *EMBO J.*, **18**(14), 3981–3989.

Kristensen, B.K., Askerlund, P., Bykova, N.V., Egsgaard, H. and Møller, I.M. (2004) Identification of oxidized proteins in the matrix of rice leaf mitochondria by immunoprecipitation and two-dimensional liquid chromatography-tandem mass spectrometry. *Phytochemistry*, **65**(12), 1839–1851.

Kruft, V., Eubel, H., Jansch, L., Werhahn, W. and Braun, H.P. (2001) Proteomic approach to identify novel mitochondrial proteins in *Arabidopsis*. *Plant Physiol.*, **127**(4), 1694–1710.

Laloi, C., Rayapuram, N., Chartier, Y., Grienenberger, J.M., Bonnard, G. and Meyer, Y. (2001) Identification and characterization of a mitochondrial thioredoxin system in plants. *Proc. Natl Acad. Sci. USA*, **98**(24), 14144–14149.

Lambers, H. (1982) Cyanide-resistant respiration: a non-phosphorylating electron transport pathway acting as an overflow. *Physiol. Plant.*, **55**(4), 478–485.

Leterme, S. and Boutry, M. (1993) Purification and preliminary characterization of mitochondrial complex I (NADH: ubiquinone reductase) from broad bean (*Vicia faba* L.). *Plant Physiol.*, **102**(2), 435–443.

Lill, R. and Kispal, G. (2001) Mitochondrial ABC transporters. *Res. Microbiol.*, **152**(3–4), 331–340.

Lister, R., Chew, O., Lee, M.N., Heazlewood, J.L., Clifton, R., Parker, K.L., Millar, A.H. and Whelan, J. (2004) A transcriptomic and proteomic characterization of the *Arabidopsis* mitochondrial protein import apparatus and its response to mitochondrial dysfunction. *Plant Physiol.*, **134**(2), 777–789.

Lurin, C., Andres, C., Aubourg, S., Bellaoui, M., Bitton, F., Bruyere, C., Caboche, M., Debast, C., Gualberto, J., Hoffmann, B., Lecharny, A., Le Ret, M., Martin-Magniette, M.L., Mireau, H., Peeters, N., Renou, J.P., Szurek, B., Taconnat, L. and Small, I. (2004) Genome-wide analysis of *Arabidopsis* pentatricopeptide repeat proteins reveals their essential role in organelle biogenesis. *Plant Cell*, **16**(8), 2089–2103.

Lutziger, I. and Oliver, D.J. (2001) Characterization of two cDNAs encoding mitochondrial lipoamide dehydrogenase from *Arabidopsis*. *Plant Physiol.*, **127**(2), 615–623.

Martinou, J.C. and Green, D.R. (2001) Breaking the mitochondrial barrier. *Nat. Rev. Mol. Cell Biol.*, **2**(1), 63–67.

Masterson, C. and Wood, C. (2001) Mitochondrial and peroxisomal beta-oxidation capacities of organs from a non-oilseed plant. *Proc. Royal Soc. Lond., Ser. B, Biol. Sci.*, **268**(1479), 1949–1953.

Maxwell, D.P., Nickels, R. and McIntosh, L. (2002) Evidence of mitochondrial involvement in the transduction of signals required for the induction of genes associated with pathogen attack and senescence. *Plant J.*, **29**(3), 269–279.

McIntosh, L., Eichler, T., Gray, G., Maxwell, D., Nickels, R. and Wang, Y. (1998) Biochemical and genetic controls exerted by plant mitochondria. *Biochim. Biophys. Acta*, **1365**(1–2), 278–284.

Miernyk, J.A., Thomas, D.R. and Wood, C. (1991) Partial purification and characterization of the mitochondrial and peroxisomal isozymes of enoyl-Coenzyme A hydratase from germinating pea seedlings. *Plant Physiol.*, **95**, 564–569.

Mihr, C., Baumgärtner, M., Dieterich, J.-H., Schmitz, U.K. and Braun, H.P. (2001) Proteomic approach for investigation of cytoplasmic male sterility (CMS) in *Brassica*. *J. Plant Physiol.*, **158**(6), 787–794.

Millar, A.H. and Heazlewood, J.L. (2003) Genomic and proteomic analysis of mitochondrial carrier proteins in *Arabidopsis*. *Plant Physiol.*, **131**(2), 443–453.

Millar, A.H., Sweetlove, L.J., Giege, P. and Leaver, C.J. (2001) Analysis of the *Arabidopsis* mitochondrial proteome. *Plant Physiol.*, **127**(4), 1711–1727.

Millar, A.H., Mittova, V., Kiddle, G., Heazlewood, J.L., Bartoli, C.G., Theodoulou, F.L. and Foyer, C.H. (2003) Control of ascorbate synthesis by respiration and its implications for stress responses. *Plant Physiol.*, **133**(2), 443–447.

Millar, A.H., Eubel, H., Jansch, L., Kruft, V., Heazlewood, J.L. and Braun, H.P. (2004a) Mitochondrial cytochrome *c* oxidase and succinate dehydrogenase complexes contain plant specific subunits. *Plant Mol. Biol.*, **56**(1), 77–90.

Millar, A.H., Trend, A.E. and Heazlewood, J.L. (2004b) Changes in the mitochondrial proteome during the anoxia to air transition in rice focus around cytochrome-containing respiratory complexes. *J. Biol. Chem.*, **279**(38), 39471–39478.

Millar, A.H., Heazlewood, J.L., Kristensen, B.K., Braun, H.P. and Møller, I.M. (2005) The plant mitochondrial proteome. *Trends Plant Sci.*, **10**(1), 36–43.

Mittler, R. (2002) Oxidative stress, antioxidants and stress tolerance. *Trends Plant Sci.*, **7**(9), 405–410.

Møller, I.M. (2001a) Plant mitochondria and oxidative stress: electron transport, NADPH turnover, and metabolism of reactive oxygen species. *Annu. Rev. Plant Physiol. Plant Mol. Biol.*, **52**, 561–591.

Møller, I.M. (2001b) A more general mechanism of cytoplasmic male fertility? *Trends Plant Sci.*, **6**(12), 560.

Møller, I.M. and Kristensen, B.K. (2004) Protein oxidation in plant mitochondria as a stress indicator. *Photochem. Photobiol. Sci.*, **3**(8), 730–735.

Møller, I.M. and Kristensen, B.K. (2006) Protein oxidation in plant mitochondria detected as oxidised tryptophan. *Free Radical, Biol. Med.*, **40**: 430–435.

Navarre, D.A., Wendehenne, D., Durner, J., Noad, R. and Klessig, D.F. (2000) Nitric oxide modulates the activity of tobacco aconitase. *Plant Physiol.*, **122**(2), 573–582.

Oliver, D.J. (1994) The glycine decarboxylase complex from plant mitochondria. *Annu. Rev. Plant Physiol. Plant Mol. Biol.*, **45**, 323–337.

Ovadi, J. and Srere, P.A. (2000) Macromolecular compartmentation and channelling. *Internatl. Rev. Cytol.*, **192**, 255–280.

Papenbrock, J. and Schmidt, A. (2000a) Characterization of a sulfurtransferase from *Arabidopsis thaliana*. *Eur. J. Biochem.*, **267**(1), 145–154.

Papenbrock, J. and Schmidt, A. (2000b) Characterization of two sulfurtransferase isozymes from *Arabidopsis thaliana*. *Eur. J. Biochem.*, **267**(17), 5571–5579.

Parisi, G., Perales, M., Fornasari, M.S., Colaneri, A., Gonzalez-Schain, N., Gomez-Casati, D., Zimmermann, S., Brennicke, A., Araya, A., Ferry, J.G., Echave, J. and Zabaleta, E. (2004) Gamma carbonic anhydrases in plant mitochondria. *Plant Mol. Biol.*, **55**(2), 193–207.

Peeters, N. and Small, I. (2001) Dual targeting to mitochondria and chloroplasts. *Biochim. Biophys. Acta*, **1541**(1–2), 54–63.

Petit, P.X., Sommarin, M., Pical, C. and Møller, I.M. (1990) Modulation of endogenous protein phosphorylation in plant mitochondria by respiratory substrates. *Physiol. Plant.*, **80**(4), 493–499.

Petronilli, V., Costantini, P., Scorrano, L., Colonna, R., Passamonti, S. and Bernardi, P. (1994) The voltage sensor of the mitochondrial permeability transition pore is tuned by the oxidation–reduction state of vicinal thiols. Increase of the gating potential by oxidants and its reversal by reducing agents. *J. Biol. Chem.*, **269**(24), 16638–16642.

Pical, C., Fredlund, K.M., Petit, P.X., Sommarin, M. and Møller, I.M. (1993) The outer membrane of plant mitochondria contains a calcium-dependent protein kinase and multiple phosphoproteins. *FEBS Lett.*, **336**(2), 347–351.

Picault, N., Hodges, M., Palmieri, L. and Palmieri, F. (2004) The growing family of mitochondrial carriers in *Arabidopsis*. *Trends Plant Sci.*, **9**(3), 138–146.

Pike, C.S., Kopecek, K.K., Russel, J.P. and Sceppa, E.A. (1991) Regulation of phosphorylation in mitochondria of etiolated oat shoots. *Plant Physiol. Biochem.*, **29**(6), 565–572.

Rasmusson, A.G., Mendel-Hartvig, J., Møller, I.M. and Wiskich, J.T. (1994) Isolation of the rotenone-sensitive NADH-ubiquinone reductase (Complex I) from red beet mitochondria. *Physiol. Plant.*, **90**(3), 607–615.

Rasmusson, A.G., Soole K.L. and Elthon, T.E. (2004) Alternative NAD(P)H dehydrogenases of plant mitochondria. *Annu. Rev. Plant. Biol.*, **55**, 23–39.

Rebeillé, F., Macherel, D., Mouillon, J.M., Garin, J. and Douce, R. (1997) Folate biosynthesis in higher plants: purification and molecular cloning of a bifunctional 6-hydroxymethyl-7,8-dihydropterin pyrophosphokinase/7,8-dihydropteroate synthase localized in mitochondria. *EMBO J.*, **16**(5), 947–957.

Reichert, A.S. and Neupert, W. (2004) Mitochondriomics or what makes us breathe. *Trends Genet.*, **20**(11), 555–562.

Richly, E., Chinnery, P.F. and Leister, D. (2003) Evolutionary diversification of mitochondrial proteomes: implications for human disease. *Trends Genet.*, **19**(7), 356–362.

Schägger, H. and Pfeiffer, K. (2000) Supercomplexes in the respiratory chains of yeast and mammalian mitochondria. *EMBO J.*, **19**(8), 1777–1783.

Schägger, H., Cramer, W.A. and von Jagow, G. (1994) Analysis of molecular masses and oligomeric states of protein complexes by blue native electrophoresis and isolation of membrane protein complexes by two-dimensional native electrophoresis. *Anal. Biochem.*, **217**(2), 220–230.

Schneider, G. and Fechner, U. (2004) Advances in the prediction of protein targeting signals. *Proteomics*, **4**(6), 1571–1580.

Schürmann, P. and Jacquot, J.P. (2000) Plant thioredoxin systems revisited. *Annu. Rev. Plant Physiol. Plant Mol. Biol.*, **51**, 371–400.

Sieger, S.M., Kristensen, B.K., Robson, C.A., Amirsadeghi, S., Eng, E.W., Abdel-Mesih, A., Møller, I.M. and Vanlerberghe, G.C. (2005) The role of alternative oxidase in modulating carbon use efficiency and growth during macronutrient stress in tobacco cells. *J. Exp. Bot.*, **56**(416), 1499–1515.

Sjöling, S. and Glaser, E. (1998) Mitochondrial targeting peptides in plants. *Trends Plant Sci.*, **3**(4), 136–140.

Sommarin, M., Petit, P.X. and Møller, I.M. (1990) Endogenous protein phosphorylation in purified plant mitochondria. *Biochim. Biophys. Acta*, **1052**(1), 195–203.

Struglics, A. and Håkansson, G. (1999) Purification of a serine and histidine phosphorylated mitochondrial nucleoside diphosphate kinase from *Pisum sativum*. *Eur. J. Biochem.*, **262**(3), 765–773.

Struglics, A., Fredlund, K.M., Møller, I.M. and Allen, J.F. (1998) Two subunits of the F_0F_1-ATPase are phosphorylated in the inner mitochondrial membrane. *Biochem. Biophys. Res. Commun.*, **243**(3), 664–668.

Struglics, A., Fredlund, K.M., Møller, I.M. and Allen, J.F. (1999) Phosphoproteins and protein kinase activities intrinsic to inner membranes of potato tuber mitochondria. *Plant Cell Physiol.*, **40**(12), 1271–1279.

Struglics, A., Fredlund, K.M., Konstantinov, Y.M., Allen, J.F. and Møller, I.M. (2000) Protein phosphorylation/dephosphorylation in the inner membrane of potato tuber mitochondria. *FEBS Lett.*, **475**(3), 213–217.

Sweetlove, L.J., Heazlewood, J.L., Herald, V., Holtzapffel, R., Day, D.A., Leaver, C.J. and Millar, A.H. (2002) The impact of oxidative stress on *Arabidopsis* mitochondria. *Plant J.*, **32**(6), 891–904.

Swidzinski, J.A., Sweetlove, L.J. and Leaver C.J. (2002) A custom microarray analysis of gene expression during programmed cell death in *Arabidopsis thaliana*. *Plant J.*, **30**(4), 431–446.

Swidzinski, J.A., Leaver, C.J. and Sweetlove, L.J. (2004) A proteomic analysis of plant programmed cell death. *Phytochemistry*, **65**(12), 1829–1838.

Taylor, N.L., Day, D.A. and Millar, A.H. (2002) Environmental stress causes oxidative damage to plant mitochondria leading to inhibition of glycine decarboxylase. *J. Biol. Chem.*, **277**(45), 42663–42668.

Taylor, N.L., Day, D.A. and Millar, A.H. (2004a) Targets of stress-induced oxidative damage in plant mitochondria and their impact on cell carbon/nitrogen metabolism. *J. Exp. Bot.*, **55**(394), 1–10.

Taylor, N.L., Heazlewood, J.L., Day, D.A. and Millar, A.H. (2004b) Lipoic acid-dependent oxidative catabolism of alpha-keto acids in mitochondria provides evidence for branched-chain amino acid catabolism in *Arabidopsis*. *Plant Physiol.*, **134**(2), 838–848.

Taylor, N.L., Heazlewood, J.L., Day, D.A. and Millar, A.H. (2005) Differential impact of environmental stresses on the pea mitochondrial proteome. *Mol. Cell. Proteom.*, **4**(8), 1122–1133.

Taylor, N.L., Rudhe, C., Hulett, J.M., Lithgow, T., Glaser, E., Day, D.A., Millar, A.H. and Whelan, J. (2003b) Environmental stresses inhibit and stimulate different protein import pathways in plant mitochondria. *FEBS Lett.*, **547**(1–3), 125–130.

Taylor, S.W., Fahy, E., Zhang, B., Glenn, G.M., Warnock, D.E., Wiley, S., Murphy, A.N., Gaucher, S.P., Capaldi, R.A., Gibson, B.W. and Ghosh, S.S. (2003a) Characterization of the human heart mitochondrial proteome. *Nat. Biotechnol.*, **21**(3), 281–286.

Tiwari, B.S., Belenghi, B. and Levine, A. (2002) Oxidative stress increased respiration and generation of reactive oxygen species, resulting in ATP depletion, opening of mitochondrial permeability transition, and programmed cell death. *Plant Physiol.*, **128**(4), 1271–1281.

Vanlerberghe, G.C. and McIntosh, L. (1997) Alternative oxidase: from gene to function. *Annu. Rev. Plant Physiol. Plant Mol. Biol.*, **48**, 703–734.

Werhahn, W. and Braun, H.P. (2002) Biochemical dissection of the mitochondrial proteome from *Arabidopsis thaliana* by three-dimensional gel electrophoresis. *Electrophoresis*, **23**(4), 640–646.

Werhahn, W., Niemeyer, A., Jänsch, L., Kruft, V., Schmitz, U.K. and Braun, H. (2001) Purification and characterization of the preprotein translocase of the outer mitochondrial membrane from *Arabidopsis*. Identification of multiple forms of TOM20. *Plant Physiol.*, **125**(2), 943–954.

Yao, N., Eisfelder, B.J., Marvin, J. and Greenberg, J.T. (2004) The mitochondrion – an organelle commonly involved in programmed cell death in *Arabidopsis thaliana*. *Plant J.*, **40**(4), 596–610.

Zhang, X.P., Sjöling, S., Tanudji, M., Somogyi, L., Andreu, D., Eriksson, L.E., Gräslund, A., Whelan, J. and Glaser, E. (2001) Mutagenesis and computer modelling approach to study determinants for recognition of signal peptides by the mitochondrial processing peptidase. *Plant J.*, **27**(5), 427–438.

Index